T0305524

Differential Equations

Textbooks in Mathematics
Series editors:
Al Boggess, Kenneth H. Rosen

Linear Algebra and Its Applications with R
Ruriko Yoshida

Maple™ Projects of Differential Equations
Robert P. Gilbert, George C. Hsiao, Robert J. Ronkese

Practical Linear Algebra
A Geometry Toolbox, Fourth Edition
Gerald Farin, Dianne Hansford

An Introduction to Analysis, Third Edition
James R. Kirkwood

Student Solutions Manual for Gallian's Contemporary Abstract Algebra, Tenth Edition
Joseph A. Gallian

Elementary Number Theory
Gove Effinger, Gary L. Mullen

Philosophy of Mathematics
Classic and Contemporary Studies
Ahmet Cevik

An Introduction to Complex Analysis and the Laplace Transform
Vladimir Eiderman

An Invitation to Abstract Algebra
Steven J. Rosenberg

Numerical Analysis and Scientific Computation
Jeffery J. Leader

Introduction to Linear Algebra
Computation, Application and Theory
Mark J. DeBonis

The Elements of Advanced Mathematics, Fifth Edition
Steven G. Krantz

Differential Equations
Theory, Technique, and Practice, Third Edition
Steven G. Krantz

Real Analysis and Foundations, Fifth Edition
Steven G. Krantz

Transition to Advanced Mathematics
Danilo R. Diedrichs and Stephen Lovett

https://www.routledge.com/Textbooks-in-Mathematics/book-series/CANDHTEXBOOMTH

Differential Equations

Theory, Technique, and Practice

Third Edition

Steven G. Krantz
with the assistance of Carl A. Nolan

CRC Press is an imprint of the
Taylor & Francis Group, an **informa** business
A CHAPMAN & HALL BOOK

As the Crew-2 mission departed the International Space Station aboard SpaceX Crew Dragon Endeavour, the crew snapped this image of the station during a flyaround of the orbiting lab that took place following undocking from the Harmony module's space-facing port on Nov. 8, 2021.

NASA's SpaceX Crew-2 mission was the second operational mission of the SpaceX Crew Dragon spacecraft and Falcon 9 rocket to the International Space Station as part of the agency's Commercial Crew Program, which has worked with the U.S. aerospace industry to launch astronauts on American rockets and spacecraft from American soil to the space station.

Third edition published 2022
by CRC Press
6000 Broken Sound Parkway NW, Suite 300, Boca Raton, FL 33487-2742

and by CRC Press
4 Park Square, Milton Park, Abingdon, Oxon, OX14 4RN

First edition published by CRC Press 1995
Second edition published by CRC Press 2014

CRC Press is an imprint of Taylor & Francis Group, LLC

LCCN number: 2022934817

ISBN: 9781032102702 (hbk)
ISBN: 9781032102719 (pbk)
ISBN: 9781003214526 (ebk)

DOI: 10.1201/9781003214526

Typeset in font CMR10
by KnowledgeWorks Global Ltd.

To George F. Simmons

For the example that he set.

Contents

Preface

Background

Differential equations are one of the oldest subjects in modern mathematics. It was not long after Newton and Leibniz invented the calculus that Bernoulli and Euler and others began to consider the heat equation and the wave equation of mathematical physics. Newton, himself, solved differential equations both in the study of planetary motion and also in his consideration of optics.

Today differential equations is the centerpiece of much of engineering, of physics, of significant parts of the life sciences, and in many areas of mathematical modeling. The audience for a sophomore course in ordinary differential equations is substantial—second only perhaps to that for calculus. There is a need for a definitive text that both describes classical ideas and provides an entrée to the newer ones. Such a text should pay careful attention to advanced topics like the Laplace transform, Sturm–Liouville theory, and boundary value problems (on the traditional side), but should also pay due homage to nonlinear theory, to modeling, and to computing (on the modern side).

This book provides a cogent and accessible introduction to all the traditional topics. It is a pleasure to have this opportunity to bring this text up to date and to add some more timely material. We have streamlined some of the exposition, and augmented other parts. There is now computer work based not only on number crunching but also on computer algebra systems such as `Mathematica` and `Maple`. Certainly a study of flows and vector fields, and of the beautiful Poincaré–Bendixson theory built thereon, is essential for any modern treatment.

Prerequisites

We assume that the student studying this book has a solid background in calculus of one and several variables. This prerequisite is inescapable. We do *not* assume that the student knows linear algebra. In fact we provide an APPENDIX that gives a quick background in that topic.

Modeling

And all of the above is a basis for *modeling.* Modeling is what brings the subject to life, and makes the ideas real for the students. Differential equations can model real life *questions*, and computer calculations and graphics can then provide real life *answers.* The symbiosis of the synthetic and the calculational provides a rich educational experience for students, and prepares them for more concrete, applied work in future courses. The *Anatomy of an Application* sections in this edition showcase some rich applications from engineering, physics, and applied science.

The *Anatomy of an Application* sections are an important feature of the book. We have worked on the presentation of these important examples, and we hope thereby to have made them more valuable and incisive. The *Historical Notes* continue to be a unique feature of this text. *Math Nuggets* are brief perspectives on mathematical lives or other features of our discipline that will enhance the reading experience.

Problems

Another special feature of this book is the Problems for Review and Discovery. These give students some open-ended material for exploration and further learning. They are an important means of extending the reach of the text, and for anticipating future work.

Organization

An important feature of this new edition is that we have re-organized the material to make it more useful and more accessible. The most frequently taught topics are now up front. And the major applications are isolated in their own chapters. This will make it easier for instructors to find and teach the material that they want. It will also make the text more plausible for students.

Of course we continue to stress applications, and we have added some new ones. But no other text has anything like our *Anatomy of an Application* sections. These set the book apart.

Structuring a Course

Chapters 1–8 plus Chapter 11 provide a solid introduction to the theory of ordinary differential equations. At schools where there are time issues one might omit Chapter 11 and Chapter 8.

The other chapters are all fundamental material and instructors may dip into those as taste and interest suggest.

Acknowledgments

It is a pleasure to thank Carl A. Nolan for working through both the text and the problems and correcting many slips and errors. He put in hundreds of hours on this job, and has been a huge help.

Thanks also to my editor Robert Ross for his advice and encouragement.

Concluding Remarks

We look forward to setting a new standard of the modern textbook on ordinary differential equations, a standard to which other texts may aspire. This will be a book that students read, and internalize, and in the end apply to other subjects and disciplines. It will lay the foundation for future studies in analytical thinking. It will be a touchstone for in-depth studies of growth and change.

Steven G. Krantz
St. Louis, Missouri

Author

Steven G. Krantz is a professor of mathematics at Washington University in St. Louis. He has previously taught at UCLA, Princeton University, and Pennsylvania State University. He has written more than 130 books and more than 250 scholarly papers and is the founding editor of the *Journal of Geometric Analysis.* An AMS Fellow, Dr. Krantz has been a recipient of the Chauvenet Prize, Beckenbach Book Award, and Kemper Prize. He received a PhD from Princeton University.

1

What is a Differential Equation?

- The concept of a differential equation
- Characteristics of a solution
- Finding a solution

1.1 Introductory Remarks

A *differential equation* is an equation relating some function f to one or more of its derivatives. An example is

$$\frac{d^2 f}{dx^2}(x) + 2x\frac{df}{dx}(x) + f^2(x) = \sin x. \tag{1.1}$$

Observe that this particular equation involves a function f together with its first and second derivatives. Any given differential equation may or may not involve f or any particular derivative of f. But, for an equation to be a *differential* equation, at least some derivative of f must appear. The objective in solving an equation like (1) is to *find the function* f. Thus we already perceive a fundamental new paradigm: When we solve an algebraic equation, we seek a number or perhaps a collection of numbers, but when we solve a differential equation, we seek one or more *functions.*

As a simple example, consider the differential equation

$$y' = y\,.$$

It is easy to determine that any function of the form $y = Ce^x$ is a solution of this equation. Because the derivative of the function is equal to itself, we see that the solution set of this particular differential equation is an infinite *family* of functions parameterized by a parameter C. This phenomenon is quite typical of what we will see when we solve differential equations in the chapters that follow.

Many of the laws of nature—in physics, in engineering, in chemistry, in biology, and in astronomy—find their most natural expression in the language of differential equations. Put in other words, differential equations are the

DOI: 10.1201/9781003214526-1

language of nature. Applications of differential equations are also abound in mathematics itself, especially in geometry and harmonic analysis and modeling. Differential equations occur in economics and systems science and other fields of mathematical science.

It is not difficult to perceive why differential equations arise so readily in the sciences. If $y = f(x)$ is a given function, then the derivative df/dx can be interpreted as the rate of change of f with respect to x. In any process of nature, the variables involved are related to their rates of change by the basic scientific principles that govern the process—that is, by the laws of nature. When this relationship is expressed in mathematical notation, the result is usually a differential equation.

Certainly Newton's law of universal gravitation, Maxwell's field equations, the motions of the planets, and the refraction of light are important examples, which can be expressed using differential equations. Much of our understanding of nature comes from our ability to solve differential equations. The purpose of this book is to introduce you to some of these techniques.

The following example will illustrate some of these ideas. According to Newton's second law of motion, the acceleration $\mathbf{a}$ of a body of mass m is proportional to the total force $\mathbf{F}$ acting on the body. The standard expression of this relationship is

$$\mathbf{F} = m \cdot \mathbf{a} \,. \tag{1.2}$$

Suppose in particular that we are analyzing a falling body. Express the height of the body from the surface of the Earth as $y(t)$ feet at time t. The only force acting on the body is due to gravity. If g is the acceleration due to gravity (about -32 ft./sec.^{2} near the surface of the Earth), then the force exerted on the body has magnitude $m \cdot g$. And of course the acceleration is d^2y/dt^2. Thus Newton's law (2) becomes

$$m \cdot g = m \cdot \frac{d^2y}{dt^2} \tag{1.3}$$

or

$$g = \frac{d^2y}{dt^2} \,.$$

We may make the problem a little more interesting by supposing that air exerts a resisting force proportional to the velocity. If the constant of proportionality is k, then the total force acting on the body is $mg - k \cdot (dy/dt)$. Then equation (3) becomes

$$m \cdot g - k \cdot \frac{dy}{dt} = m \cdot \frac{d^2y}{dt^2} \,. \tag{1.4}$$

The differential equations (3) and (4) express the essential attributes of this physical system.

A few additional examples of differential equations are these:

$$(1 - x^2)\frac{d^2y}{dx^2} - 2x\frac{dy}{dx} + p(p+1)y = 0 \,; \tag{1.5}$$

$$x^2\frac{d^2y}{dx^2}+x\frac{dy}{dx}+(x^2-p^2)y=0; \tag{1.6}$$

$$\frac{d^2y}{dx^2}+xy=0\,; \tag{1.7}$$

$$(1-x^2)y''-xy'+p^2y=0\,; \tag{1.8}$$

$$y''-2xy'+2py=0\,; \tag{1.9}$$

$$\frac{dy}{dx}=k\cdot y\,; \tag{1.10}$$

$$\frac{d^3y}{dx^3}+\left(\frac{dy}{dx}\right)^2=y^3+\sin x\,. \tag{1.11}$$

Equations (5)–(9) are called Legendre's equation, Bessel's equation, Airy's equation, Chebyshev's equation, and Hermite's equation, respectively. Each has a vast literature and a history reaching back hundreds of years. We shall touch on each of these equations later in the book. Equation (10) is the equation of exponential decay (or of biological growth).

Math Nugget

Adrien Marie Legendre (1752–1833) invented Legendre polynomials (the artifact for which he is best remembered) in the context of gravitational attraction of ellipsoids. Legendre was a fine French mathematician who suffered the misfortune of seeing most of his best work—in elliptic integrals, number theory, and the method of least squares—superseded by the achievements of younger and abler men. For instance, he devoted 40 years to the study of elliptic integrals, and his two-volume treatise on the subject had scarcely appeared in print before the discoveries of Abel and Jacobi revolutionized the field. Legendre was remarkable for the generous spirit with which he repeatedly welcomed newer and better work that made his own obsolete.

Each of equations (5)–(9) is of second order, meaning that the highest derivative that appears is the second. Equation (10) is of first order. Equation (11) is of third order. Each equation is an *ordinary differential equation*, meaning that it involves a function of a single variable and the *ordinary derivatives* of that function.

A *partial differential equation* is one involving a function of two or more variables, and in which the derivatives are *partial derivatives.* These equations are more subtle, and more difficult, than ordinary differential equations. We shall say something about partial differential equations in Chapter 11.

Math Nugget

Friedrich Wilhelm Bessel (1784–1846) was a distinguished German astronomer and an intimate friend of Gauss. The two corresponded for many years. Bessel was the first man to determine accurately the distance of a fixed star (the star 61 Cygni). In 1844 he discovered the binary (or twin) star Sirius. The companion star to Sirius has the size of a planet but the mass of a star; its density is many thousands of times the density of water. It was the first dead star to be discovered, and occupies a special place in the modern theory of stellar evolution.

1.2 A Taste of Ordinary Differential Equations

In this section we look at two *very simple* examples to get a notion of what solutions to ordinary differential equations look like.

EXAMPLE 1.2.1 Let us solve the differential equation

$$y' = x\,.$$

Solution: This is certainly an equation involving a function and some of its derivatives. It is plain to see, just intuitively, that a solution is given by

$$y = \frac{x^2}{2}\,.$$

But that is not the only solution. In fact the general solution to this differential equation is

$$y = \frac{x^2}{2} + C\,,$$

for an arbitrary real constant C.

So we see that the solution to this differential equation is an *infinite family of functions*. This is quite different from the situation when we were solving polynomials—when the solution set was a finite list of numbers.

The differential equation that we are studying here is what we call *first order*—simply meaning that the highest derivative that appears in the equation is one. So we expect there to be one free parameter in the solution, and indeed there is. □

EXAMPLE 1.2.2 Let us solve the differential equation

$$y' = y \,.$$

Solution: Just by intuition, we see that $y = e^x$ is a solution of this ODE. But that is not the only solution. In fact

$$y = Ce^x$$

is the general solution.

It is curious that the arbitrary constant in the last example occurred additively, while the constant in this solution occurred multiplicatively. Why is that?

Here is another way that we might have discovered the solution to this new equation. Write the problem as

$$\frac{dy}{dx} = y \,.$$

Now manipulate the symbols to write this as

$$dy = ydx$$

or

$$\frac{dy}{y} = dx \,.$$

(At first it may seem odd to manipulate dy/dx as though it were a fraction. But we are simply using shorthand for $dy = (dy/dx)dx$.)

We can integrate both sides of the last equality to obtain

$$\ln y = x + C$$

or

$$y = e^C \cdot e^x = D \cdot e^x \,.$$

Thus we have rediscovered the solution to this ODE, and we have rather naturally discovered that the constant occurs multiplicatively. □

Exercises

1. Verify that the following functions (explicit or implicit) are solutions of the corresponding differential equations.

(a) $y = x^2 + c$ $\quad y' = 2x$
(b) $y = cx^2$ $\quad xy' = 2y$
(c) $y^2 = e^{2x} + c$ $\quad yy' = e^{2x}$
(d) $y = ce^{kx}$ $\quad y' = ky$
(e) $y = c_1 \sin 2x + c_2 \cos 2x$ $\quad y'' + 4y = 0$
(f) $y = c_1 e^{2x} + c_2 e^{-2x}$ $\quad y'' - 4y = 0$
(g) $y = c_1 \sinh 2x + c_2 \cosh 2x$ $\quad y'' - 4y = 0$

2. Find the general solution of each of the following differential equations.

(a) $y' = e^{3x} - x$
(b) $y' = xe^{x^2}$
(c) $(1 + x)y' = x$
(d) $(1 + x^2)y' = x$
(e) $(1 + x^2)y' = \arctan x$

3. For each of the following differential equations, find the particular solution that satisfies the given initial condition.

(a) $y' = xe^x$ $\quad y = 3$ when $x = 1$
(b) $y' = 2 \sin x \cos x$ $\quad y = 1$ when $x = 0$
(c) $y' = \ln x$ $\quad y = 0$ when $x = e$

1.3 The Nature of Solutions

An ordinary differential equation of order n is an equation involving an unknown function f together with its derivatives

$$\frac{df}{dx}, \frac{d^2f}{dx^2}, \ldots, \frac{d^nf}{dx^n}.$$

We might, in a more formal manner, express such an equation as

$$F\left(x, y, \frac{df}{dx}, \frac{d^2f}{dx^2}, \ldots, \frac{d^nf}{dx^n}\right) = 0.$$

How do we verify that a given function f is actually the solution of such an equation?

The answer to this question is best understood in the context of concrete examples.

EXAMPLE 1.3.1 Consider the differential equation

$$y'' - 5y' + 6y = 0\,.$$

Without saying how the solutions are actually *found*, verify that $y_1(x) = e^{2x}$ and $y_2(x) = e^{3x}$ are both solutions.

Solution: To verify this assertion, we note that

$$\begin{aligned} y_1'' - 5y_1' + 6y_1 &= 2 \cdot 2 \cdot e^{2x} - 5 \cdot 2 \cdot e^{2x} + 6 \cdot e^{2x} \\ &= [4 - 10 + 6] \cdot e^{2x} \\ &\equiv 0 \end{aligned}$$

and

$$\begin{aligned} y_2'' - 5y_2' + 6y_2 &= 3 \cdot 3 \cdot e^{3x} - 5 \cdot 3 \cdot e^{3x} + 6 \cdot e^{3x} \\ &= [9 - 15 + 6] \cdot e^{3x} \\ &\equiv 0\,. \end{aligned}$$

□

This process, of verifying that a given *function* is a solution of the given differential equation, is entirely new. The reader will want to practice and become accustomed to it. In the present instance, the reader may check that any function of the form

$$y(x) = c_1 e^{2x} + c_2 e^{3x} \tag{1.12}$$

(where c_1 and c_2 are arbitrary constants) is also a solution of the differential equation.

An important obverse consideration is this: When you are going through the procedure to solve a differential equation, how do you know when you are finished? The answer is that the solution process is complete when all derivatives have been eliminated from the equation. For then you will have y expressed in terms of x, at least implicitly. Thus you will have found the sought-after function.

For a large class of equations that we shall study in detail in the present book, we shall find a number of "independent" solutions equal to the order of the differential equation. Then we shall be able to form a so-called general solution by combining them as in (12). Picard's existence and uniqueness theorem, covered in detail in Section 13.1, will help us to see that our general solution is complete—there are no other solutions. Of course we shall provide all the details in the development below.

Sometimes the solution of a differential equation will be expressed as an *implicitly defined function*. An example is the equation

$$\frac{dy}{dx} = \frac{y^2}{1 - xy}\,, \tag{1.13}$$

which has solution

$$xy = \ln y + c\,. \tag{1.14}$$

Note here that the hallmark of what we call a *solution* is that it has no derivatives in it: it is a direct formula, relating y (the dependent variable) to x (the independent variable). To verify that (14) is indeed a solution of (13), let us differentiate:

$$\frac{d}{dx}[xy] = \frac{d}{dx}[\ln y + c]$$

hence

$$1 \cdot y + x \cdot \frac{dy}{dx} = \frac{dy/dx}{y}$$

or

$$\frac{dy}{dx}\left(\frac{1}{y} - x\right) = y\,.$$

In conclusion,

$$\frac{dy}{dx} = \frac{y^2}{1 - xy}\,,$$

as desired.

One unifying feature of the two examples that we have now seen of verifying solutions is this: When we solve an equation of order n, we expect n "independent solutions" (we shall have to say later just what this word "independent" means) and we expect n undetermined constants. In the first example, the equation was of order 2 and the undetermined constants were c_1 and c_2. In the second example, the equation was of order 1 and the undetermined constant was c.

Math Nugget

Sir George Biddell Airy (1801–1892) was Astronomer Royal of England for many years. He was a hard-working, systematic plodder whose sense of decorum almost deprived John Couch Adams of credit for discovering the planet Neptune. As a boy, Airy was notorious for his skill in designing peashooters. Although this may have been considered to be a notable start, and in spite of his later contributions to the theory of light (he was one of the first to identify the medical condition known as astigmatism), Airy seems to have developed into an excessively practical sort of scientist who was obsessed with elaborate numerical calculations. He had little use for abstract scientific ideas. Nonetheless, Airy functions still play a prominent role in differential equations, special function theory, and mathematical physics.

EXAMPLE 1.3.2 Verify that, for any choice of the constants A and B, the function

$$y = x^2 + Ae^x + Be^{-x}$$

is a solution of the differential equation

$$y'' - y = 2 - x^2 .$$

Solution: This solution set is typical of what we shall learn to find for a second-order linear equation. There are two free parameters in the solution (corresponding to the degree 2 of the equation). Now, if $y = x^2 + Ae^x + Be^{-x}$, then

$$y' = 2x + Ae^x - Be^{-x}$$

and

$$y'' = 2 + Ae^x + Be^{-x} .$$

Hence

$$y'' - y = \left[2 + Ae^x + Be^{-x}\right] - \left[x^2 + Ae^x + Be^{-x}\right] = 2 - x^2$$

as required. □

EXAMPLE 1.3.3 One of the useful things that we do in this subject is to use an "initial condition" to specify a particular solution. For example, let us solve the (very simple) problem

$$y' = 2y, \quad y(0) = 1 .$$

Solution: Of course the general solution of this differential equation is $y = C \cdot e^{2x}$. We seek the solution that has value 1 when $x = 0$. So we set

$$1 = y(0) = C \cdot e^{2 \cdot 0} .$$

This gives

$$1 = C .$$

So we find the particular solution

$$y = e^{2x} .$$

□

Remark 1.3.4 One of the powerful and fascinating features of the study of differential equations is the geometric interpretation that we can often place on the solution of a problem. This connection is most apparent when we consider a first-order equation. Consider the equation

$$\frac{dy}{dx} = F(x, y) . \tag{1.15}$$

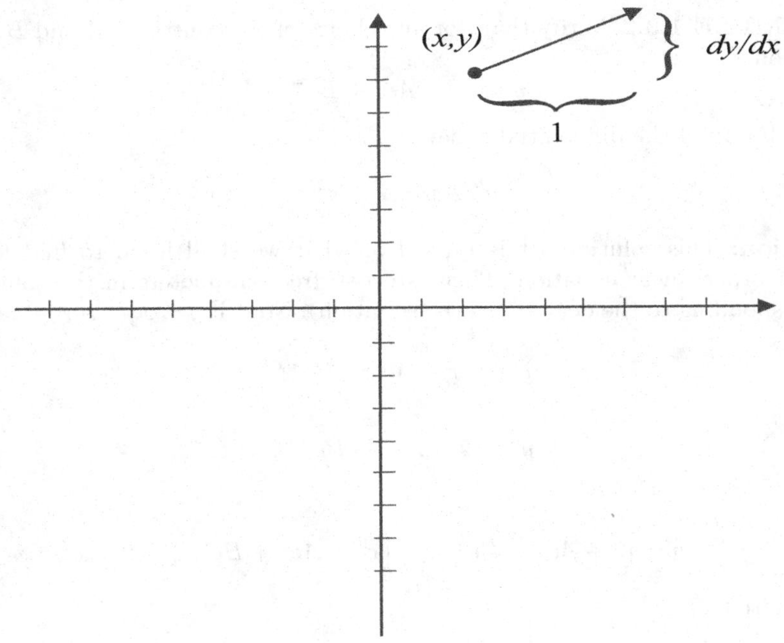

FIGURE 1.1
A vector field.

We may think of equation (15) as assigning to each point (x, y) in the plane a slope dy/dx. For the purposes of drawing a picture, it is more convenient to think of the equation as assigning to the point (x, y) the *vector* $\langle 1, dy/dx \rangle$. See Figure 1.1. Figure 1.2 illustrates how the differential equation

$$\frac{dy}{dx} = x$$

assigns such a vector to each point in the plane. Figure 1.3 illustrates how the differential equation

$$\frac{dy}{dx} = -y$$

assigns such a vector to each point in the plane.

Chapter 12 will explore in greater detail how the geometric analysis of these so-called vector fields can lead to an understanding of the solutions of a variety of differential equations.

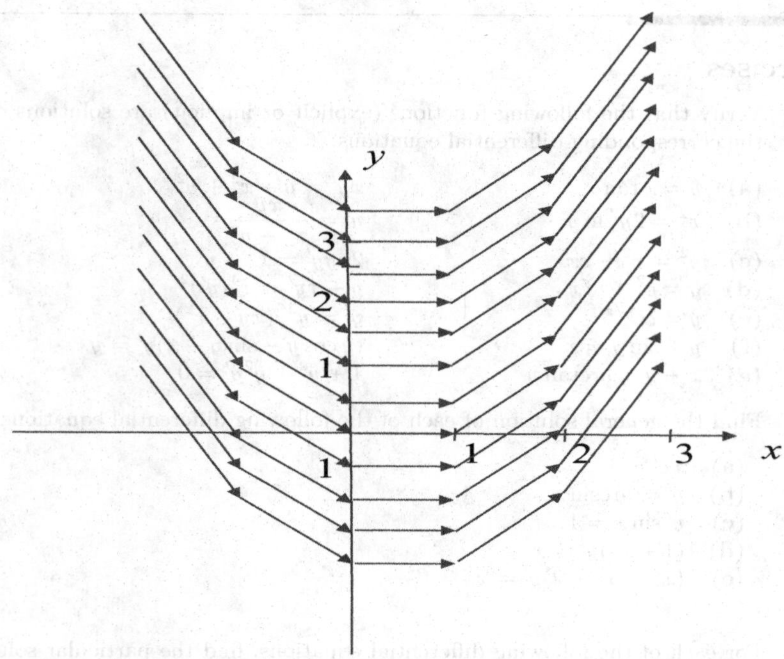

FIGURE 1.2
The vector field for $dy/dx = x$.

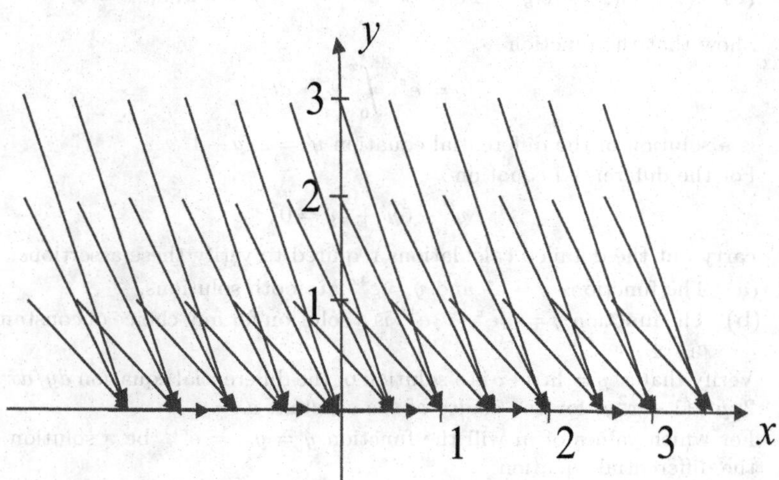

FIGURE 1.3
The vector field for $dy/dx = -y$.

Exercises

1. Verify that the following functions (explicit or implicit) are solutions of the corresponding differential equations.

(a)	$y = x \tan x$	$xy' = y + x^2 + y^2$
(b)	$x^2 = 2y^2 \ln y$	$y' = \dfrac{xy}{x^2 + y^2}$
(c)	$y^2 = x^2 - cx$	$2xyy' = x^2 + y^2$
(d)	$y = c^2 + c/x$	$y + xy' = x^4(y')^2$
(e)	$y = ce^{y/x}$	$y' = y^2/(xy - x^2)$
(f)	$y + \sin y = x$	$(y \cos y - \sin y + x)y' = y$
(g)	$x + y = \arctan y$	$1 + y^2 + y^2 y' = 0$

2. Find the general solution of each of the following differential equations.

(a) $xy' = 1$
(b) $y' = \arcsin x$
(c) $y' \sin x = 1$
(d) $(1 + x^3)y' = x$
(e) $(x^2 - 3x + 2)y' = x$

3. For each of the following differential equations, find the particular solution that satisfies the given initial condition.

(a)	$(x^2 - 1)y' = 1$	$y = 0$ when $x = 2$
(b)	$x(x^2 - 4)y' = 1$	$y = 0$ when $x = 1$
(c)	$(x + 1)(x^2 + 1)y' = 2x^2 + x$	$y = 1$ when $x = 0$

4. Show that the function

$$y = e^{x^2} \int_0^x e^{-t^2}\, dt$$

is a solution of the differential equation $y' = 2xy + 1$.

5. For the differential equation

$$y'' - 5y' + 4y = 0\,,$$

carry out the detailed calculations required to verify these assertions:

(a) The functions $y = e^x$ and $y = e^{4x}$ are both solutions.

(b) The function $y = c_1 e^x + c_2 e^{4x}$ is a solution for any choice of constants c_1, c_2.

6. Verify that $x^2 y = \ln y + c$ is a solution of the differential equation $dy/dx = 2xy^2/(1 - x^2 y)$ for any choice of the constant c.

7. For which values of m will the function $y = y_m = e^{mx}$ be a solution of the differential equation

$$2y''' + y'' - 5y' + 2y = 0\,?$$

Find three such values m. Use the ideas in Exercise 5 to find a solution containing three arbitrary constants c_1, c_2, c_3.

2

Solving First-Order Equations

- Separable equations
- First-order linear equations
- Exact equations
- Orthogonal trajectories
- Homogeneous equations
- Integrating factors
- Reduction of order

2.1 Separable Equations

In this section we shall encounter our first general class of equations with the properties that

(i) We can immediately recognize members of this class of equations.

(ii) We have a simple and direct method for (in principle) solving such equations.

This is the class of *separable equations.*

A first-order ordinary differential equation is *separable* if it is possible, by elementary algebraic manipulation, to arrange the equation so that all the dependent variables (usually the y variable) are on one side and all the independent variables (usually the x variable) are on the other side. Let us learn the method by way of some examples.

EXAMPLE 2.1.1 Solve the ordinary differential equation

$$y' = 2xy\,.$$

Solution: In the method of separation of variables—which is a method for *first-order* equations only—it is useful to write the derivative using Leibniz notation. Thus we have

$$\frac{dy}{dx} = 2xy.$$

DOI: 10.1201/9781003214526-2

We rearrange this equation as

$$\frac{dy}{y} = 2x\,dx\,.$$

(It should be noted here that we use the shorthand dy to stand for $\frac{dy}{dx}\,dx$ and we of course assume that $y \neq 0$.) You can see that we have separated the variables.

Now we can integrate both sides of the last displayed equation to obtain

$$\int \frac{dy}{y} = \int 2x\,dx\,.$$

We are fortunate in that both integrals are easily evaluated. We obtain

$$\ln|y| = x^2 + c\,.$$

(It is important here that we include the constant of integration.) Thus

$$y = e^{x^2+c}\,.$$

We may rewrite this as

$$y = De^{x^2}\,. \tag{2.1}$$

Notice two important features of our final representation for the solution:

(i) We have re-expressed the constant e^c as the positive constant D. We will even allow D to be negative, so we no longer need to worry about the absolute values around y.

(ii) Our solution contains one free constant, as we may have anticipated since the differential equation is of order 1.

We invite the reader to verify that the solution in equation (1) actually satisfies the original differential equation. □

Remark 2.1.2 Of course it would be foolish to expect that all first-order differential equations will be separable. For example, the equation

$$\frac{dy}{dx} = x^2 + y^2$$

certainly is not separable. The property of being separable is quite special. But it is rather surprising that quite a few of the equations of mathematical physics turn out to be separable (as we shall see later in the book).

EXAMPLE 2.1.3 Solve the differential equation

$$xy' = (1 - 2x^2)\tan y\,.$$

Solution: We first write the equation in Leibniz notation. Thus

$$x \cdot \frac{dy}{dx} = (1 - 2x^2) \tan y \, .$$

Separating variables, we find that

$$\cot y \, dy = \left(\frac{1}{x} - 2x \right) \, dx \, .$$

Applying the integral to both sides gives

$$\int \cot y \, dy = \int \left(\frac{1}{x} - 2x \right) \, dx$$

or

$$\ln |\sin y| = \ln |x| - x^2 + C \, .$$

Again note that we were careful to include a constant of integration.

We may express our solution as

$$\sin y = e^{\ln x - x^2 + C}$$

or

$$\sin y = D \cdot x \cdot e^{-x^2} \, .$$

The result is

$$\sin y = D \cdot x \cdot e^{-x^2}$$

or

$$y = \sin^{-1} \left(D \cdot x \cdot e^{-x^2} \right) .$$

We invite the reader to verify that this is indeed a solution to the given differential equation. □

Remark 2.1.4 Of course the technique of separable equations is one that is specifically designed for first-order equations. It makes no sense for second-order equations. Later in the book we shall learn techniques for reducing a second-order equation to a first-order equation; then it may happen that the separation-of-variables technique applies.

Exercises

1. Use the method of separation of variables to solve each of these ordinary differential equations.

(a) $x^5y' + y^5 = 0$
(b) $y' = 4xy$
(c) $y' + y\tan x = 0$
(d) $(1 + x^2)\,dy + (1 + y^2)\,dx = 0$
(e) $y\ln y\,dx - x\,dy = 0$
(f) $xy' = (1 - 4x^2)\tan y$
(g) $y'\sin y = x^2$
(h) $y' - y\tan x = 0$
(i) $xyy' = y - 1$
(j) $xy^2 - y'x^2 = 0$

2. For each of the following differential equations, find the particular solution that satisfies the additional given property (called an *initial condition*).

(a) $y'y = x + 1$ $\qquad y = 3$ when $x = 1$
(b) $(dy/dx)x^2 = y$ $\qquad y = 1$ when $x = 0$
(c) $\dfrac{y'}{1 + x^2} = \dfrac{x}{y}$ $\qquad y = 3$ when $x = 1$
(d) $y^2y' = x + 2$ $\qquad y = 4$ when $x = 0$
(e) $y' = x^2y^2$ $\qquad y = 2$ when $x = -1$
(f) $y'(1 + y) = 1 - x^2$ $\qquad y = -2$ when $x = -1$

3. For the differential equation

$$\frac{y''}{y'} = x^2 ,$$

make the substitution $y' = p$, $y'' = p'$ to reduce the order. Then solve the new equation by separation of variables. Now resubstitute and find the solution y of the original equation.

4. Use the method of Exercise 3 to solve the equation

$$y'' \cdot y' = x(1 + x)$$

subject to the initial conditions $y(0) = 1$, $y'(0) = 0$.

2.2 First-Order Linear Equations

Another class of differential equations that is easily recognized and readily solved (at least in principle)[1] is that of first-order linear equations.

An equation is said to be *first-order linear* if it has the form

$$y' + a(x)y = b(x) . \tag{2.2}$$

The "first-order" aspect is obvious: only first derivatives appear in the equation. The "linear" aspect depends on the fact that the left-hand side involves a

[1]We throw in this caveat because it can happen, and frequently does happen, that we can write down integrals that represent solutions of our differential equation, but *we are unable to evaluate those integrals.* This is annoying, but we shall later learn numerical techniques that will address such an impasse.

differential operator that acts linearly on the space of differentiable functions. Roughly speaking, a differential equation is linear if y and its derivatives are not multiplied together, not raised to powers, and do not occur as the arguments of functions. This is an advanced idea that we explicate in the appendix. If you have not had a course in linear algebra, or if your linear algebra is rusty, then you may find it useful to review this appendix. For now, the reader should simply accept that an equation of the form (2) is first-order linear, and that we will soon have a recipe for solving it.

As usual, we learn this new method by proceeding directly to the examples.

EXAMPLE 2.2.1 Consider the differential equation

$$y' + 2xy = x\,.$$

Find a complete solution.

Solution: First note that the equation definitely has the form (2) of a linear differential equation of first order. So we may proceed.

We endeavor to multiply both sides of the equation by some function that will make each side readily integrable. It turns out that there is a trick that always works: You multiply both sides by $e^{\int a(x)\,dx}$.

Like many tricks, this one may seem unmotivated. But let us try it out and see how it works. Now

$$\int a(x)\,dx = \int 2x\,dx = x^2\,.$$

(At this point we *could* include a constant of integration, but it is not necessary.) Thus $e^{\int a(x)\,dx} = e^{x^2}$. Multiplying both sides of our equation by this factor gives

$$e^{x^2} \cdot y' + e^{x^2} \cdot 2xy = e^{x^2} \cdot x$$

or

$$\left(e^{x^2} \cdot y\right)' = x \cdot e^{x^2}.$$

It is the last step that is a bit tricky. For a first-order linear equation, it is *guaranteed* that, if we multiply through by $e^{\int a(x)\,dx}$, then the left-hand side of the equation will end up being the derivative of $[e^{\int a(x)\,dx} \cdot y]$. Now of course we integrate both sides of the equation:

$$\int \left(e^{x^2} \cdot y\right)' dx = \int x \cdot e^{x^2}\,dx.$$

We can perform both the integrations: on the left-hand side, we simply apply the fundamental theorem of calculus; on the right-hand side, we do the integration. The result is

$$e^{x^2} \cdot y = \frac{1}{2} \cdot e^{x^2} + C$$

or

$$y = \frac{1}{2} + Ce^{-x^2} .$$

Observe that, as we usually expect, the solution has one free constant (because the original differential equation was of order 1). We invite the reader to check in detail that this solution actually satisfies the original differential equation. □

Remark 2.2.2 The last example illustrates a phenomenon that we shall encounter repeatedly in this book. That is the idea of a "general solution." Generally speaking, a differential equation will have an entire family of solutions. And, especially when the problem comes from physical considerations, we shall often have initial conditions that must be met by the solution. The family of solutions will depend on one or more parameters (in the last example there was one parameter C), and those parameters will often be determined by the initial conditions.

We shall see as the book develops that the amount of freedom built into the family of solutions—that is, the number of degrees of freedom provided by the parameters—meshes very nicely with the number of initial conditions that fit the problem (in the last example, one initial condition would be appropriate). Thus we shall generally be able to solve uniquely for numerical values of the parameters. Picard's Existence and Uniqueness Theorem—treated in Section 13.1—gives a precise mathematical framework for the informal discussion in the present remark.

Summary of the Method

To solve a first-order linear equation

$$y' + a(x)y = b(x) ,$$

multiply both sides of the equation by the "integrating factor" $e^{\int a(x)\,dx}$ and then integrate.

EXAMPLE 2.2.3 Solve the differential equation

$$x^2y' + xy = x^3 .$$

Solution: First observe that this equation is not in the standard form (equation (2)) for first-order linear. We render it so by multiplying through by a factor of $1/x^2$. Thus the equation becomes

$$y' + \frac{1}{x}y = x .$$

Now $a(x) = 1/x$, $\int a(x)\,dx = \ln|x|$, and $e^{\int a(x)\,dx} = |x|$. We multiply the differential equation by this factor. In fact, in order to simplify the calculus, we shall restrict attention to $x > 0$. Thus we may eliminate the absolute value signs.

Thus

$$xy' + y = x^2 .$$

Now, as is guaranteed by the theory, we may rewrite this equation as

$$\left(x \cdot y\right)' = x^2 .$$

Applying the integral to both sides gives

$$\int \left(x \cdot y\right)' \, dx = \int x^2 \, dx .$$

Now, as usual, we may use the fundamental theorem of calculus on the left; and we may simply integrate on the right. The result is

$$x \cdot y = \frac{x^3}{3} + C .$$

We finally find that our solution is

$$y = \frac{x^2}{3} + \frac{C}{x} .$$

You should plug this answer into the differential equation and check that it works. □

Exercises

1. Find the general solution of each of the following first-order, linear ordinary differential equations.

   (a) $y' - xy = 0$
   (b) $y' + xy = x$
   (c) $y' + y = \dfrac{1}{1 + e^{2x}}$
   (d) $y' + y = 2xe^{-x} + x^2$
   (e) $(2y - x^3)\,dx = x\,dy$
   (f) $y' + 2xy = 0$
   (g) $xy' - 3y = x^4$
   (h) $(1 + x^2)\,dy + 2xy\,dx = \cot x\,dx$
   (i) $y' + y \cot x = 2x \csc x$
   (j) $y - x + xy \cot x + xy' = 0$

2. For each of the following differential equations, find the particular solution that satisfies the given initial data.

(a) $y' - xy = 0$ $\quad y = 3$ when $x = 1$

(b) $y' - 2xy = 6xe^{x^2}$ $\quad y = 1$ when $x = 1$

(c) $xy' + y = 3x^2$ $\quad y = 2$ when $x = 2$

(d) $y' - (1/x)y = x^2$ $\quad y = 3$ when $x = 1$

(e) $y' + 4y = e^{-x}$ $\quad y = 0$ when $x = 0$

(f) $x^2y' + xy = 2x$ $\quad y = 1$ when $x = 1$

3. The equation

$$\frac{dy}{dx} + P(x)y = Q(x)y^n$$

is known as Bernoulli's equation. It is linear when $n = 0$ or 1, otherwise not. In fact the equation can be reduced to a linear equation when $n > 1$ by the change of variables $z = y^{1-n}$. Use this method to solve each of the following equations.

(a) $xy' + y = x^4y^3$ $\qquad$ **(c)** $x\,dy + y\,dx = xy^2\,dx$

(b) $xy^2y' + y^3 = x\cos x$ $\qquad$ **(d)** $y' + xy = xy^4$

4. The usual Leibniz notation dy/dx implies that x is the independent variable and y is the dependent variable. In solving a differential equation, it is sometimes useful to reverse the roles of the two variables. Treat each of the following equations by reversing the roles of y and x:

(a) $(e^y - 2xy)y' = y^2$ $\qquad$ **(c)** $xy' + 2 = x^3(y-1)y'$

(b) $y - xy' = y'y^2e^y$ $\qquad$ **(d)** $f(y)^2\dfrac{dx}{dy} + 3f(y)f'(y)x = f'(y)$

5. We know from our solution technique that the general solution of a first-order linear equation is a family of curves of the form

$$y = c \cdot f(x) + g(x)\,.$$

Show, conversely, that the differential equation of any such family is linear and first order.

6. Show that the differential equation $y' + Py = Qy\ln y$ can be solved by the change of variables $z = \ln y$. Apply this method to solve the equation

$$xy' = 2x^2y + y\ln y\,.$$

7. One solution of the differential equation $y'\sin 2x = 2y + 2\cos x$ remains bounded as $x \to \pi/2$. Find this solution.

8. A tank contains 10 gallons of brine in which 2 pounds of salt are dissolved. New brine containing 1 pound of salt per gallon is pumped into the tank at the rate of 3 gallons per minute. The mixture is stirred and drained off at the rate of 4 gallons per minute. Find the amount $x = x(t)$ of salt in the tank at any time t.

9. A tank contains 40 gallons of pure water. Brine with 3 pounds of salt per gallon flows in at the rate of 2 gallons per minute. The thoroughly stirred mixture then flows out at the rate of 3 gallons per minute.

(a) Find the amount of salt in the tank when the brine in it has been reduced to 20 gallons.

(b) When is the amount of salt in the tank greatest?

2.3 Exact Equations

A great many first-order equations may be written in the form

$$M(x, y)\,dx + N(x, y)\,dy = 0\,. \tag{2.3}$$

This particular format is quite suggestive, for it brings to mind a family of curves. Namely, if it happens that there is a function $f(x, y)$ so that

$$\frac{\partial f}{\partial x} = M \qquad \text{and} \qquad \frac{\partial f}{\partial y} = N\,, \tag{2.4}$$

then we can rewrite the differential equation as

$$\frac{\partial f}{\partial x}\,dx + \frac{\partial f}{\partial y}\,dy = 0\,. \tag{2.5}$$

Of course the only way that such an equation can hold is if

$$\frac{\partial f}{\partial x} \equiv 0 \qquad \text{and} \qquad \frac{\partial f}{\partial y} \equiv 0\,.$$

And this entails that the function f be identically constant. In other words,

$$f(x, y) \equiv c\,.$$

This last equation describes a family of curves: for each fixed value of c, the equation expresses y implicitly as a function of x, and hence gives a curve. Refer to Figure 2.1 for an example. In later parts of this book we shall learn much from thinking of the set of solutions of a differential equation as a smoothly varying family of curves in the plane.

The method of solution just outlined is called the *method of exact equations.* It depends critically on being able to tell when an equation of the form (3) can be written in the form (5). This in turn begs the question of when (4) will hold.

Fortunately, we learned in calculus a complete answer to this question. Let us review the key points. First note that, if it is the case that

$$\frac{\partial f}{\partial x} = M \qquad \text{and} \qquad \frac{\partial f}{\partial y} = N\,, \tag{2.6}$$

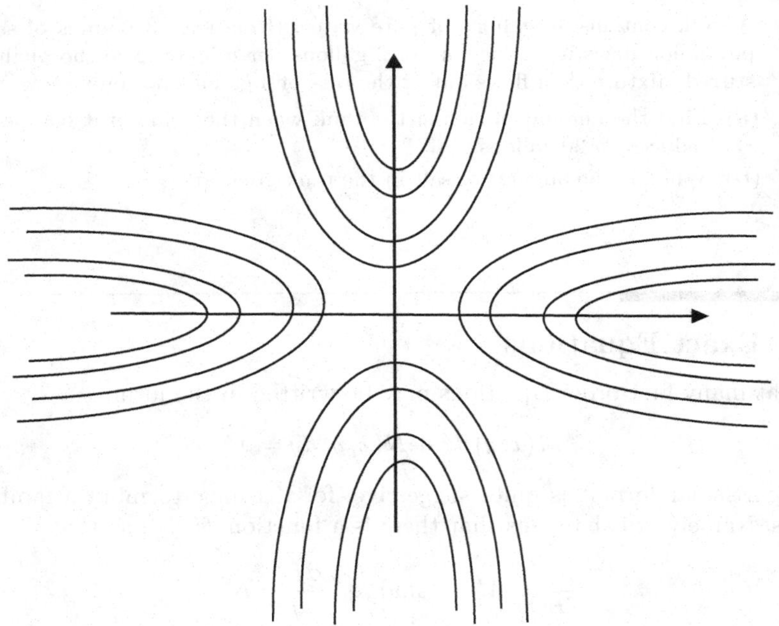

FIGURE 2.1
A family of curves defined by $x^2 - y^2 = c$.

then we see (by differentiation) that

$$\frac{\partial^2 f}{\partial y \partial x} = \frac{\partial M}{\partial y} \qquad \text{and} \qquad \frac{\partial^2 f}{\partial x \partial y} = \frac{\partial N}{\partial x} .$$

Since mixed partials of a smooth function may be taken in any order, we find that a *necessary condition* for condition (6) to hold is that

$$\frac{\partial M}{\partial y} = \frac{\partial N}{\partial x} . \tag{2.7}$$

We call (7) the *exactness condition.* This provides us with a useful test for when the method of exact equations will apply.

It turns out that condition (7) is also sufficient—at least on a domain with no holes. We refer the reader to any good calculus book (see, for instance, [BLK]) for the details of this assertion. We shall use our worked examples to illustrate the point.

EXAMPLE 2.3.1 Use the method of exact equations to solve

$$\frac{x^2}{2} \cdot \cot y \cdot \frac{dy}{dx} = -x .$$

Solution: First, we rearrange the equation as

$$2x \sin y \, dx + x^2 \cos y \, dy = 0 .$$

Observe that the role of $M(x, y)$ is played by $2x \sin y$ and the role of $N(x, y)$ is played by $x^2 \cos y$. Next we see that

$$\frac{\partial M}{\partial y} = 2x \cos y = \frac{\partial N}{\partial x} .$$

Thus our necessary condition (exactness) for the method of exact equations to work is satisfied. We shall soon see explicitly from our calculations that it is also sufficient.

We seek a function f such that $\partial f/\partial x = M(x, y) = 2x \sin y$ and $\partial f/\partial y = N(x, y) = x^2 \cos y$. Let us begin by concentrating on the first of these:

$$\frac{\partial f}{\partial x} = 2x \sin y ,$$

hence

$$\int \frac{\partial f}{\partial x} \, dx = \int 2x \sin y \, dx .$$

The left-hand side of this equation may be evaluated with the fundamental theorem of calculus. Treating x and y as independent variables (which is part of this method), we can also compute the integral on the right. The result is

$$f(x, y) = x^2 \sin y + \phi(y) . \tag{2.8}$$

Now there is an important point that must be stressed. The reader should by now have expected a constant of integration to show up. But in fact our "constant of integration" is $\phi(y)$. This is because our integral was with respect to x, and therefore our constant of integration should be the most general possible expression *that does not depend on* x. That, of course, would be a function of y.

Now we differentiate both sides of (8) with respect to y to obtain

$$N(x, y) = \frac{\partial f}{\partial y} = x^2 \cos y + \phi'(y) .$$

But of course we already know that $N(x, y) = x^2 \cos y$. The upshot is that

$$\phi'(y) = 0$$

or

$$\phi(y) = d ,$$

an ordinary constant.

Plugging this information into equation (8) now yields that

$$f(x, y) = x^2 \sin y + d .$$

We stress that *this is not the solution of the differential equation.* Before you proceed, please review the outline of the method of exact equations that preceded this example. Our job now is to set

$$f(x, y) = c.$$

So

$$x^2 \cdot \sin y = \widetilde{c},$$

where $\widetilde{c} = c - d$.

This is in fact the solution of our differential equation, expressed implicitly. If we wish, we can solve for y in terms of x to obtain

$$y = \sin^{-1} \frac{\widetilde{c}}{x^2}. \qquad \square$$

EXAMPLE 2.3.2 Use the method of exact equations to solve the differential equation

$$y^2 \, dx - x^2 \, dy = 0.$$

Solution: We first test the exactness condition:

$$\frac{\partial M}{\partial y} = 2y \neq -2x = \frac{\partial N}{\partial x}.$$

The exactness condition fails. As a result, this ordinary differential equation cannot be solved by the method of exact equations. □

It is a fact that, even when a differential equation fails the "exact equations test," it is always possible to multiply the equation through by an "integrating factor" so that it *will* pass the exact equations test. Unfortunately, it can be quite difficult to discover explicitly what that integrating factor might be. We shall learn more about the method of integrating factors later.

EXAMPLE 2.3.3 Use the method of exact equations to solve

$$e^y \, dx + (xe^y + 2y) \, dy = 0.$$

Solution: First we check for exactness:

$$\frac{\partial M}{\partial y} = \frac{\partial}{\partial y}[e^y] = e^y = \frac{\partial}{\partial x}[xe^y + 2y] = \frac{\partial N}{\partial x}.$$

This condition is verified, so we can proceed to solve for f:

$$\frac{\partial f}{\partial x} = M = e^y$$

hence

$$f(x,y) = x \cdot e^y + \phi(y).$$

But then

$$\frac{\partial}{\partial y} f(x,y) = \frac{\partial}{\partial y}(x \cdot e^y + \phi(y)) = x \cdot e^y + \phi'(y).$$

And this last expression must equal $N(x,y) = xe^y + 2y$. It follows that

$$\phi'(y) = 2y$$

or

$$\phi(y) = y^2 + d.$$

Altogether, then, we conclude that

$$f(x,y) = x \cdot e^y + y^2 + d.$$

We must not forget the final step. The solution of the differential equation is

$$f(x,y) = c$$

or

$$x \cdot e^y + y^2 = c - d = \widetilde{c}.$$

This time we must content ourselves with the solution expressed implicitly, since it is not feasible to solve for y in terms of x (at least not in an elementary, closed form). □

Exercises

Determine which of the following equations, in Exercises 1–19, is exact. Solve those that *are* exact.

1. $\left(x + \frac{2}{y}\right) dy + y\,dx = 0$

2. $(\sin x \tan y + 1)\,dx + \cos x \sec^2 y\,dy = 0$

3. $(y - x^3)\,dx + (x + y^3)\,dy = 0$

4. $(2y^2 - 4x + 5)\,dx = (4 - 2y + 4xy)\,dy$

5. $(y + y\cos xy)\,dx + (x + x\cos xy)\,dy = 0$

6. $\cos x \cos^2 y\,dx + 2\sin x \sin y \cos y\,dy = 0$

7. $(\sin x \sin y - xe^y)\,dy = (e^y + \cos x \cos y)\,dx$

8. $-\frac{1}{y}\sin\frac{x}{y}\,dx + \frac{x}{y^2}\sin\frac{x}{y}\,dy = 0$

9. $(1+y)\,dx + (1-x)\,dy = 0$

10. $(2xy^3 + y\cos x)\,dx + (3x^2y^2 + \sin x)\,dy = 0$

11. $dx = \dfrac{y}{1-x^2y^2}\,dx + \dfrac{x}{1-x^2y^2}\,dy$

12. $(2xy^4 + \sin y)\,dx + (4x^2y^3 + x\cos y)\,dy = 0$

13. $\dfrac{y\,dx + x\,dy}{1-x^2y^2} + x\,dx = 0$

14. $2x(1+\sqrt{x^2-y})\,dx = \sqrt{x^2-y}\,dy$

15. $(x\ln y + xy)\,dx + (y\ln x + xy)\,dy = 0$

16. $(e^{y^2} - \csc y\csc^2 x)\,dx + (2xye^{y^2} - \csc y\cot y\cot x)\,dy = 0$

17. $(1+y^2\sin 2x)\,dx - 2y\cos^2 x\,dy = 0$

18. $\dfrac{x\,dx}{(x^2+y^2)^{3/2}} + \dfrac{y\,dy}{(x^2+y^2)^{3/2}} = 0$

19. $3x^2(1+\ln y)\,dx + \left(\dfrac{x^3}{y} - 2y\right)\,dy = 0$

20. Solve

$$\frac{y\,dx - x\,dy}{(x+y)^2} + dy = dx$$

as an exact equation by two different methods. Now reconcile the results.

21. Solve

$$\frac{4y^2 - 2x^2}{4xy^2 - x^3}\,dx + \frac{8y^2 - x^2}{4y^3 - x^2y}\,dy = 0$$

as an exact equation. Later on (Section 1.8) we shall learn that we may also solve this equation as a homogeneous equation.

22. For each of the following equations, find the value of n for which the equation is exact. Then solve the equation for that value of n.

(a) $(xy^2 + nx^2y)\,dx + (x^3 + x^2y)\,dy = 0$

(b) $(x + ye^{2xy})\,dx + nxe^{2xy}\,dy = 0$

2.4 Orthogonal Trajectories and Families of Curves

We have already noted that it is useful to think of the collection of solutions of a first-order differential equation as a family of curves. Refer, for instance, to the last example of the preceding section. We solved the differential equation

$$e^y\,dx + (xe^y + 2y)\,dy = 0$$

and found the solution set

$$x \cdot e^y + y^2 = c\,. \tag{2.9}$$

For each value of c, the equation describes a curve in the plane.

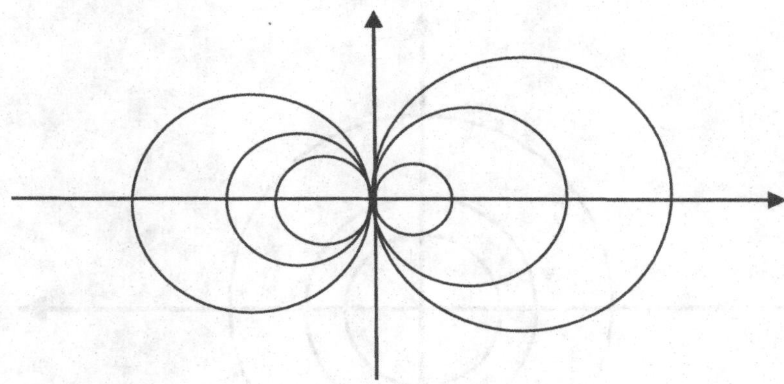

FIGURE 2.2
A family of circles.

Conversely, if we are given a family of curves in the plane, then we can produce a differential equation from which the curves all come. Consider the example of the family

$$x^2 + y^2 = 2cx \,. \tag{2.10}$$

The reader can readily see that this is the family of all circles tangent to the y-axis at the origin (Figure 2.2).

We may differentiate the equation with respect to x, thinking of y as a function of x, to obtain

$$2x + 2y \cdot \frac{dy}{dx} = 2c \,.$$

Now the original equation (10) tells us that

$$x + \frac{y^2}{x} = 2c \,,$$

and we may equate the two expressions for the quantity $2c$ (the point being to eliminate the constant c). The result is

$$2x + 2y \cdot \frac{dy}{dx} = x + \frac{y^2}{x}$$

or

$$\frac{dy}{dx} = \frac{y^2 - x^2}{2xy} \,. \tag{2.11}$$

In summary, we see that we can pass back and forth between a differential equation and its family of solution curves.

There is considerable interest, given a family $\mathcal{F}$ of curves, to find the corresponding family $\mathcal{G}$ of curves that are orthogonal (or perpendicular) to those of $\mathcal{F}$. For instance, if $\mathcal{F}$ represents the flow curves of an electric current, then $\mathcal{G}$

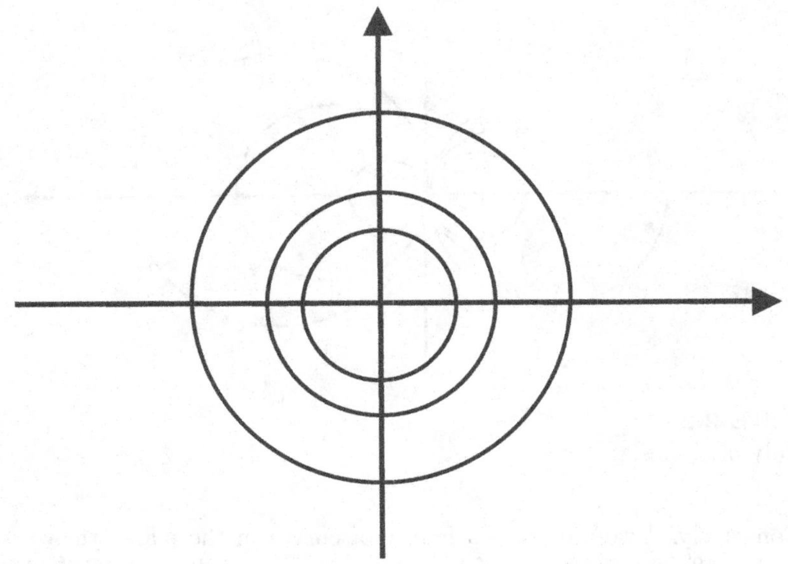

FIGURE 2.3
Circles centered at the origin.

will be the equipotential curves for the flow. If we bear in mind that orthogonality of curves means orthogonality of their tangents, and that orthogonality of the tangent lines means simply that their slopes are negative reciprocals, then it becomes clear what we must do.

EXAMPLE 2.4.1 Find the orthogonal trajectories to the family of curves

$$x^2 + y^2 = c\,.$$

Solution: First observe that we can differentiate the given equation to obtain

$$2x + 2y \cdot \frac{dy}{dx} = 0\,.$$

The constant c has disappeared, and we can take this to be the differential equation for the given family of curves (which in fact are all the circles centered at the origin—see Figure 2.3).

We rewrite the differential equation as

$$\frac{dy}{dx} = -\frac{x}{y}\,.$$

Now taking negative reciprocals, as indicated in the discussion right before this example, we obtain the new differential equation

$$\frac{dy}{dx} = \frac{y}{x}\,.$$

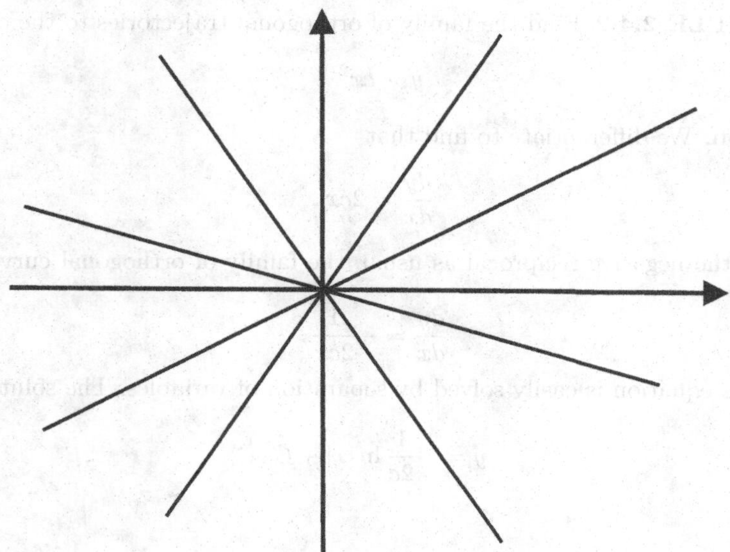

FIGURE 2.4
Lines through the origin.

We may easily separate variables to obtain

$$\frac{1}{y}\,dy = \frac{1}{x}\,dx\,.$$

Applying the integral to both sides yields

$$\int \frac{1}{y}\,dy = \int \frac{1}{x}\,dx$$

or

$$\ln|y| = \ln|x| + C\,.$$

With some algebra, this simplifies to

$$|y| = D|x|$$

or

$$y = \pm Dx\,.$$

The solution that we have found comes as no surprise: the orthogonal trajectories to the family of circles centered at the origin is the family of lines through the origin. See Figure 2.4. □

EXAMPLE 2.4.2 Find the family of orthogonal trajectories to the curves

$$y = cx^2 .$$

Solution: We differentiate to find that

$$\frac{dy}{dx} = 2cx .$$

Taking the negative reciprocal as usual, the family of orthogonal curves will satisfy

$$\frac{dy}{dx} = -\frac{1}{2cx} .$$

This equation is easily solved by separation of variables. The solution is

$$y = -\frac{1}{2c} \ln |x| + C .$$

□

Exercises

1. Sketch each of the following families of curves. In each case, find the family of orthogonal trajectories, and add those to your sketch.

   (a) $xy = c$
   (b) $y = cx^2$
   (c) $x + y = c$
   (d) $r = c(1 + \cos\theta)$
   (e) $y = ce^x$
   (f) $x - y^2 = c$

2. What are the orthogonal trajectories of the family of curves $y = cx^4$? What are the orthogonal trajectories of the family of curves $y = cx^n$ for n a positive integer? Sketch both families of curves. How does the family of orthogonal trajectories change when n is increased?

3. Sketch the family $y^2 = 4c(x + c)$ of all parabolas with axis the x-axis and focus at the origin. Find the differential equation of this family. Show that this differential equation is unaltered if dy/dx is replaced by $-dx/dy$. What conclusion can be drawn from this fact?

4. In each of parts **(a)** through **(f)**, find the family of curves that satisfy the given geometric condition (you should have six different answers for the six different parts of the problem):

   (a) The part of the tangent cut off by the axes is bisected by the point of tangency.

   (b) The projection on the x-axis of the part of the normal between (x, y) and the x-axis has length 1.

(c) The projection on the x-axis of the part of the tangent between (x, y) and the x-axis has length 1.

(d) The part of the tangent between (x, y) and the x-axis is bisected by the y-axis.

(e) The part of the normal between (x, y) and the y-axis is bisected by the x-axis.

(f) The point (x, y) is equidistant from the origin and the point of intersection of the normal with the x-axis.

5. A curve rises from the origin in the x-y plane into the first quadrant. The area under the curve from $(0, 0)$ to (x, y) is one third of the area of the rectangle with these points as opposite vertices. Find the equation of the curve.

6. Find the differential equation of each of the following one-parameter families of curves:

(a) $y = x\sin(x + c)$

(b) all circles through $(1, 0)$ and $(-1, 0)$

(c) all circles with centers on the line $y = x$ and tangent to both axes

(d) all lines tangent to the parabolas $x^2 = 4y$ (*Hint:* The slope of the tangent line at $(2a, a^2)$ is a.)

(e) all lines tangent to the unit circle $x^2 + y^2 = 1$

7. Use your symbol manipulation software, such as `Maple` or `Mathematica`, to find the orthogonal trajectories to each of these families of curves:

(a) $y = \sin x + cx^2$

(b) $y = c\ln x + x, \quad x > 0$

(c) $y = \dfrac{\cos x}{cx + \ln x}, \quad x > 0$

(d) $y = \sin x + c\cos x$

2.5 Homogeneous Equations

The reader should be cautioned that the word "homogeneous" has two meanings in this subject (as mathematics is developed simultaneously by many people all over the world, and they do not always stop to cooperate on their choices of terminology).

One usage, which we shall see later, is that an ordinary differential equation is *homogeneous* when the right-hand side is zero; that is, there is no forcing term.

The other usage will be relevant to the present section. It bears on the "balance" of weight among the different variables. It turns out that a differential equation in which the x and y variables have a balanced presence is amenable to a useful change of variables. That is what we are about to learn.

First of all, a function $g(x, y)$ of two variables is said to be *homogeneous of degree* α, for α a real number, if

$$g(tx, ty) = t^{\alpha} g(x, y) \qquad \text{for all } t > 0 \,.$$

As examples, consider:

- Let $g(x,y) = x^2 + xy$. Then $g(tx,ty) = t^2 \cdot g(x,y)$, so g is homogeneous of degree 2.
- Let $g(x,y) = \sin[x/y]$. Then $g(tx,ty) = g(x,y)$, so g is homogeneous of degree 0.
- Let $g(x,y) = \sqrt{x^2+y^2}$. Then $g(tx,ty) = t \cdot g(x,y)$, so g is homogeneous of degree 1.

If a function is not homogeneous in the sense just indicated, then we call it *nonhomogeneous* or *inhomogeneous.*

In case a differential equation has the form

$$M(x,y)\,dx + N(x,y)\,dy = 0$$

and M, N have the *same degree of homogeneity*, then it is possible to perform the change of variable $z = y/x$ and make the equation separable (see Section 2.1). Of course we then have a well-understood method for solving the equation.

The next examples will illustrate the method.

EXAMPLE 2.5.1 Use the method of homogeneous equations to solve the equation

$$(x+y)\,dx - (x-y)\,dy = 0\,.$$

Solution: Observe that $M(x,y) = x+y$ and $N(x,y) = -(x-y)$ and each is homogeneous of degree 1. We thus rewrite the equation in the form

$$\frac{dy}{dx} = \frac{x+y}{x-y}\,.$$

Dividing numerator and denominator by x, we finally have

$$\frac{dy}{dx} = \frac{1+\frac{y}{x}}{1-\frac{y}{x}}\,. \tag{2.12}$$

The point of these manipulations is that the right-hand side is now plainly homogeneous of degree 0. We introduce the change of variable

$$z = \frac{y}{x} \tag{2.13}$$

hence

$$y = zx$$

and

$$\frac{dy}{dx} = z + x \cdot \frac{dz}{dx}\,. \tag{2.14}$$

Putting (13) and (14) into (12) gives

$$z + x\frac{dz}{dx} = \frac{1+z}{1-z}\,.$$

Of course this may be rewritten as

$$x\frac{dz}{dx} = \frac{1+z^2}{1-z}$$

or

$$\frac{1-z}{1+z^2}\,dz = \frac{dx}{x}\,.$$

Notice that we have separated the variables!! We apply the integral, and rewrite the left-hand side, to obtain

$$\int \frac{dz}{1+z^2} - \int \frac{z\,dz}{1+z^2} = \int \frac{dx}{x}.$$

The integrals are easily evaluated, and we find that

$$\arctan z - \frac{1}{2}\ln(1+z^2) = \ln x + C.$$

Now we return to our original notation by setting $z = y/x$. The result is

$$\arctan\frac{y}{x} - \ln\sqrt{1+\frac{y^2}{x^2}} = \ln x + C$$

or

$$\arctan\frac{y}{x} - \left(\ln\sqrt{x^2+y^2} - \ln\sqrt{x^2}\right) = \ln x + C$$

or

$$\arctan\frac{y}{x} - \ln\sqrt{x^2+y^2} = C.$$

Thus we have expressed y implicitly as a function of x, all the derivatives are gone, and we have solved the differential equation. □

EXAMPLE 2.5.2 Solve the differential equation

$$xy' = 2x + 3y.$$

Solution: It is plain that the equation is first-order linear, and we encourage the reader to solve the equation by that method for practice and comparison purposes. Instead, developing the ideas of the present section, we shall use the method of homogeneous equations.

If we rewrite the equation as

$$-(2x+3y)\,dx + x\,dy = 0,$$

then we see that each of $M = -(2x+3y)$ and $N = x$ is homogeneous of degree 1. Thus we render the equation as

$$\frac{dy}{dx} = \frac{2x+3y}{x} = \frac{2+3(y/x)}{1} = 2 + 3\frac{y}{x}.$$

The right-hand side is homogeneous of degree 0, as we expect.

We set $z = y/x$ and $dy/dx = z + x[dz/dx]$. The result is

$$z + x\cdot\frac{dz}{dx} = 2 + 3\frac{y}{x} = 2 + 3z.$$

The equation separates, as we anticipate, into

$$\frac{dz}{2+2z} = \frac{dx}{x}.$$

This is easily integrated to yield

$$\frac{1}{2}\ln(1+z) = \ln x + C$$

or

$$\ln(1+z) = 2\ln x + D$$

or

$$\ln(1+z) = \ln x^2 + D$$

or

$$z = Ex^2 - 1\,.$$

Resubstituting $z = y/x$ gives

$$\frac{y}{x} = Ex^2 - 1$$

hence

$$y = Ex^3 - x\,.$$ □

Exercises

1. Verify that each of the following differential equations is homogeneous, and then solve it.

   (a) $(x^2 - 2y^2)\,dx + xy\,dy = 0$
   (b) $xy' - 3xy - 2y^2 = 0$
   (c) $x^2y' = 3(x^2 + y^2) \cdot \arctan \frac{y}{x} + xy$
   (d) $x\left(\sin \frac{y}{x}\right)\frac{dy}{dx} = y \sin \frac{y}{x} + x$
   (e) $xy' = y + 2xe^{-y/x}$
   (f) $(x - y)\,dx - (x + y)\,dy = 0$
   (g) $xy' = 2x - 6y$
   (h) $xy' = \sqrt{x^2 + y^2}$
   (i) $x^2y' = y^2 + 2xy$
   (j) $(x^3 + y^3)\,dx - xy^2\,dy = 0$

2. Use rectangular coordinates to find the orthogonal trajectories of the family of all circles tangent to the y-axis at the origin.

3. **(a)** If $ae \neq bd$ then show that h and k can be chosen so that the substitution $x = z - h$, $y = w - k$ reduces the equation

   $$\frac{dy}{dx} = F\left(\frac{ax + by + c}{dx + ey + f}\right)$$

   to a homogeneous equation.

   (b) If $ae = bd$ then show that there is a substitution that reduces the equation in **(a)** to one in which the variables are separable.

4. Solve each of the following differential equations.

   (a) $\frac{dy}{dx} = \frac{x + y + 4}{x - y - 6}$
   (b) $\frac{dy}{dx} = \frac{x + y + 4}{x + y - 6}$
   (c) $(2x - 2y)\,dx + (y - 1)\,dy = 0$
   (d) $\frac{dy}{dx} = \frac{x + y - 1}{x + 4y + 2}$
   (e) $(2x + 3y - 1)\,dx$
   $-4(x + 1)\,dy = 0$

5. By making the substitution $z = y/x^n$ (equivalently $y = zx^n$) and choosing a convenient value of n, show that the following differential equations can be transformed into equations with separable variables, and then solve them.

(a) $\dfrac{dy}{dx} = \dfrac{1 - xy^2}{2x^2 y}$

(b) $\dfrac{dy}{dx} = \dfrac{2 + 3xy^2}{4x^2 y}$

(c) $\dfrac{dy}{dx} = \dfrac{y - xy^2}{x + x^2 y}$

6. Show that a straight line through the origin intersects all integral curves of a homogeneous equation at the same angle.

7. Use your symbol manipulation software, such as `Maple` or `Mathematica`, to find solutions to each of the following homogeneous equations. (Note that these would be difficult to do by hand.)

(a) $y' = \sin[y/x] - \cos[y/x]$

(b) $e^{x/y}\, dx - \dfrac{y}{x}\, dy = 0$

(c) $\dfrac{dy}{dx} = \dfrac{x^2 - xy}{y^2 \cos(x/y)}$

(d) $y' = \dfrac{y}{x} \cdot \tan[y/x]$

2.6 Integrating Factors

We used a special type of integrating factor in Section 2.2 on first-order linear equations. At that time, we suggested that integrating factors may be applied in some generality to the solution of first-order differential equations. The trick is in *finding* the integrating factor.

In this section we shall discuss this matter in some detail, and indicate the uses and the limitations of the method of integrating factors.

First let us illustrate the concept of integrating factor by way of a concrete example. The differential equation

$$y\, dx + (x^2 y - x)\, dy = 0 \tag{2.15}$$

is plainly *not exact*, just because $\partial M/\partial y = 1$ while $\partial N/\partial x = 2xy - 1$, and these are unequal. However, if we multiply equation (15) through by a factor of $1/x^2$ then we obtain the equivalent equation

$$\frac{y}{x^2}\, dx + \left(y - \frac{1}{x}\right)\, dy = 0\,,$$

and this equation *is exact* (as the reader may easily verify by calculating $\partial M/\partial y$ and $\partial N/\partial x$). And of course we have a direct method (see Section 2.3) for solving such an exact equation.

We call the function $1/x^2$ in the last paragraph an *integrating factor.* It is obviously a matter of some interest to be able to find an integrating factor for any given first-order equation. So, given a differential equation

$$M(x,y)\,dx + N(x,y)\,dy = 0\,,$$

we wish to find a function $\mu(x,y)$ such that

$$\mu(x,y)\cdot M(x,y)\,dx + \mu(x,y)\cdot N(x,y)\,dy = 0$$

is exact. This entails

$$\frac{\partial(\mu\cdot M)}{\partial y} = \frac{\partial(\mu\cdot N)}{\partial x}\,.$$

Writing this condition out, we find that

$$\mu\frac{\partial M}{\partial y} + M\frac{\partial \mu}{\partial y} = \mu\frac{\partial N}{\partial x} + N\frac{\partial \mu}{\partial x}\,.$$

This last equation may be rewritten as

$$\frac{1}{\mu}\left(N\frac{\partial \mu}{\partial x} - M\frac{\partial \mu}{\partial y}\right) = \frac{\partial M}{\partial y} - \frac{\partial N}{\partial x}\,.$$

Now we use the method of wishful thinking: we suppose not only that an integrating factor μ exists, but in fact that one exists that only depends on the variable x (and not at all on y). Then the last equation reduces to

$$\frac{1}{\mu}\frac{d\mu}{dx} = \frac{\partial M/\partial y - \partial N/\partial x}{N}\,.$$

Notice that the left-hand side of this new equation is a function of x only. Hence so is the right-hand side. Call the right-hand side $g(x)$. Notice that g is something that we can always compute.

Thus

$$\frac{1}{\mu}\frac{d\mu}{dx} = g(x)$$

hence

$$\frac{d(\ln\mu)}{dx} = g(x)$$

or

$$\ln\mu = \int g(x)\,dx\,.$$

We conclude that, in case there is an integrating factor μ that depends on x only, then

$$\mu(x) = e^{\int g(x)\,dx}\,,$$

where

$$g(x) = \frac{\partial M/\partial y - \partial N/\partial x}{N}$$

can always be computed directly from the original differential equation.

Of course the best way to understand a new method like this is to look at some examples. This we now do.

EXAMPLE 2.6.1 Solve the differential equation

$$(xy - 1)\,dx + (x^2 - xy)\,dy = 0\,.$$

Solution: You may plainly check that this equation is not exact. It is also not separable. So we shall seek an integrating factor that depends only on x. Now

$$g(x) = \frac{\partial M/\partial y - \partial N/\partial x}{N} = \frac{[x] - [2x - y]}{x^2 - xy} = \frac{-(x - y)}{x(x - y)} = -\frac{1}{x}\,.$$

This g depends only on x, signaling that the methodology we just developed will actually work.

We set

$$\mu(x) = e^{\int g(x)\,dx} = e^{\int -1/x\,dx} = e^{-\ln x} = \frac{1}{x}\,.$$

This is our integrating factor. We multiply the original differential equation through by $1/x$ to obtain

$$\left(y - \frac{1}{x}\right)\,dx + (x - y)\,dy = 0\,.$$

The reader may check that *this* equation is certainly exact. We omit the details of solving this exact equation, since that technique was covered in Section 2.3. □

Of course the roles of y and x may be reversed in our reasoning for finding an integrating factor. In case the integrating factor μ depends only on y (and not at all on x) then we set

$$h(y) = -\frac{\partial M/\partial y - \partial N/\partial x}{M}$$

and define

$$\mu(y) = e^{\int h(y)\,dy}\,.$$

EXAMPLE 2.6.2 Solve the differential equation

$$y\,dx + (2x - ye^y)\,dy = 0\,.$$

Solution: First observe that the equation is not exact as it stands. Second,

$$\frac{\partial M/\partial y - \partial N/\partial x}{N} = \frac{-1}{2x - ye^y}$$

does *not* depend only on x. So instead we look at

$$-\frac{\partial M/\partial y - \partial N/\partial x}{M} = -\frac{-1}{y}\,,$$

and this expression depends only on y. So it will be our $h(y)$. We set

$$\mu(y) = e^{\int h(y)\,dy} = e^{\int 1/y\,dy} = e^{\ln y} = y\,.$$

Multiplying the differential equation through by $\mu(y) = y$, we obtain the new equation

$$y^2\,dx + (2xy - y^2 e^y)\,dy = 0\,.$$

You may easily check that this new equation is exact, and then solve it by the method of Section 2.3. □

We conclude this section by noting that the differential equation

$$xy^3\,dx + yx^2\,dy = 0$$

has the properties that

- It is not exact;
- $\dfrac{\partial M/\partial y - \partial N/\partial x}{N}$ does not depend on x only;
- $-\dfrac{\partial M/\partial y - \partial N/\partial x}{M}$ does not depend on y only.

Thus the method of the present section is not a panacea. We shall not always be able to find an integrating factor. Still, the technique has its uses.

Exercises

1. Solve each of the following differential equations by finding an integrating factor.

(a) $(3x^2 - y^2)\,dy - 2xy\,dx = 0$
(b) $(xy - 1)\,dx + (x^2 - xy)\,dy = 0$
(c) $x\,dy + y\,dx + 3x^3y^4\,dy = 0$
(d) $e^x\,dx + (e^x \cot y + 2y \csc y)\,dy = 0$
(e) $(x + 2)\sin y\,dx + x\cos y\,dy = 0$
(f) $y\,dx + (x - 2x^2y^3)\,dy = 0$
(g) $(x + 3y^2)\,dx + 2xy\,dy = 0$
(h) $y\,dx + (2x - ye^y)\,dy = 0$
(i) $(y\ln y - 2xy)\,dx + (x + y)\,dy = 0$
(j) $(y^2 + xy + 1)\,dx + (x^2 + xy + 1)\,dy = 0$
(k) $(x^3 + xy^3)\,dx + 3y^2\,dy = 0$

2. Show that if $(\partial M/\partial y - \partial N/\partial x)/(Ny - Mx)$ is a function $g(z)$ of the product $z = xy$, then
$$\mu = e^{\int g(z)\,dz}$$
is an integrating factor for the differential equation
$$M(x,y)\,dx + N(x,y)\,dy = 0\,.$$

3. Under what circumstances will the differential equation
$$M(x,y)\,dx + N(x,y)\,dy = 0$$
have an integrating factor that is a function of the sum $z = x + y$?

4. Solve the following differential equation by making the substitution $z = y/x^n$ (equivalently $y = xz^n$) and choosing a convenient value for n:
$$\frac{dy}{dx} = \frac{2y}{x} + \frac{x^3}{y} + x\tan\frac{y}{x^2}\,.$$

5. Use your symbol manipulation software, such as `Maple` or `Mathematica`, to write a routine for finding the integrating factor for a given differential equation.

2.7 Reduction of Order

Later in the book, we shall learn that virtually *any* ordinary differential equation can be transformed to a first-order *system* of equations. This is, in effect, just a notational trick, but it emphasizes the centrality of first-order equations and systems. In the present section, we shall learn how to reduce certain higher-order equations to first-order equations—ones which we can frequently solve.

We begin by concentrating on differential equations of order 2 (and thinking about how to reduce them to equations of order 1). In each differential equation in this section, x will be the independent variable and y the dependent variable. So a typical second-order equation will involve x, y, y', y''. The key to the success of each of the methods that we shall introduce in this section is that one variable must be missing from the equation.

2.7.1 Dependent Variable Missing

In case the variable y is missing from our differential equation, we make the substitution $y' = p$. This entails $y'' = p'$. Thus the differential equation is reduced to first order.

EXAMPLE 2.7.1 Solve the differential equation

$$xy'' - y' = 3x^2$$

using reduction of order.

Solution: Notice that the dependent variable y is missing from the differential equation. We set $y' = p$ and $y'' = p'$, so that the equation becomes

$$xp' - p = 3x^2 .$$

Observe that this new equation is first-order linear. We write it in standard form as

$$p' - \frac{1}{x}p = 3x .$$

We may solve this equation by using the integrating factor $\mu(x) = e^{\int -1/x\,dx} = 1/x$. Thus

$$\frac{1}{x}p' - \frac{1}{x^2}p = 3$$

so

$$\left(\frac{1}{x}p\right)' = 3$$

or

$$\int \left(\frac{1}{x}p\right)' dx = \int 3\,dx .$$

Performing the integrations, we conclude that

$$\frac{1}{x}p = 3x + C ,$$

hence

$$p(x) = 3x^2 + Cx .$$

Now we recall that $p = y'$, so we make that substitution. The result is

$$y' = 3x^2 + Cx ,$$

hence

$$y = x^3 + \frac{C}{2}x^2 + D = x^3 + Ex^2 + D .$$

We invite the reader to confirm that this is the complete and general solution to the original differential equation. □

EXAMPLE 2.7.2 Find the solution of the differential equation

$$[y']^2 = x^2y'' .$$

Solution: We note that y is missing, so we make the substitution $p = y'$, $p' = y''$. Thus the equation becomes

$$p^2 = x^2 p' .$$

This equation is amenable to separation of variables.

We begin by writing the equation as

$$p^2 = x^2 \frac{dp}{dx} .$$

Then

$$\frac{dx}{x^2} = \frac{dp}{p^2} ,$$

which integrates to

$$-\frac{1}{x} = -\frac{1}{p} + C$$

or

$$p = \frac{x}{1 + Cx}$$

for some unknown constant C. We resubstitute $p = y'$ and write the equation as

$$\frac{dy}{dx} = \frac{1/C + x}{1 + Cx} - \frac{1}{C} \cdot \frac{1}{1 + Cx} = \frac{1}{C} - \frac{1}{C} \cdot \frac{1}{1 + Cx} .$$

Now we integrate to obtain finally that

$$y(x) = \frac{x}{C} - \frac{1}{C^2} \ln(1 + Cx) + D$$

is the general solution of the original differential equation.

Note here that we have used our method to solve a nonlinear differential equation. □

2.7.2 Independent Variable Missing

In case the variable x is missing from our differential equation, we make the substitution $y' = p$. This time the corresponding substitution for y'' will be a bit different. To wit,

$$y'' = \frac{dp}{dx} = \frac{dp}{dy}\frac{dy}{dx} = \frac{dp}{dy} \cdot p .$$

This change of variable will reduce our differential equation to first order. In the reduced equation, we treat p as the dependent variable (or function) and y as the independent variable.

EXAMPLE 2.7.3 Solve the differential equation

$$y'' + k^2 y = 0$$

(where it is understood that k is an unknown real constant).

Solution: We notice that the independent variable x is missing. So we make the substitution

$$y' = p, \quad y'' = p \cdot \frac{dp}{dy}.$$

The equation then becomes

$$p \cdot \frac{dp}{dy} + k^2 y = 0.$$

In this new equation we can separate variables:

$$p\,dp = -k^2 y\,dy$$

hence

$$\frac{p^2}{2} = -k^2 \frac{y^2}{2} + C,$$

$$p = \pm\sqrt{D - k^2 y^2} = \pm k\sqrt{E - y^2}.$$

Now we resubstitute $p = dy/dx$ to obtain

$$\frac{dy}{dx} = \pm k\sqrt{E - y^2}.$$

We can separate variables to obtain

$$\frac{dy}{\sqrt{E - y^2}} = \pm k\,dx$$

hence (integrating)

$$\sin^{-1}\frac{y}{\sqrt{E}} = \pm kx + F$$

or

$$\frac{y}{\sqrt{E}} = \sin(\pm kx + F)$$

thus

$$y = \sqrt{E}\sin(\pm kx + F).$$

Now we apply the sum formula for sine to rewrite the last expression as

$$y = \sqrt{E}\cos F \sin(\pm kx) + \sqrt{E}\sin F \cos(\pm kx).$$

A moment's thought reveals that we may consolidate the constants and finally write our general solution of the differential equation as

$$y = A\sin(kx) + B\cos(kx).$$

□

We shall learn in the next chapter a different, and perhaps more expeditious, method of attacking examples of the last type. It should be noted quite plainly in the last example, and also in some of the earlier examples of the section, that the method of reduction of order basically transforms the problem of solving one second-order equation to a new problem of solving *two* first-order equations. Examine each of the examples we have presented and see whether you can say what the two new equations are.

In the next example, we shall solve a differential equation subject to an *initial condition.* This will be an important idea throughout the book. Solving a differential equation gives rise to a *family* of functions. Specifying an initial condition is a natural way to specialize down to a particular solution. In applications, these initial conditions will make good physical sense.

EXAMPLE 2.7.4 Use the method of reduction of order to solve the differential equation

$$y'' = y' \cdot e^y$$

with initial conditions $y(0) = 0$ and $y'(0) = 1$.

Solution: Noting that the independent variable x is missing, we make the substitution

$$y' = p, \quad y'' = p \cdot \frac{dp}{dy} .$$

So the equation becomes

$$p \cdot \frac{dp}{dy} = p \cdot e^y .$$

We of course may separate variables, so the equation becomes

$$dp = e^y \, dy .$$

This is easily integrated to give

$$p = e^y + C .$$

Now we resubstitute $p = y'$ to find that

$$y' = e^y + C$$

or

$$\frac{dy}{dx} = e^y + C .$$

Because of the initial conditions $[dy/dx](0) = 1$ and $y(0) = 0$, we may conclude right away

$$1 = e^0 + C$$

hence that $C = 0$. Thus our equation is

$$\frac{dy}{dx} = e^y$$

or

$$\frac{dy}{e^y} = dx .$$

This may be integrated to

$$-e^{-y} = x + D .$$

Of course we can rewrite the equation finally as

$$y = -\ln(-x + E) .$$

Since $y(0) = 0$, we conclude that

$$y(x) = -\ln(-x + 1)$$

is the solution of our initial value problem. □

Exercises

1. Solve the following differential equations using the method of reduction of order.

   (a) $yy'' + (y')^2 = 0$
   (b) $xy'' = y' + (y')^3$
   (c) $y'' - k^2y = 0$
   (d) $x^2y'' = 2xy' + (y')^2$
   (e) $2yy'' = 1 + (y')^2$
   (f) $yy'' - (y')^2 = 0$
   (g) $xy'' + y' = 4x$

2. Find the specified particular solution of each of the following equations.

   (a) $(x^2 + 2y')y'' + 2xy' = 0$, $y = 1$ and $y' = 0$ when $x = 0$
   (b) $yy'' = y^2y' + (y')^2$, $y = -1/2$ and $y' = 1$ when $x = 0$
   (c) $y'' = y'e^y$, $y = 0$ and $y' = 2$ when $x = 0$

3. Solve each of these differential equations using both methods of this section, and reconcile the results.

   (a) $y'' = 1 + (y')^2$ (b) $y'' + (y')^2 = 1$

4. Inside the Earth, the force of gravity is proportional to the distance from the center. A hole is drilled through the Earth from pole to pole and a rock is dropped into the hole. This rock will fall all the way through the hole, pause at the other end, and return to its starting point. How long will the complete round trip take?

3

Some Applications of the First-Order Theory

- Hanging chain and pursuit curves
- Electric circuits

3.1 The Hanging Chain and Pursuit Curves

3.1.1 The Hanging Chain

Imagine a flexible steel chain, attached firmly at equal height at both ends, hanging under its own weight (see Figure 3.1). What shape will it describe as it hangs?

This is a classical problem of mechanical engineering, and its analytical solution involves calculus, elementary physics, and differential equations. We describe it here.

We analyze a portion of the chain between points A and B, as shown in Figure 3.2, where A is the lowest point of the chain and $B = (x, y)$ is a variable point.

We let

- T_1 be the horizontal tension at A;
- T_2 be the component of tension *tangent* to the chain at B;
- w be the weight of the chain per unit of length.

Here T_1, T_2, and w are numbers. Figure 3.3 exhibits these quantities.

Notice that if s is the length of the chain between two given points, then sw is the downward force of gravity on this portion of the chain; this is indicated in the figure. We use the symbol θ to denote the angle that the tangent to the chain at B makes with the horizontal.

By Newton's first law, we may equate horizontal components of force to obtain

$$T_1 = T_2 \cos\theta \,. \tag{3.1}$$

Likewise, we equate vertical components of force to obtain

$$ws = T_2 \sin\theta \,. \tag{3.2}$$

DOI: 10.1201/9781003214526-3

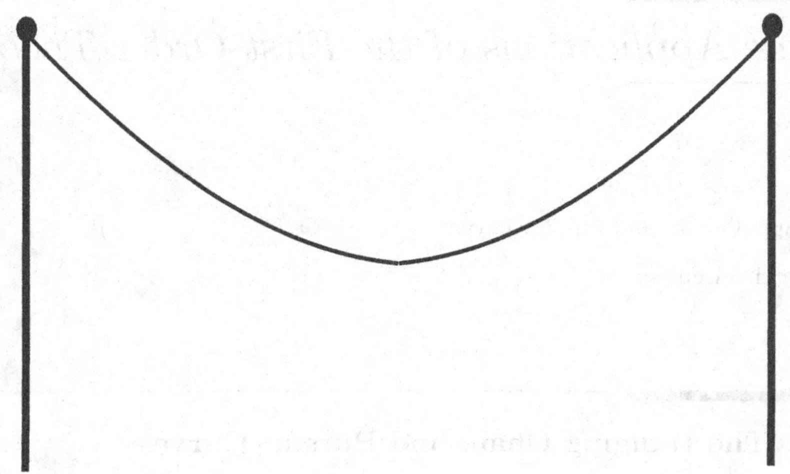

FIGURE 3.1
The hanging chain.

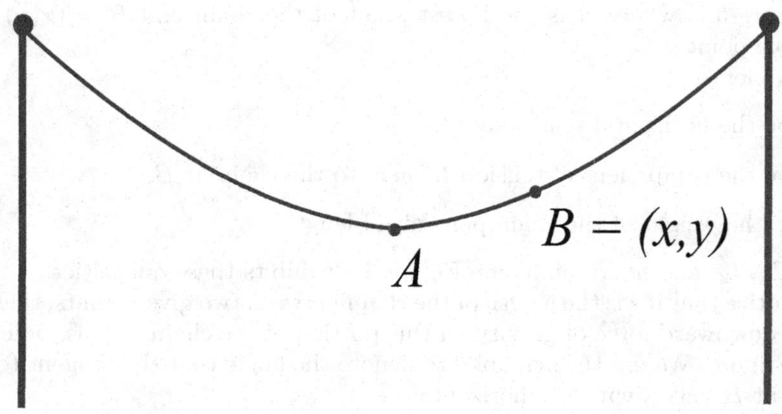

FIGURE 3.2
Analysis of the hanging chain.

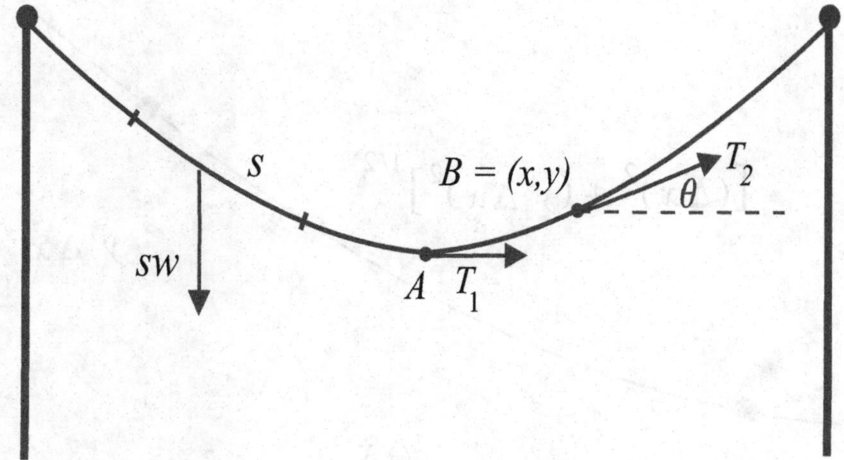

FIGURE 3.3
The quantities T_1 and T_2.

Dividing the right side of (17) by the right side of (16) and the left side of (17) by the left side of (16) and equating gives

$$\frac{ws}{T_1} = \tan\theta .$$

Think of the hanging chain as the graph of a function: y is a function of x. Then y' at B equals $\tan\theta$ so we may rewrite the last equation as

$$y' = \frac{ws}{T_1} .$$

We can simplify this equation by a change of notation: set $q = y'$. Then we have

$$q(x) = \frac{w}{T_1} s(x) . \tag{3.3}$$

If Δx is an increment of x, then $\Delta q = q(x+\Delta x) - q(x)$ is the corresponding increment of q and $\Delta s = s(x+\Delta x) - s(x)$ the increment in s. As Figure 3.4 indicates, Δs is well approximated by

$$\Delta s \approx \left((\Delta x)^2 + (y'\Delta x)^2\right)^{1/2} = \left(1 + (y')^2\right)^{1/2} \Delta x = (1+q^2)^{1/2} \Delta x .$$

Thus, from (3.3), we have

$$\Delta q = \frac{w}{T_1} \Delta s \approx \frac{w}{T_1} (1+q^2)^{1/2} \Delta x .$$

Dividing by Δx and letting Δx tend to zero gives the equation

$$\frac{dq}{dx} = \frac{w}{T_1} (1+q^2)^{1/2} . \tag{3.4}$$

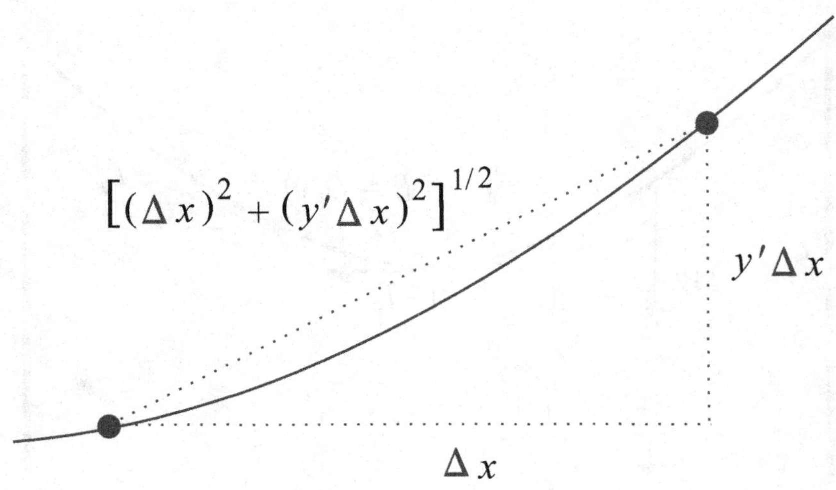

FIGURE 3.4
An increment of the chain.

This may be rewritten as

$$\int \frac{dq}{(1+q^2)^{1/2}} = \frac{w}{T_1} \int dx .$$

It is trivial to perform the integration on the right side of the equation, and a little extra effort enables us to integrate the left side (use the substitution $u = \tan\psi$, or else use inverse hyperbolic trigonometric functions). Thus we obtain

$$\sinh^{-1} q = \frac{w}{T_1} x + C .$$

We know that the chain has a horizontal tangent when $x = 0$ (this corresponds to the point A—Figures 3.2 and 3.3). Thus $q(0) = y'(0) = 0$. Substituting this into the last equation gives $C = 0$. Thus our solution is

$$\sinh^{-1} q(x) = \frac{w}{T_1} x$$

or

$$q(x) = \sinh\left(\frac{w}{T_1} x\right)$$

or

$$\frac{dy}{dx} = \sinh\left(\frac{w}{T_1} x\right) .$$

Finally, we integrate this last equation to obtain

$$y(x) = \frac{T_1}{w} \cosh\left(\frac{w}{T_1} x\right) + D ,$$

where D is a constant of integration. The constant D can be determined from the height h_0 of the point A from the x-axis:

$$h_0 = y(0) = \frac{T_1}{w}\cosh(0) + D$$

hence

$$D = h_0 - \frac{T_1}{w}\,.$$

Our hanging chain is completely described by the equation

$$y(x) = \frac{T_1}{w}\cosh\left(\frac{w}{T_1}x\right) + \left(h_0 - \frac{T_1}{w}\right).$$

This curve is called a *catenary*, from the Latin word for chain (*catena*). Catenaries arise in a number of other physical problems. The St. Louis arch is in the shape of a catenary.

Math Nugget

Many of the special curves of classical mathematics arise in problems of mechanics. The tautochrone property of the cycloid curve (that a bead sliding down the curve will reach the bottom in the same time, no matter where on the curve it begins) was discovered by the great Dutch scientist Christiaan Huygens (1629–1695). He published it in 1673 in his treatise on the theory of pendulum clocks, and it was well-known to all European mathematicians at the end of the seventeenth century. When Johann Bernoulli published his discovery of the brachistochrone (that special curve connecting two points down which a bead will slide in the least possible time) in 1696, he expressed himself in the following exuberant language (of course, as was the custom of the time, he wrote in Latin): "With justice we admire Huygens because he first discovered that a heavy particle falls down along a common cycloid in the same time no matter from what point on the cycloid it begins its motion. But you will be petrified with astonishment when I say that precisely this cycloid, the tautochrone of Huygens, is our required brachistochrone." These curves are discussed in greater detail in Section 8.3 below.

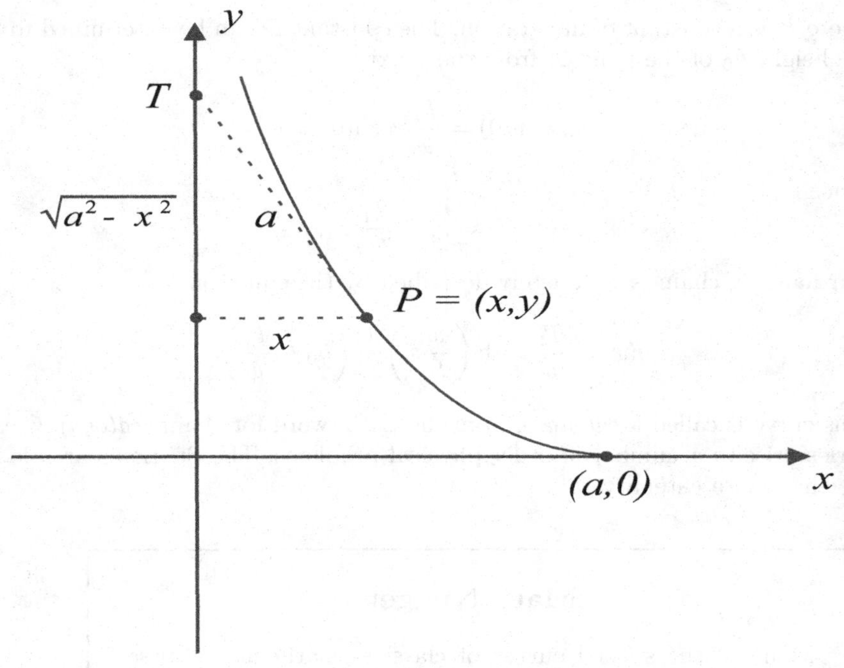

FIGURE 3.5
A tractrix.

3.1.2 Pursuit Curves

A submarine speeds across the ocean bottom in a particular path, and a destroyer at a remote location decides to engage in pursuit. What path does the destroyer follow? Problems of this type are of interest in a variety of applications. We examine a few examples. The first one is purely mathematical, and devoid of "real-world" trappings.

EXAMPLE 3.1.1 A point P is dragged along the x-y plane by a string PT of fixed length a. If T begins at the origin and moves along the positive y-axis, and if P starts at the point $(a, 0)$, then what is the path of P?

Solution: The curve described by the motion of P is called, in the classical literature, a *tractrix* (from the Latin *tractum*, meaning "drag"). Figure 3.5 exhibits the salient features of the problem.

Observe that we can calculate the slope of the pursuit curve at the point P in two ways: **(i)** as the derivative of y with respect to x and **(ii)** as the ratio of sides of the relevant triangle. This leads to the equation

$$\frac{dy}{dx} = -\frac{\sqrt{a^2 - x^2}}{x} .$$

This is a separable, first-order differential equation. We write

$$\int dy = -\int \frac{\sqrt{a^2 - x^2}}{x}\, dx\,.$$

Performing the integrations (the right-hand side requires the trigonometric substitution $x = \sin\psi$), we find that

$$y = a \ln\left(\frac{a + \sqrt{a^2 - x^2}}{x}\right) - \sqrt{a^2 - x^2}$$

is the equation of the tractrix.[1] □

EXAMPLE 3.1.2 A rabbit begins at the origin and runs up the y-axis with speed a feet per second. At the same time, a dog runs at speed b from the point $(c, 0)$ in pursuit of the rabbit. What is the path of the dog?

Solution: At time t, measured from the instant both the rabbit and the dog start, the rabbit will be at the point $R = (0, at)$ and the dog at $D = (x, y)$. We wish to solve for y as a function of x. Refer to Figure 3.6.

The premise of a pursuit analysis is that the line through D and R is tangent to the path—that is, the dog will always run straight at the rabbit. This immediately gives the differential equation

$$\frac{dy}{dx} = \frac{y - at}{x}\,.$$

This equation is a bit unusual for us, since x and y are both unknown functions of t. First, we rewrite the equation as

$$xy' - y = -at\,.$$

(Here the $'$ on y stands for differentiation in x.)

We differentiate this equation with respect to x, which gives

$$xy'' = -a\frac{dt}{dx}\,.$$

Since s is arc length along the path of the dog, it follows that $ds/dt = b$. Hence

$$\frac{dt}{dx} = \frac{dt}{ds} \cdot \frac{ds}{dx} = -\frac{1}{b} \cdot \sqrt{1 + (y')^2}\,;$$

[1] This curve is of considerable interest in other parts of mathematics. If it is rotated about the y-axis, then the result is a surface that gives a model for non-Euclidean geometry. The surface is called a *pseudosphere* in differential geometry. It is a surface of constant negative curvature (as opposed to a traditional sphere, which is a surface of constant positive curvature).

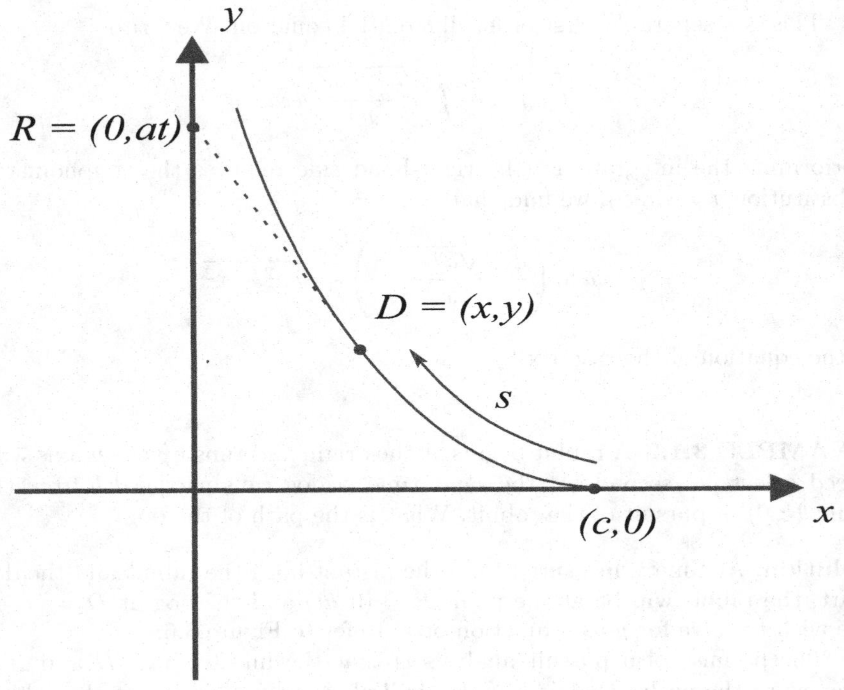

FIGURE 3.6
A pursuit curve.

here the minus sign appears because s decreases when x increases (see Figure 3.6). Of course we use the familiar expression for the derivative of arc length.

Combining the last two displayed equations gives

$$xy'' = \frac{a}{b}\sqrt{1+(y')^2}\,.$$

For convenience, we set $k = a/b$, $y' = p$, and $y'' = dp/dx$ (the latter two substitutions being one of our standard reduction of order techniques). Thus we have

$$\frac{dp}{\sqrt{1+p^2}} = k\frac{dx}{x}\,.$$

Now we may integrate, using the condition $p = 0$ when $x = c$. The result is

$$\ln\left(p+\sqrt{1+p^2}\right) = \ln\left(\frac{x}{c}\right)^k\,.$$

When we solve for p, we find that

$$\frac{dy}{dx} = p = \frac{1}{2}\left\{\left(\frac{x}{c}\right)^k - \left(\frac{c}{x}\right)^k\right\}.$$

In order to continue the analysis, we need to know something about the relative sizes of a and b. Suppose, for example, that $a < b$ (so $k < 1$), meaning that the dog will certainly catch the rabbit. Then we can integrate the last equation to obtain

$$y(x) = \frac{1}{2}\left\{\frac{c}{k+1}\left(\frac{x}{c}\right)^{k+1} - \frac{c}{(1-k)}\left(\frac{x}{c}\right)^{1-k}\right\} + D\,.$$

Since $y = 0$ when $x = c$, we find that $D = ck/(1-k^2)$. Of course the dog catches the rabbit when $x = 0$. Since both exponents on x are positive, we can set $x = 0$ and solve for y to obtain $y = ck$ as the point at which the dog and the rabbit meet. □

We invite the reader to consider what happens when $a = b$ and hence $k = 1$.

Exercises

1. Refer to our discussion of the shape of a hanging chain. Show that the tension T at an arbitrary point (x, y) on the chain is given by wy.
2. If the hanging chain supports a load of horizontal density $L(x)$, then what differential equation should be used in place of (34)?
3. What is the shape of a cable of negligible density (so that $w \equiv 0$) that supports a bridge of constant horizontal density given by $L(x) \equiv L_0$?
4. If the length of any small portion of an elastic cable of uniform density is proportional to the tension in it, then show that it assumes the shape of a parabola when hanging under its own weight.
5. A curtain is made by hanging thin rods from a cord of negligible density. If the rods are close together and equally spaced horizontally, and if the bottom of the curtain is trimmed so that it is horizontal, then what is the shape of the cord?
6. What curve lying above the x-axis has the property that the length of the arc joining any two points on it is proportional to the area under that arc?
7. Show that the tractrix discussed in Example 1.11.1 is orthogonal to the lower half of each circle with radius a and center on the positive y-axis.
8. **(a)** In Example 1.11.2, assume that $a < b$ (so that $k < 1$) and find y as a function of x. How far does the rabbit run before the dog catches him?

   (b) Assume now that $a = b$, and find y as a function of x. How close does the dog come to the rabbit?

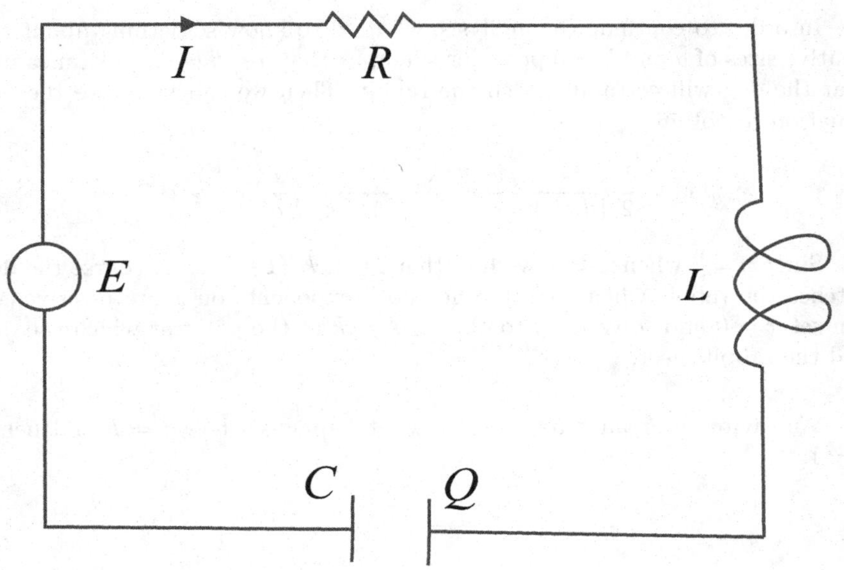

FIGURE 3.7
A simple electric circuit.

3.2 Electrical Circuits

We have alluded elsewhere in the book to the fact that our analyses of vibrating springs and other mechanical phenomena are analogous to the situation for electrical circuits. Now we shall examine this matter in some detail.

We consider the flow of electricity in the simple electrical circuit exhibited in Figure 3.7. The elements that we wish to note are these:

A. A source of electromotive force (emf) E—perhaps a battery or generator—which drives an electric charge and produces a current I. Depending on the nature of the source, E may be a constant or a function of time.

B. A resistor of resistance R, which opposes the current by producing a drop in emf of magnitude

$$E_R = RI .$$

This equation is called *Ohm's law.*

Math Nugget

Georg Simon Ohm (1787–1854) was a German physicist whose only significant contribution to science was his discovery of the law that now bears his name. When he announced it in 1827, it seemed too good to be true; sadly, it was not generally believed. Ohm was, as a consequence, deemed to be unreliable. He was subsequently so badly treated that he resigned his professorship at Cologne and lived for several years in obscurity and poverty. Ultimately, it was recognized that Ohm was right all along. So Ohm was vindicated. One of Ohm's students in Cologne was Peter Dirichlet, who later became one of the most distinguished German mathematicians of the nineteenth century.

C. An inductor of inductance L, which opposes any change in the current by producing a drop in emf of magnitude

$$E_L = L \cdot \frac{dI}{dt} .$$

D. A capacitor (or condenser) of capacitance C, which stores the charge Q. The charge accumulated by the capacitor resists the inflow of additional charge, and the drop in emf arising in this way is

$$E_C = \frac{1}{C} \cdot Q .$$

Furthermore, since the current is the rate of flow of charge, and hence the rate at which charge builds up on the capacitor, we have

$$I = \frac{dQ}{dt} .$$

Those unfamiliar with the theory of electricity may find it helpful to draw an analogy here between the current I and the rate of flow of water in a pipe. The electromotive force E plays the role of a pump producing pressure (voltage) that causes the water to flow. The resistance R is analogous to friction in the pipe—which opposes the flow by producing a drop in the pressure. The inductance L is a sort of inertia that opposes any change in flow by producing a drop in pressure if the flow is increasing and an increase in pressure if the flow is decreasing. To understand this last point, think of a cylindrical water storage tank that the liquid enters through a hole in the bottom. The deeper the water in the tank (Q), the harder it is to pump new water in; and the larger the base of the tank (C) for a given quantity of stored water, the shallower is the water in the tank and the easier to pump in new water.

These four circuit elements act together according to *Kirchhoff's Law*, which states that the algebraic sum of the electromotive forces around a closed circuit is zero. This physical principle yields

$$E - E_R - E_L - E_C = 0$$

or

$$E - RI - L\frac{dI}{dt} - \frac{1}{C}Q = 0\,,$$

which we rewrite in the form

$$L\frac{dI}{dt} + RI + \frac{1}{C}Q = E\,. \tag{3.5}$$

We may perform our analysis by regarding either the current I or the charge Q as the dependent variable (obviously time t will be the independent variable).

- In the first instance, we shall eliminate the variable Q from (20) by differentiating the equation with respect to t and replacing dQ/dt by I (since current is indeed the rate of change of charge). The result is

$$L\frac{d^2I}{dt^2} + R\frac{dI}{dt} + \frac{1}{C}I = \frac{dE}{dt}\,.$$

- In the second instance, we shall eliminate the I by replacing it by dQ/dt. The result is

$$L\frac{d^2Q}{dt^2} + R\frac{dQ}{dt} + \frac{1}{C}Q = E\,. \tag{3.6}$$

Both these ordinary differential equations are second-order, linear with constant coefficients. We shall study these in detail in Section 4.1. For now, in order to use the techniques we have already learned, we assume that our system has no capacitor present. Then the equation becomes

$$L\frac{dI}{dt} + RI = E\,. \tag{3.7}$$

EXAMPLE 3.2.1 Solve equation (22) when an initial current I_0 is flowing and a constant emf E_0 is impressed on the circuit at time $t = 0$.

Solution: For $t \geq 0$ our equation is

$$L\frac{dI}{dt} + RI = E_0\,.$$

We can separate variables to obtain

$$\frac{dI}{E_0 - RI} = \frac{1}{L}\,dt\,.$$

We integrate and use the initial condition $I(0) = I_0$ to obtain

$$\ln(E_0 - RI) = -\frac{R}{L}t + \ln(E_0 - RI_0)\,,$$

hence

$$I = \frac{E_0}{R} + \left(I_0 - \frac{E_0}{R}\right)e^{-Rt/L}.$$

We have learned that the current I consists of a *steady-state* component E_0/R and a *transient component* $(I_0 - E_0/R)e^{-Rt/L}$ that approaches zero as $t \to +\infty$. Consequently, Ohm's law $E_0 = RI$ is nearly true for t large. We also note that, if $I_0 = 0$, then

$$I = \frac{E_0}{R}(1 - e^{-Rt/L})\,;$$

if instead $E_0 = 0$, then $I = I_0 e^{-Rt/L}$. □

Exercises

1. In Example 1.12, with $I_0 = 0$ and $E_0 \neq 0$, show that the current in the circuit builds up to half its theoretical maximum in $(L \ln 2)/R$ seconds.
2. Solve equation (37) for the case in which the circuit has an initial current I_0 and the emf impressed at time $t = 0$ is given by
   (a) $E = E_0 e^{-kt}$
   (b) $E = E_0 \sin \omega t$
3. Consider a circuit described by equation (37) and show that
   (a) Ohm's law is satisfied whenever the current is at a maximum or minimum.
   (b) The emf is increasing when the current is at a minimum and decreasing when it is at a maximum.
4. If $L = 0$ in equation (21) and if $Q = 0$ when $t = 0$, then find the charge buildup $Q = Q(t)$ on the capacitor in each of the following cases:
   (a) E is a constant E_0
   (b) $E = E_0 e^{-t}$
   (c) $E = E_0 \cos \omega t$
5. Use equation (35) with $R = 0$ and $E = 0$ to find $Q = Q(t)$ and $I = I(t)$ for the discharge of a capacitor through an inductor of inductance L, with initial conditions $Q = Q_0$ and $I = 0$ when $t = 0$.

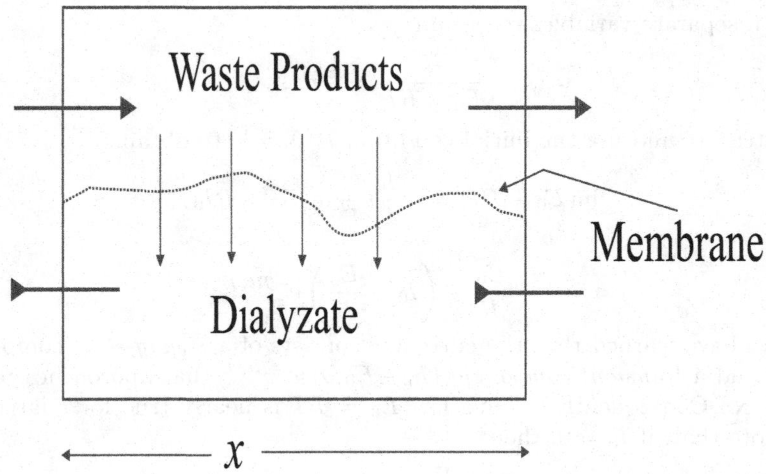

FIGURE 3.8
A dialysis machine.

Anatomy of an Application

The Design of a Dialysis Machine

The purpose of the kidneys is to filter out waste from the blood. When the kidneys malfunction, the waste material can build up to dangerous levels and be poisonous to the system. Doctors will use a kidney dialysis machine (or *dialyzer*) to assist the kidneys in the cleansing process.

How does the dialyzer work? Blood flows from the patient's body into the machine. There is a cleansing fluid, called the dialyzate, that flows through the machine in the opposite direction to the blood. The blood and the dialyzate are separated by a semi-permeable membrane. See Figure 3.8.

The membrane in the dialyzer has minute pores, which will not allow the passage of blood but *will* allow the passage of the waste matter (which has much smaller molecules). The design of the dialysis machine concerns the flow rate of the waste material through the membrane. That flow is determined by the differences in concentration (of the waste material) on either side of the membrane. Of course the flow is from high concentration to low concentration. Refer again to the figure.

Certainly the physician (and the patient) care about the rate at which the waste material is removed. This rate will depend on **(i)** the flow rate of blood through the dialyzer, **(ii)** the flow rate of dialyzate through the dialyzer, **(iii)** the capacity of the dialyzer, and **(iv)** the permeability of the membrane.

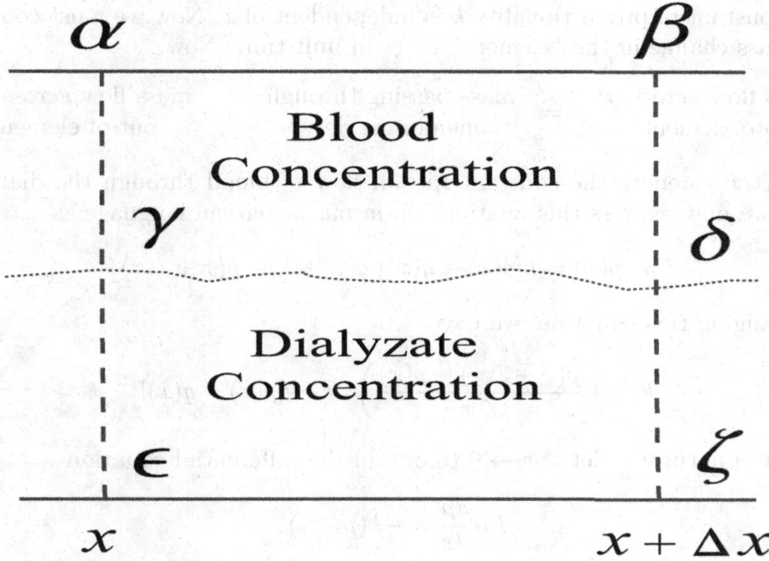

FIGURE 3.9
Cross section of the flow.

It is convenient (and plausible) for our analysis here to take the capacity of the dialyzer and the permeability of the membrane to be fixed constants. Our analysis will focus on the dependence of the removal rate (of the waste) on the flow rate. We let x denote the horizontal position in the dialyzer (Figure 3.8). Our analysis centers on a small cross section of the dialyzer from position x to position to $x + \Delta x$. We refer to the cross section of the total flow pictured in Figure 3.9 as an "element" of the flow.

Clearly, from everything we have said so far, the most important variables for our analysis are the concentration of waste in the blood (call this quantity $p(x)$) and the concentration of waste in the dialyzate (call this quantity $q(x)$). There is in fact a standard physical law governing the passage of waste material through the membrane. This is *Fick's Law.* The enunciation is

> The amount of material passing through the membrane is proportional to the difference in concentration.

Let us examine Figure 3.9 in order to understand the movement of concentration of waste. The difference in concentration across $\alpha\epsilon$ (as one moves from the upper half of the figure to the lower half) is $p(x) - q(x)$; therefore the transfer of waste mass through a section of the membrane of width 1 and length Δx from blood solution to dialyzate solution in unit time is approximately

$$k[p(x) - q(x)] \cdot \Delta x\,.$$

The constant of proportionality k is independent of x. Now we must consider the mass change in the "element" $\alpha\beta\zeta\epsilon$ in unit time. Now

$$\begin{matrix}\text{mass flow across } \alpha\gamma \\ \text{into element}\end{matrix} = \begin{matrix}\text{mass passing through} \\ \text{membrane } \gamma\delta\end{matrix} + \begin{matrix}\text{mass flow across } \beta\delta \\ \text{out of element.}\end{matrix}$$

Let F_B denote the constant rate of flow of blood through the dialyzer. Then we may express this relationship in mathematical language as

$$F_B \cdot p(x) = k[p(x) - q(x)] \triangle x + F_B \cdot p(x + \triangle x) .$$

Rearranging this equation, we have

$$F_B \cdot \left(\frac{p(x + \triangle x) - p(x)}{\triangle x} \right) = -k[p(x) - q(x)] .$$

Now it is natural to let $\triangle x \to 0$ to obtain the differential equation

$$F_B \frac{dp}{dx} = -k(p - q) . \tag{3.8}$$

This last analysis, which led to equation (23), was based on an examination of the flow of the blood. We may perform a similar study of the flow of the dialyzate to obtain

$$-F_D \frac{dq}{dx} = k(p - q) \tag{3.9}$$

(note that the presence of the minus sign comes from the fact that the blood flows in the opposite direction as the dialyzate). Of course F_D is the flow rate of dialyzate through the machine.

Now we add equations (23) and (24) to obtain

$$\frac{dp}{dx} - \frac{dq}{dz} = -\frac{k}{F_B}(p - q) + \frac{k}{F_D}(p - q) .$$

Notice that p and q occur in this equation in an antisymmetric manner. Thus it is advantageous to make the substitution $r = p - q$. We finally obtain

$$\frac{dr}{dx} = -\alpha r , \tag{3.10}$$

where $\alpha = k/F_B - k/F_D$.

This equation is easily solved with separation of variables. The result is

$$r(x) = Ae^{-\alpha x} , \tag{3.11}$$

where A is an arbitrary constant. Of course we wish to relate this solution to p and to q. Look again at equation (23). We see that

$$\frac{dp}{dx} = -\frac{k}{F_B} r = -\frac{k}{F_B} Ae^{-\alpha x} .$$

Integration yields

$$p = B + \frac{kA}{\alpha F_B} e^{-\alpha x}, \tag{3.12}$$

where B is an arbitrary constant. We now combine (26) and (27), recalling that $r = p - q = Ae^{-\alpha x}$, to obtain that

$$q = B + \frac{kA}{\alpha F_D} e^{-\alpha x}.$$

Finally, we must consider the initial conditions in the problem. We suppose that the blood has initial waste concentration p_0 and the dialyzate has initial waste concentration 0. Thus

$$\begin{aligned} p &= p_0 \quad \text{at } x = 0 \\ q &= 0 \quad \text{at } x = L. \end{aligned}$$

Here L is the length of the dialyzer machine. Some tedious algebra, applied in a by-now-familiar manner, finally tells us that

$$\begin{aligned} p(x) &= p_0 \left(\frac{(e^{-\alpha L}/F_D) - (e^{-\alpha x}/F_B)}{(e^{-\alpha L}/F_D) - (1/F_B)} \right) \\ q(x) &= \frac{p_0}{F_D} \left(\frac{e^{-\alpha L} - e^{-\alpha x}}{(e^{-\alpha L}/F_D) - (1/F_D)} \right). \end{aligned}$$

These two equations represent a definitive analysis of the concentrations of waste in the blood and in the dialyzate. To interpret these, we observe that the amount of waste removed from the blood in unit time is

$$\begin{aligned} \int_0^L k[p(x) - q(x)]\, dx &= -F_B \int_0^L \frac{dp}{dx}\, dx \qquad \text{(by (23))} \\ &= -F_B \int_{p_0}^{p(L)} dp \\ &= -F_B[p_0 - p(L)]. \end{aligned}$$

Those who design dialyzers focus their attention on the "clearance" Cl, which is defined to be

$$Cl = \frac{F_B}{p_0}[p_0 - p(L)].$$

Our equations for p and q yield, after some calculation, that

$$Cl = F_B \left(\frac{1 - e^{-\alpha L}}{1 - (F_B/F_D)e^{-\alpha L}} \right).$$

Here

$$\alpha L = \frac{kL}{F_B}(1 - F_B/F_D).$$

The actual design of a dialyzer would entail testing these theoretical results against experimental data, taking into account factors like the variation of k with x, the depth of the channels, and variations in the membrane.

Problems for Review and Discovery

A. Drill Exercises

1. Find the general solution to each of the following differential equations.

(a) $xy' + y = x$
(b) $x^2y' + y = x^2$
(c) $x^2 \dfrac{dy}{dx} = y$
(d) $\sec x \cdot \dfrac{dy}{dx} = \sec y$
(e) $\dfrac{dy}{dx} = \dfrac{x^2 + y^2}{x^2 - y^2}$
(f) $\dfrac{dy}{dx} = \dfrac{x + 2y}{2x - y}$
(g) $2xy\, dx + x^2\, dy = 0$
(h) $-\sin x \sin y\, dx + \cos x \cos y\, dy = 0$

2. Solve each of the following initial value problems.

(a) $xy' - y = 2x\,, \quad y(1) = 1$
(b) $x^2y' - 2y = 3x^2\,, \quad y(1) = 2$
(c) $y^2 \dfrac{dy}{dx} = x\,, \quad y(-1) = 3$
(d) $\csc x \cdot \dfrac{dy}{dx} = \csc y\,, \quad y(\pi/2) = 1$
(e) $\dfrac{dy}{dx} = \dfrac{x + y}{x - y}\,, \quad y(1) = 1$
(f) $\dfrac{dy}{dx} = \dfrac{x^2 + 2y^2}{x^2 - 2y^2}\,, \quad y(0) = 1$
(g) $2x \cos y\, dx - x^2 \sin y\, dy = 0\,, \quad y(1) = 1$
(h) $\dfrac{1}{y}\, dx - \dfrac{x}{y^2}\, dy = 0\,, \quad y(0) = 2$

3. Find the orthogonal trajectories to the family of curves $y = c(x^2 + 1)$.

4. Use the method of reduction of order to solve each of the following differential equations.

(a) $y \cdot y'' - (y')^2 = 0$
(b) $xy'' = y' - 2(y')^3$
(c) $yy'' + y' = 0$
(d) $xy'' - 3y' = 5x$

B. Challenge Problems

1. A tank contains 50 gallons of brine in which 25 pounds of salt are dissolved. Beginning at time $t - 0$, water runs into this tank at the rate of 2 gallons per minute; the mixture flows out at the same rate through a second tank initially containing 50 gallons of pure water. When will the second tank contain the greatest amount of salt?

2. A natural extension of the first-order linear equation

$$y' = p(x) + q(x)y$$

is the *Riccati equation*

$$y' = p(x) + q(x)y + r(x)y^2 .$$

In general, this equation cannot be explicitly solved by elementary methods. However, if a particular solution $y_1(x)$ is known, then the general solution has the form

$$y(x) = y_1(x) + z(x) ,$$

where $z(x)$ is the general solution of the associated Bernoulli equation

$$z' - (q + 2ry_1)z = rz^2 .$$

Prove this assertion, and use this set of techniques to find the general solution of the equation

$$y' = \frac{y}{x} + x^3y^2 - x^5 . \qquad (*)$$

(*Hint:* The equation $(*)$ has $y_1(x) = x$ as a particular solution.)

3. The propagation of a single act in a large population (for instance, buying a Lexus rather than a Cadillac) often depends partly on external circumstances (e.g., price, quality, and frequency-of-repair records) and partly on a human tendency to imitate other people who have already performed the same act. In this case, the rate of increase of the proportion $y(t)$ of people who have performed the act can be expressed by the formula

$$\frac{dy}{dt} = (1 - y)[s(t) + Iy] , \qquad (**)$$

where $s(t)$ measures the external stimulus and I is a constant called the *imitation coefficient.*

(a) Notice that $(**)$ is a Riccati equation (see the last exercise) and that $y \equiv 1$ is a particular solution. Use the result of the last exercise to find the Bernoulli equation satisfied by $z(t)$.

(b) Find $y(t)$ for the case in which the external stimulus increases steadily with time so that $s(t) = at$ for a positive constant a. Leave your answer in the form of an integral.

4. If Riccati equation from Exercise 2 above has a known solution $y_1(x)$, then show that the general solution has the form of the one-parameter family of curves

$$y = \frac{cf(x) + g(x)}{cF(x) + G(x)} .$$

Show, conversely, that the differential equation of *any* one-parameter family of this form is a Riccati equation.

5. It begins to snow at a steady rate sometime in the morning. A snow plow begins plowing at a steady rate at noon. The plow clears twice as much area in the first hour as it does in the second hour. When did it start snowing?

C. Problems for Discussion and Exploration

1. A rabbit starts at the origin and runs up the right branch of the parabola $y = x^2$ with speed a. At the same time a dog, running with speed b, starts at the point $(c, 0)$ and pursues the rabbit. Write a differential equation for the path of the dog.

2. Consider the initial value problem

$$\frac{dy}{dx} = \frac{\sin(xy)}{1 + x^2 + y^2} \, .$$

This equation cannot be solved by any of the methods presented in this chapter. However, we can obtain some information about solutions by using other methods.

(a) On a large sheet of graph paper draw arrows indicating the direction of the curve at a large number of points. For example, at the point $(1, 1)$, the equation tells us that

$$\frac{dy}{dx} = \frac{\sin 1}{3} \, .$$

Draw a little arrow with base at the point $(1, 1)$ indicating that the curve is moving in the indicated direction.

Do the same at many other points. Connect these arrows with "flow lines." (There will be many distinct flow lines, corresponding to different initial conditions.) Thus you obtain a family of curves, representing the different solutions of the differential equation.

(b) What can you say about the nature of the flow lines that you obtained in part **(a)**? Are they curves that you can recognize? Are they polynomial curves? Exponential curves?

(c) What does your answer to part **(b)** tell you about this problem?

3. Suppose that the function $F(x, y)$ is continuously differentiable (i.e., continuous with continuous first derivatives). Show that the initial value problem

$$\frac{dy}{dx} = F(x, y) \, , \quad y(0) = y_0$$

has at most one solution in a neighborhood of the origin.

4

Second-Order Linear Equations

- Second-order linear equations
- The nature of solutions of second-order linear equations
- General solutions
- Undetermined coefficients
- Variation of parameters
- Use of a known solution to find another

4.1 Second-Order Linear Equations with Constant Coefficients

Second-order linear equations are important because (considering Newton's second law) they arise frequently in engineering and physics. For instance, acceleration is given by the second derivative, and force is mass times acceleration.

In this section we learn about *second-order linear equations with constant coefficients.* The "linear" attribute means, just as it did in the first-order situation, that the unknown function and its derivatives are not multiplied together, are not raised to powers, and are not the arguments of other functions. So, for example,

$$y'' - 3y' + 6y = 0$$

is second-order linear while

$$\sin(y'') - y' + 5y = 0$$

and

$$y \cdot y'' + 4y' + 3y = 0$$

are not.

DOI: 10.1201/9781003214526-4

The "constant coefficient" attribute means that the coefficients in the equation are not functions—they are constants. Thus a second-order linear equation with constant coefficient will have the form

$$ay'' + by' + cy = d\,, \tag{4.1}$$

where a, b, c, and d are constants.

We in fact begin with the *homogeneous case*; this is the situation in which $d = 0$ (recall that in Section 2.5 we had a different meaning for the word "homogeneous"). We solve equation (1) by a process of organized guessing: any solution of (1) will be a function that cancels with its derivatives. Thus it is a function that is similar in form to its derivatives. Certainly exponentials fit this description. Thus we guess a solution of the form

$$y = e^{rx}\,.$$

Plugging this guess into (1) gives

$$a\left(e^{rx}\right)'' + b\left(e^{rx}\right)' + c\left(e^{rx}\right) = 0\,.$$

Calculating the derivatives, we find that

$$ar^2e^{rx} + bre^{rx} + ce^{rx} = 0$$

or

$$[ar^2 + br + c] \cdot e^{rx} = 0\,.$$

Of course the exponential never vanishes. So this last equation can only be true (for all x) if

$$ar^2 + br + c = 0\,.$$

Of course this is just a quadratic equation (called the *associated polynomial equation*),[1] and we may solve it using the quadratic formula. This process will lead to our solution set. As usual, we master these new ideas by way of some examples.

EXAMPLE 4.1.1 Solve the differential equation

$$y'' - 5y' + 4y = 0\,.$$

Solution: Following the paradigm just outlined, we guess a solution of the form $y = e^{rx}$. This leads to the quadratic equation for r given by

$$r^2 - 5r + 4 = 0\,.$$

---

[1]Some texts will call this the *characteristic polynomial*, although that terminology has other meanings in mathematics.

Of course this factors directly to

$$(r-1)(r-4)=0,$$

so $r = 1, 4$.

Thus e^x and e^{4x} are solutions to the differential equation (you should check this assertion for yourself). A *general solution* is given by

$$y = A \cdot e^x + B \cdot e^{4x}, \tag{4.2}$$

where A and B are arbitrary constants. The reader may check that any function of the form (2) solves the original differential equation. Observe that our general solution (2) has two undetermined constants, which is consistent with the fact that we are solving a second-order differential equation. □

Remark 4.1.2 Again we see that the solving of a differential equation leads to a family of solutions. In this last example, that family is indexed by two parameters A and B. As we shall see below (especially Section 5.1), a typical physical problem will give rise to two initial conditions that determine those parameters. The Picard Existence and Uniqueness Theorem (Section 13.1) gives the mathematical underpinning for these ideas.

EXAMPLE 4.1.3 Solve the differential equation

$$2y'' + 6y' + 2y = 0.$$

Solution: The associated polynomial equation is

$$2r^2 + 6r + 2 = 0.$$

This equation does not factor in any obvious way, so we use the quadratic formula:

$$r = \frac{-6 \pm \sqrt{6^2 - 4 \cdot 2 \cdot 2}}{2 \cdot 2} = \frac{-6 \pm \sqrt{20}}{4} = \frac{-6 \pm 2\sqrt{5}}{4} = \frac{-3 \pm \sqrt{5}}{2}.$$

Thus the general solution to the differential equation is

$$y = A \cdot e^{\frac{-3+\sqrt{5}}{2} \cdot x} + B \cdot e^{\frac{-3-\sqrt{5}}{2} \cdot x}.$$ □

EXAMPLE 4.1.4 Solve the differential equation

$$y'' - 6y' + 9y = 0.$$

Solution: In this case the associated polynomial is

$$r^2 - 6r + 9 = 0.$$

This algebraic equation has the single solution $r = 3$. But our differential equation is second order, and therefore we seek *two independent solutions.*

In the case that the associated polynomial has just one root, we find the other solution with an augmented guess: Our new guess is $y = x \cdot e^{3x}$. (See Section 4.4 for an explanation of where this guess comes from.) The reader may check for himself/herself that this new guess is also a solution. So the general solution of the differential equation is

$$y = A \cdot e^{3x} + B \cdot xe^{3x} \, . \qquad \square$$

As a prologue to our next example, we must review some ideas connected with complex exponentials. Recall that

$$e^x = 1 + x + \frac{x^2}{2!} + \frac{x^3}{3!} + \cdots . = \sum_{j=0}^{\infty} \frac{x^j}{j!} \, .$$

This equation persists if we replace the real variable x by a complex variable z. Thus

$$e^z = 1 + z + \frac{z^2}{2!} + \frac{z^3}{3!} + \cdots = \sum_{j=0}^{\infty} \frac{z^j}{j!} \, .$$

Now write $z = x + iy$, and let us gather together the real and imaginary parts of this last equation:

$$\begin{aligned} e^z &= e^{x+iy} \\ &= e^x \cdot e^{iy} \\ &= e^x \cdot \left(1 + iy + \frac{(iy)^2}{2!} + \frac{(iy)^3}{3!} + \frac{(iy)^4}{4!} + \cdots \right) \\ &= e^x \cdot \left\{ \left(1 - \frac{y^2}{2!} + \frac{y^4}{4!} - + \cdots \right) + i \left(y - \frac{y^3}{3!} + \frac{y^5}{5!} - + \cdots \right) \right\} \end{aligned}$$

Now the first expression in parentheses is just the power series for cosine and the second expression in parentheses is just the power series for sine. Therefore this last equals

$$e^x \left(\cos y + i \sin y \right) .$$

Taking $x = 0$ we obtain the famous identity

$$e^{iy} = \cos y + i \sin y \, .$$

This equation—much used in mathematics, engineering, and physics—is known as *Euler's formula*, in honor of Leonhard Euler (1707–1783). We shall make considerable use of the more general formula

$$e^{x+iy} = e^x \left(\cos y + i \sin y \right) .$$

In using complex numbers, the reader should of course remember that the square root of a negative number is an *imaginary number*. For instance,

$$\sqrt{-4} = \pm 2i \qquad \text{and} \qquad \sqrt{-25} = \pm 5i\,.$$

EXAMPLE 4.1.5 Solve the differential equation

$$4y'' + 4y' + 2y = 0\,.$$

Solution: The associated polynomial is

$$4r^2 + 4r + 2 = 0\,.$$

We apply the quadratic equation to solve it:

$$r = \frac{-4 \pm \sqrt{4^2 - 4 \cdot 4 \cdot 2}}{2 \cdot 4} = \frac{-4 \pm \sqrt{-16}}{8} = \frac{-4 \pm 4i}{8} = \frac{-1 \pm i}{2}\,.$$

Thus the solutions to our differential equation are

$$y = e^{\frac{-1+i}{2} \cdot x} \qquad \text{and} \qquad y = e^{\frac{-1-i}{2} \cdot x}\,.$$

A general solution is given by

$$y = A \cdot e^{\frac{-1+i}{2} \cdot x} + B \cdot e^{\frac{-1-i}{2} \cdot x}\,.$$

Using Euler's formula, we may rewrite this general solution as

$$\begin{aligned} y &= A \cdot e^{-x/2} e^{ix/2} + B \cdot e^{-x/2} e^{-ix/2} \\ &= A \cdot e^{-x/2}[\cos x/2 + i \sin x/2] + Be^{-x/2}[\cos x/2 - i \sin x/2]\,. \qquad (3) \end{aligned}$$

We shall now use some propitious choices of A and B to extract meaningful real-valued solutions. First choose $A = 1/2$, $B = 1/2$. Putting these values in equation (3) gives

$$y = \frac{1}{2} \cdot e^{-x/2}[\cos x/2 + i \sin x/2] + \frac{1}{2} e^{-x/2}[\cos x/2 - i \sin x/2] = e^{-x/2} \cos x/2\,.$$

Instead taking $A = -i/2$, $B = i/2$ gives the solution

$$y = \frac{-i}{2} \cdot e^{-x/2}[\cos x/2 + i \sin x/2] + \frac{i}{2} e^{-x/2}[\cos x/2 - i \sin x/2] = e^{-x/2} \sin x/2\,.$$

As a result of this little trick, we may rewrite our general solution as

$$y = E \cdot e^{-x/2} \cos x/2 + F \cdot e^{-x/2} \sin x/2\,.$$

As usual, we invite the reader to plug this last solution into the differential equation to verify that it really works. □

We conclude this section with a last example of a homogeneous, second-order, linear ordinary differential equation with constant coefficients, and with complex roots, just to show how straightforward the methodology really is.

EXAMPLE 4.1.6 Solve the differential equation

$$y'' - 2y' + 5y = 0 .$$

Solution: The associated polynomial is

$$r^2 - 2r + 5 = 0 .$$

According to the quadratic formula, the solutions of this equation are

$$r = \frac{2 \pm \sqrt{(-2)^2 - 4 \cdot 1 \cdot 5}}{2} = \frac{2 \pm \sqrt{-16}}{2} = \frac{2 \pm 4i}{2} = 1 \pm 2i .$$

Hence the roots of the associated polynomial are $r = 1 + 2i$ and $1 - 2i$.

According to what we have learned, two independent solutions to the differential equation are thus given by

$$y = e^x \cos 2x \qquad \text{and} \qquad y = e^x \sin 2x .$$

Therefore the general solution is given by

$$y = Ae^x \cos 2x + Be^x \sin 2x .$$

Please verify this solution for yourself. □

Exercises

1. Find the general solution of each of the following differential equations.

(a) $y'' + y' - 6y = 0$
(b) $y'' + 2y' + y = 0$
(c) $y'' + 8y = 0$
(d) $2y'' - 4y' + 8y = 0$
(e) $y'' - 4y' + 4y = 0$
(f) $y'' - 9y' + 20y = 0$
(g) $2y'' + 2y' + 3y = 0$
(h) $4y'' - 12y' + 9y = 0$
(i) $y'' + y' = 0$
(j) $y'' - 6y' + 25y = 0$
(k) $4y'' + 20y' + 25y = 0$
(l) $y'' + 2y' + 3y = 0$
(m) $y'' = 4y$
(n) $4y'' - 8y' + 7y = 0$
(o) $2y'' + y' - y = 0$
(p) $16y'' - 8y' + y = 0$
(q) $y'' + 4y' + 5y = 0$
(r) $y'' + 4y' - 5y = 0$

2. Find the solution of each of the following initial value problems:

(a) $y'' - 5y' + 6y = 0,$ $\quad y(1) = e^2$ and $y'(1) = 3e^2$
(b) $y'' - 6y' + 5y = 0,$ $\quad y(0) = 3$ and $y'(0) = 11$
(c) $y'' - 6y' + 9y = 0,$ $\quad y(0) = 0$ and $y'(0) = 5$

(d) $y'' + 4y' + 5y = 0,\quad y(0) = 1$ and $y'(0) = 0$
(e) $y'' + 4y' + 2y = 0,\quad y(0) = -1$ and $y'(0) = 2 + 3\sqrt{2}$
(f) $y'' + 8y' - 9y = 0,\quad y(1) = 2$ and $y'(1) = 0$

3. Show that the general solution of the equation

$$y'' + Py' + Qy = 0$$

(where P and Q are constant) approaches 0 as $x \to +\infty$ if and only if P and Q are both positive.

4. Show that the derivative of any solution of

$$y'' + Py' + Qy = 0$$

(where P and Q are constant) is also a solution.

5. The equation

$$x^2y'' + pxy' + qy = 0\,,$$

where p and q are constants, is known as *Euler's equidimensional equation*. Show that the change of variable $x = e^z$ transforms Euler's equation into a new equation with constant coefficients. Apply this technique to find the general solution of each of the following equations.

(a) $x^2y'' + 3xy' + 10y = 0$
(b) $2x^2y'' + 10xy' + 8y = 0$
(c) $x^2y'' + 2xy' - 12y = 0$
(d) $4x^2y'' - 3y = 0$
(e) $x^2y'' - 3xy' + 4y = 0$
(f) $x^2y'' + 2xy' - 6y = 0$
(g) $x^2y'' + 2xy' + 3y = 0$
(h) $x^2y'' + xy' - 2y = 0$
(i) $x^2y'' + xy' - 16y = 0$

6. Find the differential equation of each of the following general solution sets.

(a) $Ae^x + Be^{-2x}$
(b) $A + Be^{2x}$
(c) $Ae^{3x} + Be^{5x}$
(d) $Ae^x \cos 3x + Be^x \sin 3x$
(e) $Ae^{3x} + Be^{-x}$
(f) $Ae^{-x} + Be^{-4x}$
(g) $Ae^{2x} + Be^{-2x}$
(h) $Ae^{-4x} \cos x + Be^{-4x} \sin x$

4.2 The Method of Undetermined Coefficients

"Undetermined coefficients" is a method of organized guessing. We have already seen guessing, in one form or another, serve us well in solving first-order linear equations and also in solving homogeneous second-order linear equations with constant coefficients. Now we shall expand the technique to cover *nonhomogeneous* second-order linear equations.

We must begin by discussing what the solution to such an equation will look like. Consider an equation of the form

$$ay'' + by' + cy = f(x)\,. \tag{4.3}$$

Suppose that we can find (by guessing or by some other means) a function $y = y_0(x)$ that satisfies this equation. We call y_0 a *particular solution* of the differential equation. Notice that it will *not* be the case that a constant multiple of y will also solve the equation. In fact, if we consider $y = A \cdot y_0$ and plug this function into the equation, then we obtain

$$a[Ay_0]'' + b[Ay_0]' + c[Ay_0] = A[ay_0'' + by_0' + cy_0] = A \cdot f\,.$$

We see that, if $A \neq 1$, then $A \cdot y_0$ is *not* a solution. But we expect the solution of a second-order equation to have two free constants. Where will they come from?

The answer is that we must separately solve the associated *homogeneous equation*, which is

$$ay'' + by' + cy = 0\,.$$

If y_1 and y_2 are solutions of this equation, then of course (as we learned in the last section) $A \cdot y_1 + B \cdot y_2$ will be a general solution of *this homogeneous equation*. But then the general solution of the original differential equation (4) will be

$$y = y_0 + A \cdot y_1 + B \cdot y_2\,.$$

We invite the reader to verify that, no matter what the choice of A and B, this y will be a solution of the original differential equation (4).

These ideas are best hammered home by the examination of some examples.

EXAMPLE 4.2.1 Find the general solution of the differential equation

$$y'' + y = \sin x\,. \tag{4.4}$$

Solution: We might guess that $y = \sin x$ or $y = \cos x$ is a particular solution of this equation. But in fact these are solutions of the homogeneous equation

$$y'' + y = 0$$

(as we may check by using the techniques of the last section, or just by direct verification). In summary, the general solution of the homogeneous equation is

$$y = A\cos x + B\sin x\,.$$

If we want to find a particular solution of (5) then we must try a bit harder. Inspired by our experience with the case of repeated roots for the second-order, homogeneous linear equation with constant coefficients (as in the last section), we instead will guess

$$y_0 = \alpha \cdot x \cos x + \beta \cdot x \sin x$$

for our particular solution. Notice that we allow arbitrary constants in front of the functions $x \cos x$ and $x \sin x$. These are the "undetermined coefficients" that we seek.

Now we simply plug the guess into the differential equation and see what happens. Thus

$$[\alpha \cdot x \cos x + \beta \cdot x \sin x]'' + [\alpha \cdot x \cos x + \beta \cdot x \sin x] = \sin x$$

or

$$\alpha\big(2(-\sin x)+x(-\cos x)\big)+\beta\big(2\cos x+x(-\sin x)\big)+\big[\alpha x\cos x+\beta x\sin x\big] = \sin x$$

or

$$(-2\alpha)\sin x + (2\beta)\cos x + (-\beta+\beta)x\sin x + (-\alpha+\alpha)x\cos x = \sin x\,.$$

We see that there is considerable cancellation, and we end up with

$$-2\alpha \sin x + 2\beta \cos x = \sin x\,.$$

The only way that this can be an identity in x is if $-2\alpha = 1$ and $2\beta = 0$ or $\alpha = -1/2$ and $\beta = 0$.

Thus our particular solution is

$$y_0 = -\frac{1}{2} x \cos x\,.$$

We combine this with the solution of the homogeneous equation (described at the outset of this solution) to obtain the general solution of our inhomogeneous differential equation given by

$$y = -\frac{1}{2} x \cos x + A \cos x + B \sin x\,. \qquad \square$$

Remark 4.2.2 As usual, for a second-order equation we expect, and find, that there are two unknown parameters that parameterize the set of solutions (or the general solution). Notice that these *are not* the same as the "undetermined coefficients" that we used to find our particular solution.

EXAMPLE 4.2.3 Find the solution of

$$y'' - y' - 2y = 4x^2$$

that satisfies $y(0) = 0$ and $y'(0) = 1$.

Solution: The associated homogeneous equation is

$$y'' - y' - 2y = 0$$

and this has associated polynomial

$$r^2 - r - 2 = 0 .$$

The roots are obviously $r = 2, -1$ and so the general solution of the homogeneous equation is $y = A \cdot e^{2x} + B \cdot e^{-x}$.

For a particular solution, our guess will be a polynomial. Guessing a second-degree polynomial makes good sense, since a guess of a higher-order polynomial is going to produce terms of high degree that we do not want. Thus we guess that $y_p(x) = \alpha x^2 + \beta x + \gamma$. Plugging this guess into the differential equation gives

$$[\alpha x^2 + \beta x + \gamma]'' - [\alpha x^2 + \beta x + \gamma]' - 2[\alpha x^2 + \beta x + \gamma] = 4x^2$$

or

$$[2\alpha] - [\alpha \cdot 2x + \beta] - [2\alpha x^2 + 2\beta x + 2\gamma] = 4x^2 .$$

Grouping like terms together gives

$$-2\alpha x^2 + [-2\alpha - 2\beta]x + [2\alpha - \beta - 2\gamma] = 4x^2 .$$

As a result, we find that

$$\begin{aligned} -2\alpha &= 4 \\ -2\alpha - 2\beta &= 0 \\ 2\alpha - \beta - 2\gamma &= 0 . \end{aligned}$$

This system is easily solved to yield $\alpha = -2$, $\beta = 2$, $\gamma = -3$. So our particular solution is $y_0(x) = -2x^2 + 2x - 3$. The general solution of the original differential equation is then

$$y(x) = \left[-2x^2 + 2x - 3\right] + A \cdot e^{2x} + Be^{-x} . \tag{4.5}$$

Now we seek the solution that satisfies the initial conditions $y(0) = 0$ and $y'(0) = 1$. These translate to

$$0 = y(0) = [-2 \cdot 0^2 + 2 \cdot 0 - 3] + A \cdot e^0 + B \cdot e^0$$

and

$$1 = y'(0) = [-4 \cdot 0 + 2 - 0] + 2A \cdot e^0 - B \cdot e^0 .$$

This gives the equations

$$\begin{aligned} 0 &= -3 + A + B \\ 1 &= 2 + 2A - B . \end{aligned}$$

Of course we can solve this system quickly to find that $A = 2/3, B = 7/3$.

In conclusion, the solution to our initial boundary value problem is

$$y(x) = [-2x^2 + 2x - 3] + \frac{2}{3} \cdot e^{2x} + \frac{7}{3} \cdot e^{-x} . \qquad \square$$

Remark 4.2.4 Again notice that the undetermined coefficients α and β and γ that we used to guess the particular solution are *not* the same as the parameters that gave the two degrees of freedom in our general solution (6). Further notice that we *needed* those two degrees of freedom so that we could meet the two initial conditions.

Exercises

1. Find the general solution of each of the following equations.

(a) $y'' + 3y' - 10y = 6e^{4x}$
(b) $y'' + 4y = 3\sin x$
(c) $y'' + 10y' + 25y = 14e^{-5x}$
(d) $y'' - 2y' + 5y = 25x^2 + 12$
(e) $y'' - y' - 6y = 20e^{-2x}$
(f) $y'' - 3y' + 2y = 14\sin 2x - 18\cos 2x$
(g) $y'' + y = 2\cos x$
(h) $y'' - 2y' = 12x - 10$
(i) $y'' - 2y' + y = 6e^{x}$
(j) $y'' - 2y' + 2y = e^{x}\sin x$
(k) $y'' + y' = 10x^4 + 2$

2. Find the solution of the differential equation that satisfies the given initial conditions.

(a)	$y'' - 3y' + y = x,$	$y(0) = 1,$	$y'(0) = 0$
(b)	$y'' + 4y' + 6y = \cos x,$	$y(0) = 0,$	$y'(0) = 2$
(c)	$y'' + y' + y = \sin x,$	$y(1) = 1,$	$y'(1) = 1$
(d)	$y'' - 3y' + 2y = 0,$	$y(-1) = 0,$	$y'(-1) = 1$
(e)	$y'' - y' + y = x^2,$	$y(0) = 1,$	$y'(0) = 0$
(f)	$y'' + 2y' + y = 1,$	$y(1) = 1,$	$y'(1) = 0$

3. If k and b are positive constants, then find the general solution of

$$y'' + k^2 y = \sin bx .$$

4. If y_1 and y_2 are solutions of

$$y'' + P(x)y' + Q(x)y = R_1(x)$$

and

$$y'' + P(x)y' + Q(x)y = R_2(x)\,,$$

respectively, then show that $y = y_1 + y_2$ is a solution of

$$y'' + P(x)y' + Q(x)y = R_1(x) + R_2(x)\,.$$

This is called the *principle of superposition*. Use this idea to find the general solution of

(a) $y'' + 4y = 4\cos 2x + 6\cos x + 8x^2$

(b) $y'' + 9y = 2\sin 3x + 4\sin x - 26e^{-2x}$

5. Use your symbol manipulation software, such as `Maple` or `Mathematica`, to write a routine for solving for the undetermined coefficients in the solution of an ordinary differential equation, once an initial guess is given.

4.3 The Method of Variation of Parameters

Variation of parameters is a method for producing a *particular solution* to a nonhomogeneous equation by exploiting the (usually much simpler to find) solutions to the associated homogeneous equation.

Let us consider the differential equation

$$y'' + p(x)y' + q(x)y = r(x)\,. \tag{4.6}$$

Assume that, by some method or other, we have found the general solution of the associated homogeneous equation

$$y'' + p(x)y' + q(x)y = 0$$

to be

$$y = Ay_1(x) + By_2(x)\,.$$

What we do now is to *guess* that a particular solution to the original equation (7) is

$$y_0(x) = v_1(x)\cdot y_1(x) + v_2(x)\cdot y_2(x) \tag{4.7}$$

for some choice of functions v_1, v_2.

Now let us proceed on this guess. We calculate that

$$y_0' = [v_1'y_1 + v_1y_1'] + [v_2'y_2 + v_2y_2'] = [v_1'y_1 + v_2'y_2] + [v_1y_1' + v_2y_2']\,. \tag{4.8}$$

We also need to calculate the second derivative of y_0. But we do not want the extra complication of having second derivatives of v_1 and v_2. So we shall mandate that the first expression in brackets on the far right-hand side of (9) is identically zero. Thus we have

$$v_1'y_1 + v_2'y_2 \equiv 0\,. \tag{4.9}$$

Thus

$$y_0' = v_1y_1' + v_2y_2' \tag{4.10}$$

and we can now calculate that

$$y_0'' = [v_1'y_1' + v_1y_1''] + [v_2'y_2' + v_2y_2'']\,. \tag{4.11}$$

Now let us substitute (8), (11), and (12) into the differential equation. The result is

$$\left([v_1'y_1'+v_1y_1'']+[v_2'y_2'+v_2y_2'']\right)+p(x)\cdot\left(v_1y_1'+v_2y_2'\right)+q(x)\cdot\left(v_1y_1+v_2y_2\right)=r(x)\,.$$

After some algebraic manipulation, this becomes

$$v_1\left(y_1'' + py_1' + qy_1\right) + v_2\left(y_2'' + py_2' + qy_2\right) + v_1'y_1' + v_2'y_2' = r\,.$$

Since y_1, y_2 are solutions of the homogeneous equation, the expressions in parentheses vanish. The result is

$$v_1'y_1' + v_2'y_2' = r\,. \tag{4.12}$$

At long last we have two equations to solve in order to determine what v_1 and v_2 must be. Namely, we focus on equations (10) and (13) to obtain

$$v_1'y_1 + v_2'y_2 = 0\,,$$

$$v_1'y_1' + v_2'y_2' = r\,.$$

In practice, these can be solved for v_1', v_2', and then integration tells us what v_1, v_2 must be.

As usual, the best way to understand a new technique is by way of some examples.

EXAMPLE 4.3.1 Find the general solution of

$$y'' + y = \csc x\,.$$

Solution: Of course the general solution to the associated homogeneous equation is familiar. It is

$$y(x) = A\sin x + B\cos x\,.$$

We of course think of $y_1(x) = \sin x$ and $y_2(x) = \cos x$. In order to find a particular solution, we need to solve the equations

$$\begin{aligned} v_1'\sin x + v_2'\cos x &= 0 \\ v_1'(\cos x) + v_2'(-\sin x) &= \csc x\,. \end{aligned}$$

This is a simple algebra problem, and we find that

$$v_1'(x) = \cot x \qquad \text{and} \qquad v_2'(x) = -1\,.$$

As a result,

$$v_1(x) = \ln|\sin x| \qquad \text{and} \qquad v_2(x) = -x\,.$$

(As the reader will see, we do not need any constants of integration.)

The final result is then that a particular solution of our differential equation is

$$y_0(x) = v_1(x)y_1(x) + v_2(x)y_2(x) = [\ln|\sin x|] \cdot \sin x + [-x] \cdot \cos x\,.$$

We invite the reader to check that this solution actually works. The general solution of the original differential equation is thus

$$y(x) = \left\{ [\ln|\sin x|] \cdot \sin x + [-x] \cdot \cos x \right\} + A \sin x + B \cos x\,. \qquad \square$$

Remark 4.3.2 The method of variation of parameters has the advantage—over the method of undetermined coefficients—of not involving any guessing. It is a direct method that always leads to a solution. However, the integrals that we may need to perform to carry the technique to completion may, for some problems, be rather difficult.

EXAMPLE 4.3.3 Solve the differential equation

$$y'' - y' - 2y = 4x^2$$

using the method of variation of parameters.

Solution: The reader will note that, in the last section (Example 4.2.3), we solved this same equation using the method of undetermined coefficients (or organized guessing). Now we shall solve it a second time by our new method.

As we saw before, the homogeneous equation has the general solution

$$y = Ae^{2x} + Be^{-x}\,.$$

Hence we solve the system

$$v_1'e^{2x} + v_2'e^{-x} = 0\,,$$

$$v_1'[2e^{2x}] + v_2'[-e^{-x}] = 4x^2\,.$$

The result is

$$v_1'(x) = \frac{4}{3}x^2e^{-2x} \qquad \text{and} \qquad v_2'(x) = -\frac{4}{3}x^2e^{x}\,.$$

We may use integration by parts to then determine that

$$v_1(x) = -\frac{2x^2}{3}e^{-2x} - \frac{2x}{3}e^{-2x} - \frac{1}{3}e^{-2x}$$

and

$$v_2(x) = -\frac{4x^2}{3}e^{x} + \frac{8x}{3}e^{x} - \frac{8}{3}e^{x} .$$

We finally see that a particular solution to our differential equation is

$$\begin{aligned} y_0(x) &= v_1(x) \cdot y_1(x) + v_2(x) y_2(x) \\ &= \left(-\frac{2x^2}{3}e^{-2x} - \frac{2x}{3}e^{-2x} - \frac{1}{3}e^{-2x}\right) \cdot e^{2x} \\ &\quad + \left(-\frac{4x^2}{3}e^{x} + \frac{8x}{3}e^{x} - \frac{8}{3}e^{x}\right) \cdot e^{-x} \\ &= \left(-\frac{2x^2}{3} - \frac{2x}{3} - \frac{1}{3}\right) + \left(-\frac{4x^2}{3} + \frac{8x}{3} - \frac{8}{3}\right) \\ &= -2x^2 + 2x - 3 . \end{aligned}$$

In conclusion, the general solution of the original differential equation is

$$y(x) = \left\{-2x^2 + 2x - 3\right\} + Ae^{2x} + Be^{-x} .$$

As you can see, this is the same answer that we obtained in Section 4.2, Example 4.2.3, by the method of undetermined coefficients. □

Remark 4.3.4 Of course the method of variation of parameters is a technique for finding a *particular solution* of a differential equation. The general solution of the associated homogeneous equation must be found by a different technique.

Exercises

1. Find a particular solution of each of the following differential equations.

(a) $y'' + 4y = \tan 2x$
(b) $y'' + 2y' + y = e^{-x} \ln x$
(c) $y'' - 2y' - 3y = 64xe^{-x}$
(d) $y'' + 2y' + 5y = e^{-x} \sec 2x$
(e) $2y'' + 3y' + y = e^{-3x}$
(f) $y'' - 3y' + 2y = (1 + e^{-x})^{-1}$

2. Find a particular solution of each of the following differential equations.

(a) $y'' + y = \sec x$
(b) $y'' + y = \cot^2 x$
(c) $y'' + y = \cot 2x$
(d) $y'' + y = x \cos x$
(e) $y'' + y = \tan x$
(f) $y'' + y = \sec x \tan x$
(g) $y'' + y = \sec x \csc x$

3. Find a particular solution of

$$y'' - 2y' + y = 2x\,,$$

first by inspection and then by variation of parameters.

4. Find a particular solution of

$$y'' - y' - 6y = e^{-x}\,,$$

first by undetermined coefficients and then by variation of parameters.

5. Find the general solution of each of the following equations.

(a) $(x^2 - 1)y'' - 2xy' + 2y = (x^2 - 1)^2$
(b) $(x^2 + x)y'' + (2 - x^2)y' - (2 + x)y = x(x + 1)^2$
(c) $(1 - x)y'' + xy' - y = (1 - x)^2$
(d) $xy'' - (1 + x)y' + y = x^2 e^{2x}$
(e) $x^2 y'' - 2xy' + 2y = xe^{-x}$

6. Use your symbol manipulation software, such as `Maple` or `Mathematica`, to find a particular solution to an ordinary differential equation, once the solutions to the homogeneous equation are given.

4.4 The Use of a Known Solution to Find Another

Consider a general second-order linear equation of the form

$$y'' + p(x)y' + q(x)y = 0\,. \tag{4.13}$$

It often happens—and we have seen this in our earlier work—that one can either guess or elicit one solution to the equation. But finding the second independent solution is more difficult. In this section we introduce a method for finding that second solution.

In fact we exploit a notational trick that served us well in Section 4.3 on variation of parameters. Namely, we shall assume that we have found the one solution y_1 and we shall suppose that the second solution we seek is $y_2 = v \cdot y_1$ for some undetermined function v. Our job, then, is to find v.

Assuming, then, that y_1 is a solution of (14), we shall substitute $y_2 = v \cdot y_1$ into (14) and see what this tells us about calculating v. We see that

$$[v \cdot y_1]'' + p(x) \cdot [v \cdot y_1]' + q(x) \cdot [v \cdot y_1] = 0$$

or

$$[v'' \cdot y_1 + 2v' \cdot y_1' + v \cdot y_1''] + p(x) \cdot [v' \cdot y_1 + v \cdot y_1'] + q(x) \cdot [v \cdot y_1] = 0\,.$$

We rearrange this identity to find that

$$v \cdot [y_1'' + p(x) \cdot y_1' + q(x) y_1] + [v'' \cdot y_1] + [v' \cdot (2y_1' + p(x) \cdot y_1)] = 0\,.$$

Now we are *assuming* that y_1 is a solution of the differential equation (14), so the first expression in brackets must vanish. As a result,

$$[v'' \cdot y_1] + [v' \cdot (2y_1' + p \cdot y_1)] = 0\,.$$

In the spirit of separation of variables, we may rewrite this equation as

$$\frac{v''}{v'} = -2\frac{y_1'}{y_1} - p\,.$$

Integrating once, we find that

$$\ln v' = -2 \ln y_1 - \int p(x)\, dx$$

or

$$v' = \frac{1}{y_1^2} e^{-\int p(x)\, dx}\,.$$

Applying the integral one last time yields

$$v = \int \frac{1}{y_1^2} e^{-\int p(x)\, dx}\, dx\,.$$

What does this mean? It tells us that a *second solution* to our differential equation is given by

$$y_2(x) = v(x) \cdot y_1(x) = \left[\int \frac{1}{y_1^2} e^{-\int p(x)\, dx}\, dx \right] \cdot y_1(x)\,. \tag{4.14}$$

In order to really understand what this means, let us apply the method to some particular differential equations.

EXAMPLE 4.4.1 Find the general solution of the differential equation

$$y'' - 4y' + 4y = 0\,.$$

Solution: When we first encountered this type of equation in Section 4.1, we learned to study the associated polynomial

$$r^2 - 4r + 4 = 0\,.$$

Unfortunately, the polynomial has only the repeated root $r = 2$, so we find just the one solution $y_1(x) = e^{2x}$. Where do we find another?

In Section 4.1, we found the second solution by guessing. Now we have a more systematic way of finding that second solution, and we use it now to test out the new methodology. Observe that $p(x) = -4$ and $q(x) = 4$. According to formula (15), we can find a second solution $y_2 = v \cdot y_1$ with

$$\begin{aligned} v &= \int \frac{1}{y_1^2} e^{-\int p(x)\,dx}\,dx \\ &= \int \frac{1}{[e^{2x}]^2} e^{-\int -4\,dx}\,dx \\ &= \int e^{-4x} \cdot e^{4x}\,dx \\ &= \int 1\,dx = x\,. \end{aligned}$$

Thus the second solution to our differential equation is $y_2 = v \cdot y_1 = x \cdot e^{2x}$ and the general solution is therefore

$$y = A \cdot e^{2x} + B \cdot xe^{2x}\,.$$

This reaffirms what we learned in Section 4.1 by a different and more elementary technique. □

Next we turn to an example of a nonconstant coefficient equation.

EXAMPLE 4.4.2 Find the general solution of the differential equation

$$x^2y'' + xy' - y = 0\,.$$

Solution: Differentiating a monomial once lowers the degree by 1 and differentiating it twice lowers the degree by 2. So it is natural to guess that this differential equation has a power of x as a solution. And $y_1(x) = x$ works.

We use formula (15) to find a second solution of the form $y_2 = v \cdot y_1$. First we rewrite the equation in the standard form as

$$y'' + \frac{1}{x}y' - \frac{1}{x^2}y = 0$$

and we note then that $p(x) = 1/x$ and $q(x) = -1/x^2$. Thus

$$\begin{aligned} v(x) &= \int \frac{1}{y_1^2} e^{-\int p(x)\,dx}\,dx \\ &= \int \frac{1}{x^2} e^{-\int 1/x\,dx}\,dx \\ &= \int \frac{1}{x^2} e^{-\ln x}\,dx \\ &= \int \frac{1}{x^2}\frac{1}{x}\,dx \\ &= -\frac{1}{2x^2}\,. \end{aligned}$$

In conclusion, $y_2 = v \cdot y_1 = [-1/(2x^2)] \cdot x = -1/(2x)$ and the general solution is

$$y(x) = A \cdot x + B \cdot \left(-\frac{1}{2x}\right) . \qquad \square$$

Exercises

1. Use the method of this section to find y_2 and the general solution of each of the following equations from the given solution y_1.

   (a) $y'' + y = 0 , \quad y_1(x) = \sin x$ (b) $y'' - y = 0 , \quad y_1(x) = e^x$

2. The equation $xy'' + 3y' = 0$ has the obvious solution $y_1 \equiv 1$. Find y_2 and find the general solution.
3. Verify that $y_1 = x^2$ is one solution of $x^2y'' + xy' - 4y = 0$, and then find y_2 and the general solution.
4. The equation

   $$(1 - x^2)y'' - 2xy' + 2y = 0 \qquad (*)$$

   is a special case, corresponding to $p = 1$, of the Legendre equation

   $$(1 - x^2)y'' - 2xy' + p(p + 1)y = 0 .$$

   Equation $(*)$ has $y_1 = x$ as a solution. Find the general solution.
5. The equation

   $$x^2y'' + xy' + \left(x^2 - \frac{1}{4}\right) y = 0 \qquad (\star)$$

   is a special case, corresponding to $p = 1/2$, of the Bessel equation

   $$x^2y'' + xy' + (x^2 - p^2)y = 0 .$$

   Verify that $y_1(x) = x^{-1/2} \sin x$ is a solution of $(\star)$ for $x > 0$ and find the general solution.
6. For each of the following equations, $y_1(x) = x$ is one solution. In each case, find the general solution.

   (a) $y'' - \dfrac{x}{x-1}y' + \dfrac{1}{x-1}y = 0$

   (b) $x^2y'' + 2xy' - 2y = 0$

   (c) $x^2y'' - x(x + 2)y' + (x + 2)y = 0$
7. Find the general solution of the differential equation

   $$y'' - xf(x)y' + f(x)y = 0 .$$

8. Verify that one solution of the equation

   $$xy'' - (2x + 1)y' + (x + 1)y = 0$$

   is $y_1(x) = e^x$. Find the general solution.

9. If y_1 is a nonzero solution of the differential equation

$$y'' + P(x)y' + Q(x)y = 0$$

and if $y_2 = v \cdot y_1$, with

$$v(x) = \int \frac{1}{y_1^2} \cdot e^{-\int p dx} dx \, ,$$

then show that y_1 and y_2 are linearly independent.

4.5 Higher-Order Equations

We treat here some aspects of higher-order equations that bear a similarity to what we learned about second-order examples. We shall concentrate primarily on linear equations with constant coefficients. As usual, we illustrate the ideas with a few key examples.

We consider an equation of the form

$$y^{(n)} + a_{n-1}y^{(n-1)} + \cdots + a_1 y^{(1)} + a_0 y = f \, . \tag{4.15}$$

Here a superscript $^{(j)}$ denotes a jth derivative and f is some continuous function. This is a linear, ordinary differential equation of order n with constant coefficients.

Following what we learned about second-order equations, we expect the general solution of (16) to have the form

$$y = y_p + y_g \, ,$$

where y_p is a particular solution of (16) and y_g is the general solution of the associated homogeneous equation

$$y^{(n)} + a_{n-1}y^{(n-1)} + \cdots + a_1 y^{(1)} + a_0 y = 0 \, . \tag{4.16}$$

Furthermore, we expect that y_g will have the form

$$y_g = A_1 y_1 + A_2 y_2 + \cdots + A_{n-1} y_{n-1} + A_n y_n \, ,$$

where the y_j are "independent" solutions of (17) (to repeat, we shall later use linear algebra to say precisely what "independent" really means).

We begin by studying the homogeneous equation (17) and seeking the general solution y_g. Again following the paradigm that we developed for second-order equations, we guess a solution of the form $y = e^{rx}$. Substituting this guess into (17), we find that

$$e^{rx} \cdot \left(r^n + a_{n-1} r^{n-1} + \cdots + a_1 r + a_0 \right) = 0 \, .$$

Thus we are led to solving the *associated polynomial*

$$r^n + a_{n-1}r^{n-1} + \cdots + a_1 r + a_0 = 0\,.$$

The fundamental theorem of algebra tells us that every polynomial of degree n has a total of n complex roots $r_1, r_2, \ldots, r_n$ (there may be repetitions in this list). Thus the polynomial factors as

$$(r - r_1)\cdot(r - r_2)\cdot\cdots\cdot(r - r_{n-1})\cdot(r - r_n)\,.$$

In practice there may be some difficulty in *actually finding* the complete set of roots of a given polynomial. For instance, it is known that for polynomials of degree 5 and greater there is no elementary formula for the roots. Let us pass over this sticky point for the moment, and continue to comment on the theoretical setup.

I. Distinct Real Roots: For a given associated polynomial, if the polynomial roots $r_1, r_2, \ldots, r_n$ are distinct and real, then we can be sure that

$$e^{r_1 x}, e^{r_2 x}, \ldots, e^{r_n x}$$

are n distinct solutions to the differential equation (17). It then follows, just as in the order-2 case, that

$$y_g = A_1 e^{r_1 x} + A_2 e^{r_2 x} + \cdots + A_n e^{r_n x}$$

is the general solution to (17) that we seek.

II. Repeated Real Roots: If the roots are real, but two of them are equal (say that $r_1 = r_2$), then of course $e^{r_1 x}$ and $e^{r_2 x}$ are *not* distinct solutions of the differential equation. Just as in the case of order-2 equations, what we do in this case is manufacture two distinct solutions of the form $e^{r_1 x}$ and $x \cdot e^{r_1 x}$.

More generally, if several of the roots are equal, say $r_1 = r_2 = \cdots = r_k$, then we manufacture distinct solutions of the form $e^{r_1 x}, x \cdot e^{r_1 x}, x^2 e^{r_1 x}, \ldots,$ $x^{k-1} \cdot e^{r_1 x}$.

III. Complex Roots: We have been assuming that the coefficients of the original differential equation ((16) or (17)) are all real. This being the case, any complex roots of the associated polynomial will occur in conjugate pairs $a + ib$ and $a - ib$. Then we have distinct solutions $e^{(a+ib)x}$ and $e^{(a-ib)x}$. Then we can use Euler's formula and a little algebra, just as we did in the order-2 case, to produce distinct real solutions $e^{ax}\cos bx$ and $e^{ax}\sin bx$.

In the case that complex roots are repeated to order k, then we take

$$e^{ax}\cos bx, xe^{ax}\cos bx, \ldots, x^{k-1}e^{ax}\cos bx$$

and

$$e^{ax}\sin bx, xe^{ax}\sin bx, \ldots, x^{k-1}e^{ax}\sin bx$$

as solutions of the ordinary differential equation.

EXAMPLE 4.5.1 Find the general solution of the differential equation

$$y^{(4)} - 5y^{(2)} + 4y = 0\,.$$

Solution: The associated polynomial is

$$r^4 - 5r^2 + 4 = 0\,.$$

Of course we may factor this as $(r^2 - 4)(r^2 - 1) = 0$ and then as

$$(r-2)(r+2)(r-1)(r+1) = 0\,.$$

We find, therefore, that the general solution of our differential equation is

$$y(x) = A_1e^{2x} + A_2e^{-2x} + A_3e^{x} + A_4e^{-x}\,. \qquad \square$$

EXAMPLE 4.5.2 Find the general solution of the differential equation

$$y^{(4)} - 8y^{(2)} + 16y = 0\,.$$

Solution: The associated polynomial is

$$r^4 - 8r^2 + 16 = 0\,.$$

This factors readily as $(r^2 - 4)(r^2 - 4) = 0$, and then as

$$(r-2)^2(r+2)^2 = 0\,.$$

According to our discussion in part **II**, the general solution of the differential equation is then

$$y(x) = A_1e^{2x} + A_2xe^{2x} + A_3e^{-2x} + A_4xe^{-2x}\,. \qquad \square$$

EXAMPLE 4.5.3 Find the general solution of the differential equation

$$y^{(4)} - 2y^{(3)} + 2y^{(2)} - 2y^{(1)} + y = 0\,.$$

Solution: The associated polynomial is

$$r^4 - 2r^3 + 2r^2 - 2r + 1 = 0\,.$$

We notice, just by inspection, that $r_1 = 1$ is a solution of this polynomial equation. Thus $r - 1$ divides the polynomial. In fact

$$r^4 - 2r^3 + 2r^2 - 2r + 1 = (r-1)\cdot(r^3 - r^2 + r - 1)\,.$$

But we again see that $r_2 = 1$ is a root of the new third-degree polynomial. Dividing out $r - 1$ again, we obtain a quadratic polynomial that we can solve directly.

The end result is

$$r^4 - 2r^3 + 2r^2 - 2r + 1 = (r-1)^2 \cdot (r^2 + 1) = 0$$

or

$$(r-1)^2(r-i)(r+i) = 0\,.$$

The roots are 1 (repeated), i, and $-i$. As a result, we find that the general solution of the differential equation is

$$y(x) = A_1 e^x + A_2 x e^x + A_3 \cos x + A_4 \sin x\,.$$

□

EXAMPLE 4.5.4 Find the general solution of the equation

$$y^{(4)} - 5y^{(2)} + 4y = \sin x\,. \tag{4.17}$$

Solution: In fact we found the general solution of the associated homogeneous equation in Example 4.5.1. To find a particular solution of (18), we use undetermined coefficients and guess a solution of the form $y = \alpha \cos x + \beta \sin x$. A little calculation reveals then that $y_p(x) = (1/10)\sin x$ is the particular solution that we seek. As a result,

$$y(x) = \frac{1}{10}\sin x + A_1 e^{2x} + A_2 e^{-2x} + A_3 e^x + A_4 e^{-x}$$

is the general solution of (18). □

EXAMPLE 4.5.5 (Coupled Harmonic Oscillators) Linear equations of order greater than 2 arise in physics by the elimination of variables from simultaneous systems of second-order equations. We give here an example that arises from coupled harmonic oscillators. Accordingly, let two carts of masses m_1, m_2 be attached to left and right walls as in Figure 4.1 with springs having spring constants k_1, k_2. If there is no damping and the carts are unattached, then of course when the carts are perturbed we have two separate harmonic oscillators.

But if we connect the carts, with a spring having spring constant k_3, then we obtain *coupled harmonic oscillators.* In fact Newton's second law of motion can now be used to show that the motions of the coupled carts will satisfy these differential equations:

$$\begin{aligned} m_1 \frac{d^2 x_1}{dt^2} &= -k_1 x_1 + k_3(x_2 - x_1)\,. \\ m_2 \frac{d^2 x_2}{dt^2} &= -k_2 x_2 - k_3(x_2 - x_1)\,. \end{aligned}$$

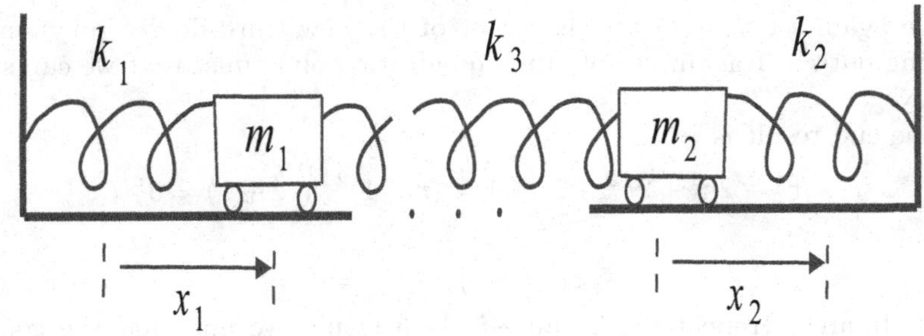

FIGURE 4.1
Coupled harmonic oscillators.

We can solve the first equation for x_2,

$$x_2 = \frac{1}{k_3}\left(x_1[k_1 + k_3] + m_1\frac{d^2x_1}{dt^2}\right),$$

and then substitute into the second equation. The result is a fourth order equation for x_1. ∎

Exercises

In each of Exercises 1–15, find the general solution of the given differential equation.

1. $y''' - 3y'' + 2y' = 0$
2. $y''' - 3y'' + 4y' - 2y = 0$
3. $y''' - y = 0$
4. $y''' + y = 0$
5. $y''' + 3y'' + 3y' + y = 0$
6. $y^{(4)} + 4y''' + 6y'' + 4y' + y = 0$
7. $y^{(4)} - y = 0$
8. $y^{(4)} + 5y'' + 4y = 0$
9. $y^{(4)} - 2a^2y'' + a^4y = 0$
10. $y^{(4)} + 2a^2y'' + a^4y = 0$
11. $y^{(4)} + 2y''' + 2y'' + 2y' + y = 0$
12. $y^{(4)} + 2y''' - 2y'' - 6y' + 5y = 0$
13. $y''' - 6y'' + 11y' - 6y = 0$
14. $y^{(4)} + y''' - 3y'' - 5y' - 2y = 0$
15. $y^{(5)} - 6y^{(4)} - 8y''' + 48y'' + 16y' - 96y = 0$

16. Find the general solution of $y^{(4)} = 0$. Now find the general solution of $y^{(4)} = \sin x + 24$.

17. Find the general solution of $y''' - 3y'' + 2y' = 10 + 42e^{3x}$.

18. Find the solution of $y''' - y' = 1$ that satisfies the initial conditions $y(0) = y'(0) = y''(0) = 4$.

19. Show that the change of independent variable given by $x = e^z$ transforms the third-order Euler equidimensional equation

$$x^3y''' + a_2x^2y'' + a_1xy' + a_0y = 0$$

into a third-order linear equation with constant coefficients. Solve the following equations by this method.

(a) $x^3y''' + 3x^2y'' = 0$

(b) $x^3y''' + x^2y'' - 2xy' + 2y = 0$

(c) $x^3y''' + 2x^2y''$ $+xy' - y = 0$

20. In determining the drag on a small sphere moving at a constant speed through a viscous fluid, it is necessary to solve the differential equation

$$x^3y^{(4)} + 8x^2y''' + 8xy'' - 8y' = 0\,.$$

If we make the substitution $w = y'$, then this becomes a third-order Euler equation that we can solve by the method of Exercise 19. Do so, and show that the general solution is

$$y(x) = c_1x^2 + c_2x^{-1} + c_3x^{-3} + c_4\,.$$

(These ideas are part of the mathematical foundation of the work of Robert Millikan in his famous oil-drop experiment of 1909 or measuring the charge of an electron. He won the Nobel Prize for this work in 1923.)

21. In Example 2.7.5, find the fourth-order differential equation for x_1 by eliminating x_2, as described at the end of the example.

22. In Exercise 21, solve the fourth-order equation for x_1 if the masses are equal and the spring constants are equal so that $m_1 = m_2 = m$ and $k_1 = k_2 = k_3 = k$. In this special case, show directly that x_2 satisfies the same differential equation as x_1. The two frequencies associated with these coupled harmonic oscillators are called the *normal frequencies* of the system. What are they?

5

Applications of the Second-Order Theory

- Vibrations and oscillations
- Electrical current
- Newton's law of gravitation
- Kepler's laws
- Higher-order equations

5.1 Vibrations and Oscillations

When a physical system in stable equilibrium is disturbed, then it is subject to forces that tend to restore the equilibrium. The result can lead to oscillations or vibrations. It is described by an ordinary differential equation of the form

$$\frac{d^2x}{dt^2} + p(t) \cdot \frac{dx}{dt} + q(t)x = r(t) .$$

In this section we shall learn how and why such an equation models the physical system we have described, and we shall see how its solution sheds light on the physics of the situation.

5.1.1 Undamped Simple Harmonic Motion

Our basic example will be a cart of mass M attached to a nearby wall by means of a spring. See Figure 5.1.

The spring exerts no force when the cart is at its rest position $x = 0$ (notice that, contrary to custom, we are locating the origin to the right of the wall). According to Hooke's law, if the cart is displaced a distance x, then the spring exerts a proportional force $F_s = -kx$, where k is a positive constant known as Hooke's constant. Observe that, if $x > 0$, then the cart is moved to the right and the spring pulls to the left; so the force is negative. Obversely, if $x < 0$ then the cart is moved to the left and the spring resists with a force to the right; so the force is positive.

Newton's second law of motion says that the mass of the cart times its acceleration equals the force acting on the cart. Thus

$$M \cdot \frac{d^2x}{dt^2} = F_s = -k \cdot x .$$

DOI: 10.1201/9781003214526-5

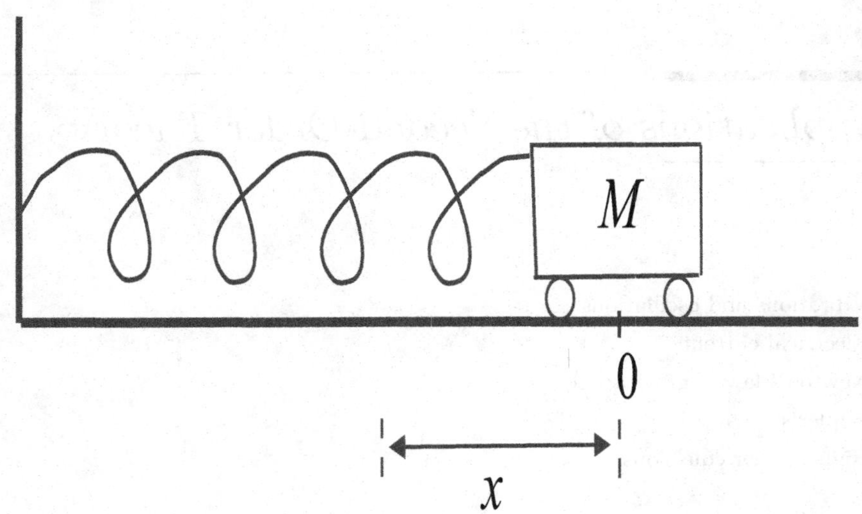

FIGURE 5.1
Hooke's law.

As a result,

$$\frac{d^2x}{dt^2} + \frac{k}{M}x = 0 .$$

It is useful to let $a = \sqrt{k/M}$ (both k and M are positive) and thus to write the equation as

$$\frac{d^2x}{dt^2} + a^2 x = 0 .$$

Of course this is a familiar differential equation for us, and we can write its general solution immediately:

$$x(t) = A \sin at + B \cos at .$$

Now suppose that the cart is pulled to the right to an initial position of $x = x_0 > 0$ and then is simply released (with initial velocity 0). Then we have the initial conditions

$$x(0) = x_0 \qquad \text{and} \qquad \frac{dx}{dt}(0) = 0 .$$

Thus

$$\begin{aligned} x_0 &= A \sin(a \cdot 0) + B \cos(a \cdot 0) \\ 0 &= Aa \cos(a \cdot 0) - Ba \sin(a \cdot 0) \end{aligned}$$

or

$$\begin{aligned} x_0 &= B \\ 0 &= A \cdot a . \end{aligned}$$

We conclude that $B = x_0$, $A = 0$, and we find the solution of the system to be

$$x(t) = x_0 \cos at .$$

In other words, if the cart is displaced a distance x_0 and released, then the result is a simple harmonic motion (described by the cosine function) with *amplitude* x_0 (i.e., the cart glides back and forth, x_0 units to the left of the origin and then x_0 units to the right) and with period $T = 2\pi/a$ (which means that the motion repeats itself every $2\pi/a$ units of time).

The *frequency* f of the motion is the number of cycles per unit of time, hence $f \cdot T = 1$, or $f = 1/T = a/(2\pi)$. It is useful to substitute back in the actual value of a so that we can analyze the physics of the system. Thus

$$\text{amplitude} = x_0$$

$$\text{period} = T = \frac{2\pi\sqrt{M}}{\sqrt{k}}$$

$$\text{frequency} = f = \frac{\sqrt{k}}{2\pi\sqrt{M}} .$$

We see that, if the stiffness k of the spring is increased, then the period becomes smaller and the frequency increases. Likewise, if the mass M of the cart is increased then the period increases and the frequency decreases.

5.1.2 Damped Vibrations

It probably has occurred to the reader that the physical model in the last subsection is not realistic. Typically, a cart that is attached to a spring and released, just as we have described, will enter a harmonic motion that *dies out over time.* In other words, resistance and friction will cause the system to be damped. Let us add that information to the system.

Physical considerations make it plausible to postulate that the resistance is proportional to the velocity of the moving cart. Thus

$$F_d = -c\frac{dx}{dt} ,$$

where F_d denotes damping force and $c > 0$ is a positive constant that measures the resistance of the medium (air or water or oil, etc.). Notice, therefore, that when the cart is traveling to the right then $dx/dt > 0$ and therefore the force of resistance is negative (i.e., in the other direction). Likewise, when the cart is traveling to the left then $dx/dt < 0$ and the force of resistance is positive.

Since the total of all the forces acting on the cart equals the mass times the acceleration, we now have

$$M \cdot \frac{d^2x}{dt^2} = F_s + F_d .$$

In other words,

$$\frac{d^2x}{dt^2} + \frac{c}{M} \cdot \frac{dx}{dt} + \frac{k}{M} \cdot x = 0 .$$

Because of convenience and tradition, we again take $a = \sqrt{k/M}$ and we set $b = c/(2M)$. Thus the differential equation takes the form

$$\frac{d^2x}{dt^2} + 2b \cdot \frac{dx}{dt} + a^2 \cdot x = 0 .$$

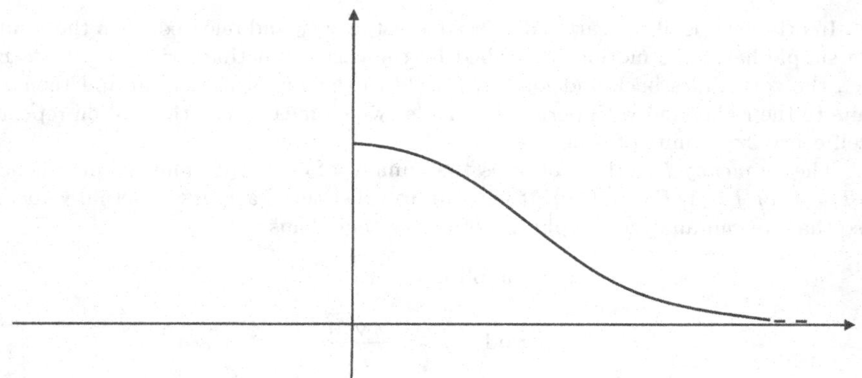

FIGURE 5.2
The motion dies out.

This is a second-order, linear, homogeneous ordinary differential equation with constant coefficients. The associated polynomial is

$$r^2 + 2br + a^2 = 0\,,$$

and it has roots

$$r_1, r_2 = \frac{-2b \pm \sqrt{4b^2 - 4a^2}}{2} = \frac{-2b \pm 2\sqrt{b^2 - a^2}}{2} = -b \pm \sqrt{b^2 - a^2}\,.$$

Now we must consider three cases.

CASE A: $c^2 - 4kM > 0$ In other words, $b^2 - a^2 > 0$. We are assuming that the frictional force (which depends on c) is significantly larger than the stiffness of the spring (which depends on k). Thus we would expect the system to damp heavily. In any event, the calculation of r_1, r_2 involves the square root of a *positive real number*, and thus r_1, r_2 are distinct real (and negative) roots of the associated polynomial equation.

Thus the general solution of our system in this case is

$$x = Ae^{r_1 t} + Be^{r_2 t}\,,$$

where (we repeat) r_1, r_2 are negative real numbers. We apply the initial conditions $x(0) = x_0$, $dx/dt(0) = 0$, just as in the last section (details are left to the reader). The result is the particular solution

$$x(t) = \frac{x_0}{r_1 - r_2}\left(r_1 e^{r_2 t} - r_2 e^{r_1 t}\right). \tag{5.1}$$

Notice that, in this heavily damped system, no oscillation occurs (i.e., there are no sines or cosines in the expression for $x(t)$). The system simply dies out. Figure 5.2 exhibits the graph of the function in (1).

CASE B: $c^2 - 4kM = 0$ In other words, $b^2 - a^2 = 0$. This is the critical case, where the resistance balances the force of the spring. We see that $b = a$ (both are known to be positive) and $r_1 = r_2 = -b = -a$. We know, then, that the general solution to our differential equation is

$$x(t) = Ae^{-at} + Bte^{-at}.$$

When the standard initial conditions are imposed, we find the particular solution

$$x(t) = x_0 \cdot e^{-at}(1 + at).$$

We see that this differs from the situation in **CASE A** by the factor $(1 + at)$. That factor of course attenuates the damping, but there is *still no oscillatory motion.* We call this the *critical case.* The graph of our new $x(t)$ is quite similar to the graph already shown in Figure 5.2.

If there is any small decrease in the viscosity, however slight, then the system will begin to vibrate (as one would expect). That is the next, and last, case that we examine.

CASE C: $c^2 - 4kM < 0$ This says that $b^2 - a^2 < 0$. Now $0 < b < a$ and the calculation of r_1, r_2 entails taking the square root of a negative number. Thus r_1, r_2 are the conjugate complex numbers $-b \pm i\sqrt{a^2 - b^2}$. We set $\alpha = \sqrt{a^2 - b^2}$.

Now the general solution of our system, as we well know, is

$$x(t) = e^{-bt}\left(A\sin\alpha t + B\cos\alpha t\right).$$

If we evaluate A, B according to our usual initial conditions, then we find the particular solution

$$x(t) = \frac{x_0}{\alpha}e^{-bt}\left(b\sin\alpha t + \alpha\cos\alpha t\right).$$

It is traditional and convenient to set $\theta = \arctan(b/\alpha)$. With this notation, we can express the last equation in the form

$$x(t) = \frac{x_0\sqrt{\alpha^2 + b^2}}{\alpha}e^{-bt}\cos(\alpha t - \theta). \tag{5.2}$$

As you can see, there is oscillation because of the presence of the cosine function. The amplitude (the expression that appears in front of cosine) clearly falls off—rather rapidly—with t because of the presence of the exponential. The graph of this function is exhibited in Figure 5.3.

Of course this function is *not* periodic—it is dying off, and not repeating itself. What *is true*, however, is that the graph crosses the t-axis (the equilibrium position $x = 0$) at regular intervals. If we consider this interval T (which is not a "period," strictly speaking) as the time required for one complete cycle, then $\alpha T = 2\pi$ so

$$T = \frac{2\pi}{\alpha} = \frac{2\pi}{\sqrt{k/M - c^2/(4M^2)}}. \tag{5.3}$$

We define the number f, which plays the role of "frequency" with respect to the indicated time interval, to be

$$f = \frac{1}{T} = \frac{1}{2\pi}\sqrt{\frac{k}{M} - \frac{c^2}{4M^2}}.$$

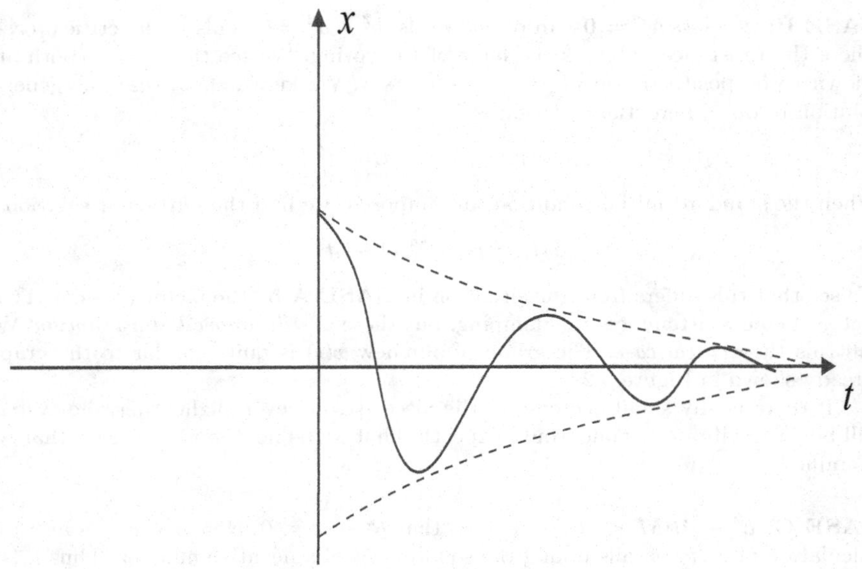

FIGURE 5.3
A damped vibration.

This number is commonly called the *natural frequency* of the system. When the viscosity vanishes, then our solution clearly reduces to the one we found earlier when there was no viscosity present. We also see that the frequency of the vibration is reduced by the presence of damping; increasing the viscosity further reduces the frequency.

5.1.3 Forced Vibrations

The vibrations that we have considered so far are called *free vibrations* because all the forces acting on the system are internal to the system itself. We now consider the situation in which there is an *external force* $F_e = f(t)$ acting on the system. This force could be an external magnetic field (acting on the steel cart) or vibration of the wall, or perhaps a stiff wind blowing. Again setting mass times acceleration equal to the resultant of all the forces acting on the system, we have

$$M \cdot \frac{d^2x}{dt^2} = F_s + F_d + F_e \,.$$

Taking into account the definitions of the various forces, we may write the differential equation as

$$M\frac{d^2x}{dt^2} + c\frac{dx}{dt} + kx = f(t)\,.$$

So we see that the equation describing the physical system is second-order linear, and that the external force gives rise to an nonhomogeneous term on the right. An interesting special case occurs when $f(t) = F_0 \cdot \cos\omega t$, in other words when that external force is periodic. Thus our equation becomes

$$M\frac{d^2x}{dt^2} + c\frac{dx}{dt} + kx = F_0 \cdot \cos \omega t\,. \tag{5.4}$$

If we can find a particular solution of this equation, then we can combine it with the information about the solution of the associated homogeneous equation in the last subsection and then come up with the general solution of the differential equation. We shall use the method of undetermined coefficients. Considering the form of the right-hand side, our guess will be

$$x(t) = A\sin \omega t + B\cos \omega t\,.$$

Substituting this guess into the differential equation gives

$$\begin{aligned} M\frac{d^2}{dt^2}[A\sin \omega t + B\cos \omega t] + c\frac{d}{dt}[A\sin \omega t + B\cos \omega t] \\ +k[A\sin \omega t + B\cos \omega t] = F_0 \cdot \cos \omega t\,. \end{aligned}$$

With a little calculus and a little algebra we are led to the algebraic equations

$$\begin{aligned} \omega c A + (k - \omega^2 M)B &= F_0 \\ (k - \omega^2 M)A - \omega c B &= 0\,. \end{aligned}$$

We solve for A and B to obtain

$$A = \frac{\omega c F_0}{(k-\omega^2 M)^2 + \omega^2 c^2} \qquad \text{and} \qquad B = \frac{(k-\omega^2 M)F_0}{(k-\omega^2 M)^2 + \omega^2 c^2}\,.$$

Thus we have found the particular solution

$$x_0(t) = \frac{F_0}{(k-\omega^2 M)^2 + \omega^2 c^2}\left(\omega c \sin \omega t + (k - \omega^2 M)\cos \omega t\right).$$

We may write this in a more useful form with the notation $\phi = \arctan[\omega c/(k-\omega^2 M)]$. Thus

$$x_0(t) = \frac{F_0}{\sqrt{(k-\omega^2 M)^2 + \omega^2 c^2}} \cdot \cos(\omega t - \phi)\,. \tag{5.5}$$

If we assume that we are dealing with the underdamped system, which is **CASE C** of the last subsection, we find that the general solution of our differential equation with a periodic external forcing term is

$$\begin{aligned} x(t) \quad &= \quad e^{-bt}\left(A\cos \alpha t + B\sin \alpha t\right) \\ &\quad + \frac{F_0}{\sqrt{(k-\omega^2 M)^2 + \omega^2 c^2}} \cdot \cos(\omega t - \phi)\,. \end{aligned}$$

We see that, as long as some damping is present in the system (that is, b is nonzero and positive), then the first term in the definition of $x(t)$ is clearly transient (i.e., it dies as $t \to \infty$ because of the exponential term). Thus, as time goes on, the motion assumes the character of the second term in $x(t)$, which is the *steady state* term. So we can say that, for large t, the physical nature of the general solution to our system is more or less like that of the particular solution $x_0(t)$ that we found.

The frequency of this forced vibration equals the impressed frequency (originating with the external forcing term) $\omega/2\pi$. The amplitude is the coefficient

$$\frac{F_0}{\sqrt{(k-\omega^2 M)^2+\omega^2 c^2}}. \tag{5.6}$$

This expression for the amplitude depends on all the relevant physical constants, and it is enlightening to analyze it a bit. Observe, for instance, that if the viscosity c is very small and if ω is close to $\sqrt{k/M}$ (so that $k-\omega^2 M$ is very small) then the motion is lightly damped and the external (impressed) frequency $\omega/2\pi$ is close to the natural frequency

$$\frac{1}{2\pi}\sqrt{\frac{k}{M}-\frac{c^2}{4M^2}}.$$

Then the amplitude is very large (because we are dividing by a number close to 0). This phenomenon is known as *resonance*. There are classical examples of resonance.[1] For instance, several years ago there was a celebration of the anniversary of the Golden Gate Bridge (built in 1937), and many thousands of people marched in unison across the bridge. The frequency of their footfalls was so close to the natural frequency of the bridge (thought of as a suspended string under tension) that the bridge nearly fell apart. A famous incident at the Tacoma Narrows Bridge has been attributed to resonance, although more recent studies suggest a more complicated combination of effects (see the movie of this disaster at `http://www.ketchum.org/bridgecollapse.html`).

5.1.4 A Few Remarks about Electricity

It is known that if a periodic electromotive force, $E = E_0$, acts in a simple circuit containing a resistor, an inductor, and a capacitor, then the charge Q on the capacitor is governed by the differential equation

$$L\frac{d^2Q}{dt^2}+R\frac{dQ}{dt}+\frac{1}{C}Q=E_0\cos\omega t\,.$$

This equation is of course quite similar to equation (20) in Section 3.2 for the oscillating cart with external force. In particular, the following correspondences (or analogies) are suggested:

$$\begin{aligned}
\text{mass } M &\quad\longleftrightarrow\quad \text{inductance } L\,;\\
\text{viscosity } c &\quad\longleftrightarrow\quad \text{resistance } R\,;\\
\text{stiffness of spring } k &\quad\longleftrightarrow\quad \text{reciprocal of capacitance } \frac{1}{C}\,;\\
\text{displacement } x &\quad\longleftrightarrow\quad \text{charge } Q \text{ on capacitor}\,.
\end{aligned}$$

The analogy between the mechanical and electrical systems renders identical the mathematical analysis of the two systems, and enables us to carry over at once all mathematical conclusions from the first to the second. In the given electric circuit, we therefore have a critical resistance below which the free behavior of the circuit will be vibratory with a certain natural frequency, a forced steady-state vibration

[1] One of the basic ideas behind filter design is resonance.

of the charge Q, and resonance phenomena that appear when the circumstances are favorable.

Math Nugget

Charles Proteus Steinmetz (1865–1923) was a mathematician, inventor, and electrical engineer. He pioneered the use of complex numbers in the study of electrical circuits. After he left Germany (on account of his socialist political activities) and emigrated to America, he was employed by the General Electric Company. He soon solved some of GE's biggest problems—to design a method to mass-produce electric motors, and to find a way to transmit electricity more than 3 miles. With these contributions alone Steinmetz had a massive impact on mankind.

Steinmetz was a dwarf, crippled by a congenital deformity. He lived in pain, but was well liked for his humanity and his sense of humor, and certainly admired for his scientific prowess. The following Steinmetz story comes from the Letters section of *Life* Magazine (May 14, 1965):

> Sirs: In your article on Steinmetz (April 23) you mentioned a consultation with Henry Ford. My father, Burt Scott, who was an employee of Henry Ford for many years, related to me the story behind the meeting. Technical troubles developed with a huge new generator at Ford's River Rouge plant. His electrical engineers were unable to locate the difficulty so Ford solicited the aid of Steinmetz. When "the little giant" arrived at the plant, he rejected all assistance, asking only for a notebook, pencil and cot. For two straight days and nights he listened to the generator and made countless computations. Then he asked for a ladder, a measuring tape, and a piece of chalk.
>
> He laboriously ascended the ladder, made careful measurements, and put a chalk mark on the side of the generator. He descended and told his skeptical audience to remove a plate from the side of the generator [at the marked spot] and take out 16 windings from the field coil at that location. The corrections were made and the generator then functioned perfectly. Subsequently Ford received a bill for $10,000 signed by Steinmetz for G.E. Ford returned the bill acknowledging the good job done by Steinmetz but respectfully requesting an itemized statement. Steinmetz replied as follows: Making chalk mark on generator $1. Knowing where to make mark $9,999. Total due $10,000.

Exercises

1. Consider the forced vibration (5) in the underdamped case, and find the impressed frequency for which the amplitude (6) attains a maximum. Will such an impressed frequency necessarily exist? This value of the impressed frequency, when it exists, is called the *resonance frequency*. Show that the resonance frequency is always less than the natural frequency.
2. Consider the underdamped free vibration described by formula (2). Show that x assumes maximum values for $t = 0, T, 2T, \ldots$, where T is the "period," as given in formula (3). If x_1 and x_2 are any two successive maxima, then show that $x_1/x_2 = e^{bT}$. The logarithm of this quantity, or bT, is known as the *logarithmic decrement* of the vibration.
3. A spherical buoy of radius r floats half-submerged in water. If it is depressed slightly, then a restoring force equal to the weight of the displaced water presses it upward; and if it is then released, it will bob up and down. Find the period of oscillation if the friction of the water is negligible.
4. A cylindrical buoy 2 feet in diameter floats with its axis vertical in fresh water of density 62.4 lb./ft.^{3} When depressed slightly and released, its period of oscillation is observed to be 1.9 seconds. What is the weight of the buoy?
5. Suppose that a straight tunnel is drilled through the Earth between two points on its surface. If tracks are laid, then—neglecting friction—a train placed in the tunnel at one end will roll through the Earth under its own weight, stop at the other end, and return. Show that the time required for a single, complete round trip is the same for all such tunnels (no matter what the beginning and ending points), and estimate its value. If the tunnel is $2L$ miles long, then what is the greatest speed attained by the train on its journey?
6. The cart in Figure 5.1 weighs 128 pounds and is attached to the wall by a spring with spring constant $k = 64$ lb./ft. The cart is pulled 6 inches in the direction away from the wall and released with no initial velocity. Simultaneously, a periodic external force $F_e = f(t) = 32\sin 4t$ lb is applied to the cart. Assuming that there is no air resistance, find the position $x = x(t)$ of the cart at time t. Note particularly that $|x(t)|$ assumes arbitrarily large values as $t \to +\infty$. This phenomenon is known as *pure resonance* and is caused by the fact that the forcing function has the same period as the free vibrations of the unforced system.
7. Use your symbol manipulation software, such as `Maple` or `Mathematica`, to solve the ordinary differential equation with the given damping term and forcing term. In each instance you should assume that both the damping and the forcing terms occur on the right-hand side of the differential equation and that $t > 0$.

   (a) damping $= -e^t\, dx/dt, \quad f = \sin t + \cos 2t$

   (b) damping $= -\ln t\, dx/dt, \quad f = e^t$

   (c) damping $= -[e^t] \cdot \ln t\, dx/dt, \quad f = \cos 2t$

   (d) damping $= -t^3\, dx/dt, \quad f = e^{-t}$

5.2 Newton's Law of Gravitation and Kepler's Laws

Newton's law of universal gravitation is one of the great ideas of modern physics. It underlies so many important physical phenomena that it is part of the bedrock of science. In this section we show how Kepler's laws of planetary motion can be derived from Newton's gravitation law. It might be noted that Johannes Kepler himself (1571–1630) used thousands of astronomical observations (made by his teacher Tycho Brahe (1546–1601)) in order to formulate his laws. Kepler was a follower of Copernicus, who postulated that the planets orbited about the sun, but Brahe held the more traditional view that the Earth was the center of the orbits. Brahe did not want to let Kepler use his data, for he feared that Kepler would promote the Copernican theory. As luck would have it, Brahe died from a burst bladder after a night of excessive beer drinking at a social function. So Kepler was able to get the valuable numbers from Tycho Brahe's family.

Interestingly, Copernicus believed that the orbits were circles (rather than ellipses, as we now know them to be). Newton determined how to derive the laws of motion analytically, and he was able to *prove* that the orbits must be ellipses (although it should be noted that the ellipses are very nearly circular). Furthermore, the eccentricity of an elliptical orbit has an important physical interpretation. The present section explores all these ideas.

KEPLER'S LAWS OF PLANETARY MOTION

I. The orbit of each planet is an ellipse with the sun at one focus (Figure 5.4).

II. The segment from the center of the sun to the center of an orbiting planet sweeps out area at a constant rate (Figure 5.5).

III. The square of the period of revolution of a planet is proportional to the cube of the length of the major axis of its elliptical orbit, with the same constant of proportionality for any planet. (Figure 5.6).

It turns out that the eccentricities of the ellipses that arise in the orbits of the planets are very small so that the orbits are nearly circles, but they are definitely *not* circles. That is the importance of Kepler's first law.

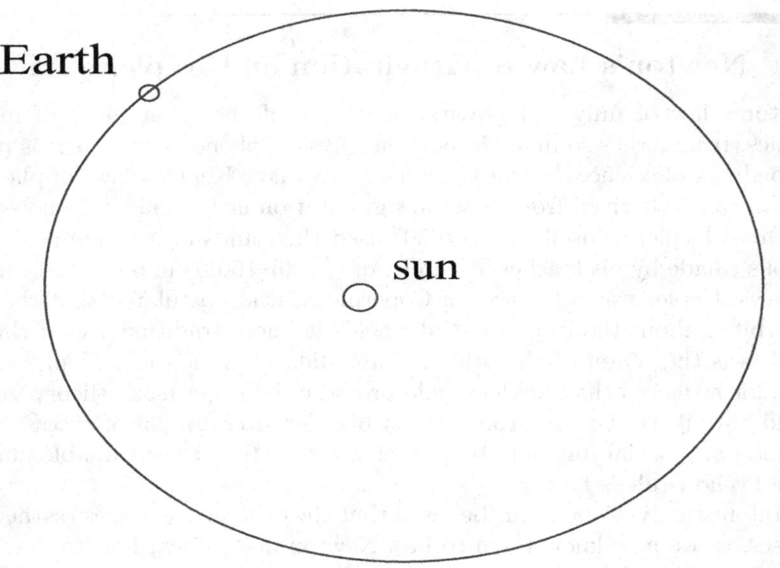

FIGURE 5.4
The elliptical orbit of the Earth about the sun.

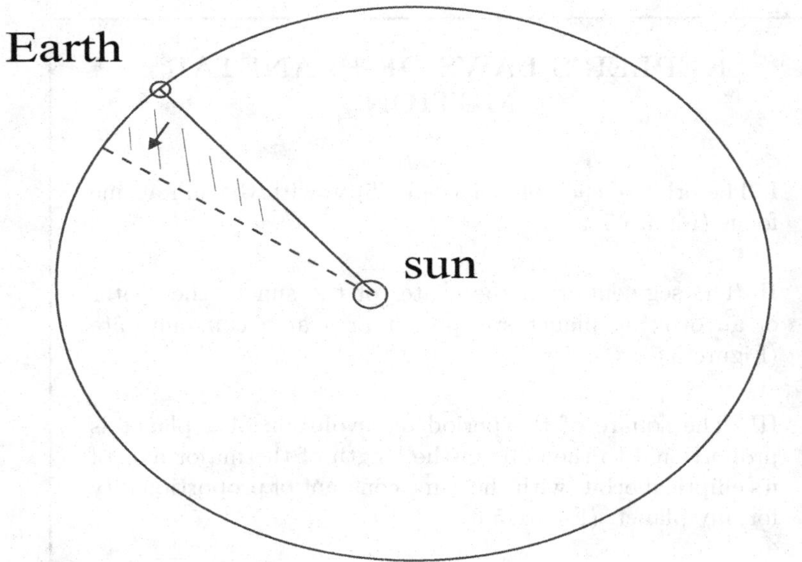

FIGURE 5.5
Area is swept out at a constant rate.

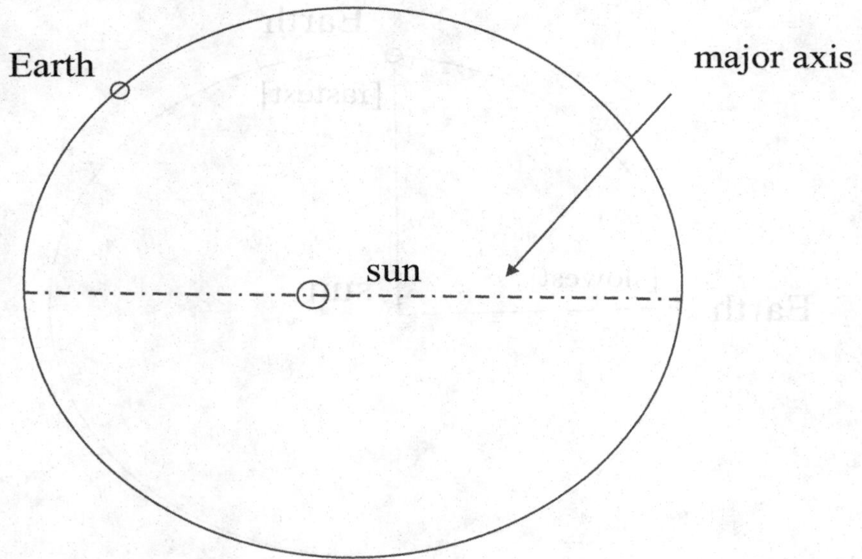

FIGURE 5.6
The square of the period is proportional to the cube of the major axis.

The second law tells us that when the planet is at its apogee (furthest from the sun) then it is traveling relatively slowly, whereas at its perigee (nearest point to the sun), it is traveling relatively rapidly—Figure 5.7. In fact the second law is valid for any central force, and Newton knew this important fact.

The third law allows us to calculate the length of a year on any given planet from knowledge of the shape of its orbit.

In this section we shall learn how to derive Kepler's three laws from Newton's inverse square law of gravitational attraction. To keep matters as simple as possible, we shall assume that our solar system contains a fixed sun and just one planet (the Earth, for instance). The problem of analyzing the gravitation influence of three or more planets on each other is incredibly complicated and is still not thoroughly understood.

The argument that we present is due to S. Kochen and is used with his permission.

5.2.1 Kepler's Second Law

It is convenient to derive the second law first. We use a polar coordinate system with the origin at the center of the sun. We analyze a single planet which orbits the sun, and we denote the position of that planet at time t by the vector $\mathcal{R}(t)$. The only physical facts that we shall use in this portion of

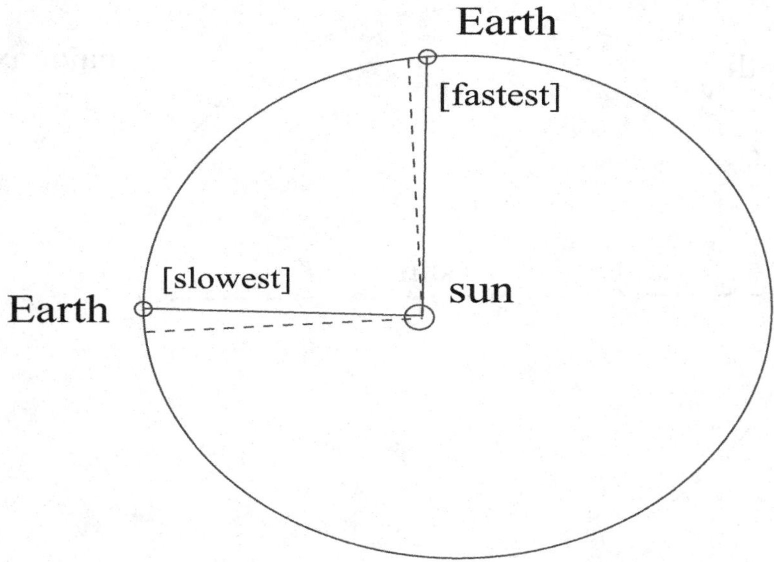

FIGURE 5.7
Apogee motion vs. perigee motion.

the argument are Newton's second law and the self-evident assertion that the gravitational force exerted by the sun on a planet is a vector parallel to $\mathcal{R}(t)$. See Figure 5.8.

If **F** is force, m is the mass of the planet (Earth), and **a** is its acceleration then Newton's second law says that

$$\mathbf{F} = m\mathbf{a} = m\mathcal{R}''(t).$$

We conclude that $\mathcal{R}(t)$ is parallel to $\mathcal{R}''(t)$ for every value of t.

Now

$$\frac{d}{dt}\left(\mathcal{R}(t) \times \mathcal{R}'(t)\right) = \left[\mathcal{R}'(t) \times \mathcal{R}'(t)\right] + \left[\mathcal{R}(t) \times \mathcal{R}''(t)\right].$$

Note that the first of these terms is zero because the cross product of any vector with itself is zero. The second is zero because $\mathcal{R}(t)$ is parallel to $\mathcal{R}''(t)$ for every t. We conclude that

$$\mathcal{R}(t) \times \mathcal{R}'(t) = \mathbf{C}, \tag{5.7}$$

where **C** is a constant vector. Notice that this already guarantees that $\mathcal{R}(t)$ and $\mathcal{R}'(t)$ always lie in the same plane, hence that the orbit takes place in a plane.

Now let Δt be an increment of time, $\Delta\mathcal{R}$ the corresponding increment of position, and ΔA the increment of area swept out. Look at Figure 5.9.

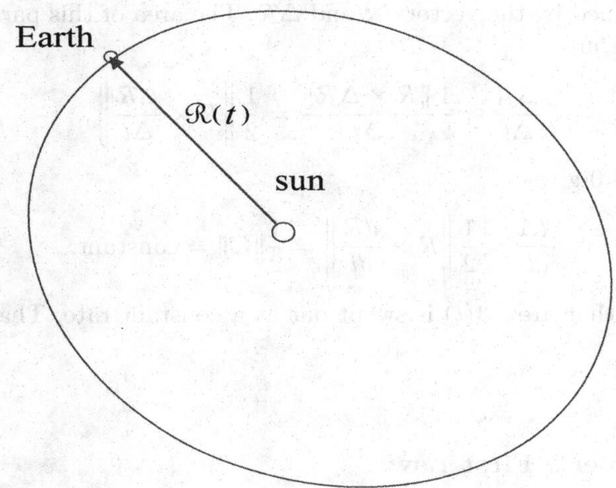

FIGURE 5.8
Polar coordinate system for the motion of the Earth about the sun.

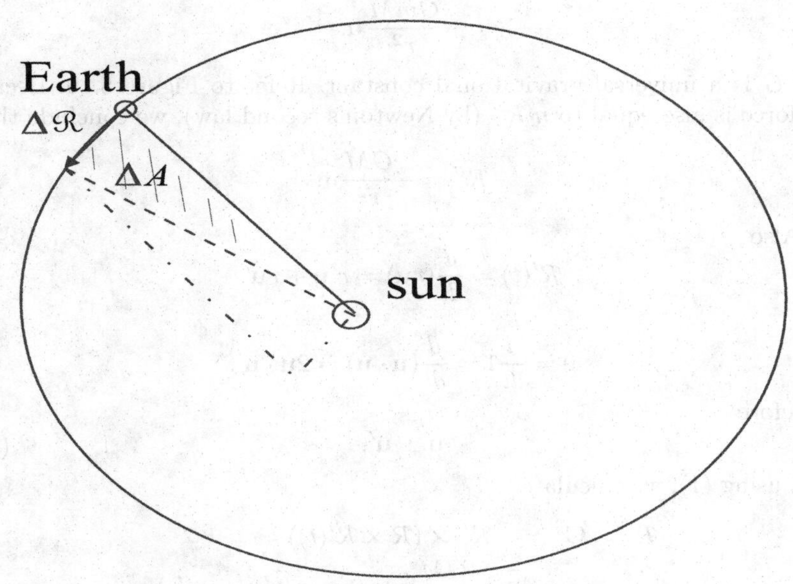

FIGURE 5.9
The increment of area.

We see that ΔA is approximately equal to half the area of the parallelogram determined by the vectors $\mathcal{R}$ and $\Delta\mathcal{R}$. The area of this parallelogram is $\|\mathcal{R} \times \Delta\mathcal{R}\|$. Thus

$$\frac{\Delta A}{\Delta t} \approx \frac{1}{2}\frac{\|\mathcal{R} \times \Delta\mathcal{R}\|}{\Delta t} = \frac{1}{2}\left\|\mathcal{R} \times \frac{\Delta\mathcal{R}}{\Delta t}\right\|.$$

Letting $\Delta t \to 0$ gives

$$\frac{dA}{dt} = \frac{1}{2}\left\|\mathcal{R} \times \frac{d\mathcal{R}}{dt}\right\| = \frac{1}{2}\|\mathbf{C}\| = \text{constant}.$$

We conclude that area $A(t)$ is swept out at a constant rate. That is Kepler's second law.

5.2.2 Kepler's First Law

Now we write $\mathcal{R}(t) = r(t)\mathbf{u}(t)$, where $\mathbf{u}$ is a unit vector pointing in the same direction as $\mathcal{R}$ and r is a positive, scalar-valued function representing the length of $\mathcal{R}$. We use Newton's inverse square law for the attraction of two bodies. If one body (the sun) has mass M and the other (the planet) has mass m then Newton says that the force exerted by gravity on the planet is

$$-\frac{GmM}{r^2}\mathbf{u}.$$

Here G is a universal gravitational constant. Refer to Figure 5.10. Because this force is also equal to $m\mathcal{R}''$ (by Newton's second law), we conclude that

$$\mathcal{R}'' = -\frac{GM}{r^2}\mathbf{u}.$$

Also

$$\mathcal{R}'(t) = \frac{d}{dt}(r\mathbf{u}) = r'\mathbf{u} + r\mathbf{u}'$$

and

$$0 = \frac{d}{dt}1 = \frac{d}{dt}(\mathbf{u}\cdot\mathbf{u}) = 2\mathbf{u}\cdot\mathbf{u}'.$$

Therefore

$$\mathbf{u} \perp \mathbf{u}'. \tag{5.8}$$

Now, using (1), we calculate

$$\begin{aligned}
\mathcal{R}'' \times \mathbf{C} &= \mathcal{R}'' \times (\mathcal{R} \times \mathcal{R}'(t)) \\
&= -\frac{GM}{r^2}\mathbf{u} \times (r\mathbf{u} \times (r'\mathbf{u} + r\mathbf{u}')) \\
&= -\frac{GM}{r^2}\mathbf{u} \times (r\mathbf{u} \times r\mathbf{u}') \\
&= -GM\left(\mathbf{u} \times (\mathbf{u} \times \mathbf{u}')\right).
\end{aligned}$$

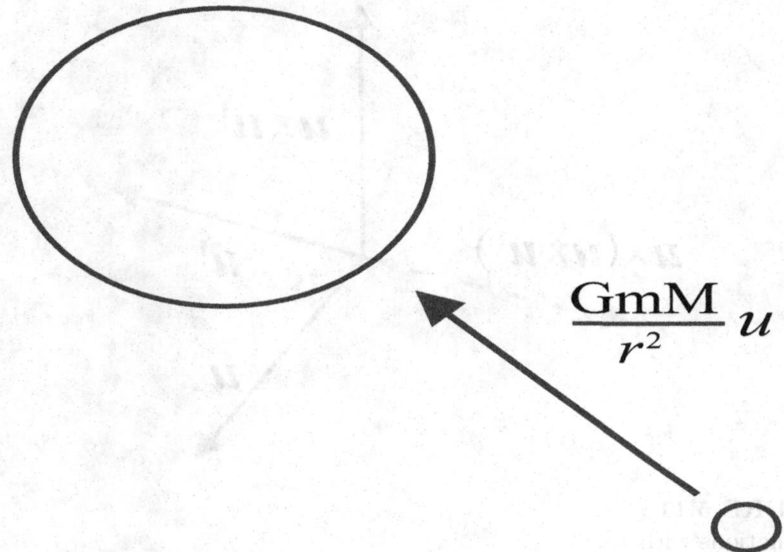

FIGURE 5.10
Newton's universal law of gravitation.

We can determine the vector $\mathbf{u} \times (\mathbf{u} \times \mathbf{u}')$. For, with formula (8), we see that $\mathbf{u}$ and $\mathbf{u}'$ are perpendicular and that $\mathbf{u} \times \mathbf{u}'$ is perpendicular to both of these. Because $\mathbf{u} \times (\mathbf{u} \times \mathbf{u}')$ is perpendicular to the first and last of these three, it must therefore be parallel to $\mathbf{u}'$. It also has the same length as $\mathbf{u}'$ and, by the right-hand rule, points in the opposite direction. Look at Figure 5.11. We conclude that $\mathbf{u} \times (\mathbf{u} \times \mathbf{u}') = -\mathbf{u}'$, hence that

$$\mathcal{R}'' \times \mathbf{C} = GM\mathbf{u}'.$$

If we antidifferentiate this last equality we obtain

$$\mathcal{R}'(t) \times \mathbf{C} = GM(\mathbf{u} + \mathbf{K}),$$

where $\mathbf{K}$ is a constant vector of integration.

Thus we have

$$\mathcal{R} \cdot (\mathcal{R}'(t) \times \mathbf{C}) = r\mathbf{u}(t) \cdot GM(\mathbf{u}(t) + \mathbf{K}) = GMr(1 + \mathbf{u}(t) \cdot \mathbf{K}),$$

because $\mathbf{u}(t)$ is a unit vector. If $\theta(t)$ is the angle between $\mathbf{u}(t)$ and $\mathbf{K}$ then we may rewrite our equality as

$$\mathcal{R} \cdot (\mathcal{R}' \times \mathbf{C}) = GMr(1 + \|\mathbf{K}\| \cos\theta)\,.$$

By a standard triple product formula,

$$\mathcal{R} \cdot (\mathcal{R}'(t) \times \mathbf{C}) = (\mathcal{R} \times \mathcal{R}'(t)) \cdot \mathbf{C}\,,$$

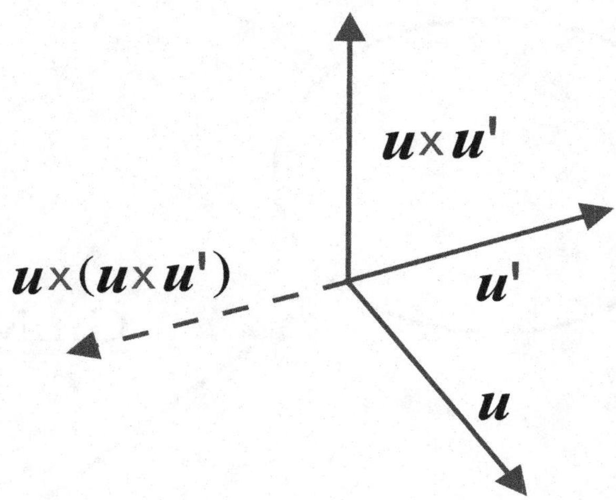

FIGURE 5.11
Calculations with $\boldsymbol{u}'$.

which in turn equals

$$\mathbf{C} \cdot \mathbf{C} = \|\mathbf{C}\|^2.$$

(Here we have used the fact, which we derived in the proof of Kepler's second law, that $\mathcal{R} \times \mathcal{R}' = \mathbf{C}$.)

Thus

$$\|\mathbf{C}\|^2 = GMr(1 + \|\mathbf{K}\| \cos\theta).$$

(Notice that this equation can be true only if $\|\mathbf{K}\| \leq 1$. This fact will come up again below.)

We conclude that

$$r = \frac{\|\mathbf{C}\|^2}{GM} \cdot \left(\frac{1}{1 + \|\mathbf{K}\| \cos\theta} \right).$$

This is the polar equation for an ellipse of eccentricity $\|\mathbf{K}\|$. (Exercises 4 and 5 will say a bit more about such polar equations.)

We have verified Kepler's first law.

5.2.3 Kepler's Third Law

Look at Figure 5.12. The length $2a$ of the major axis of our elliptical orbit is equal to the maximum value of r plus the minimum value of r. From the equation for the ellipse we see that these occur, respectively, when $\cos\theta$ is $+1$ and when $\cos\theta$ is -1. Thus

$$2a = \frac{\|\mathbf{C}\|^2}{GM} \frac{1}{1 - \|\mathbf{K}\|} + \frac{\|\mathbf{C}\|^2}{GM} \frac{1}{1 + \|\mathbf{K}\|} = \frac{2\|\mathbf{C}\|^2}{GM(1 - \|\mathbf{K}\|^2)}.$$

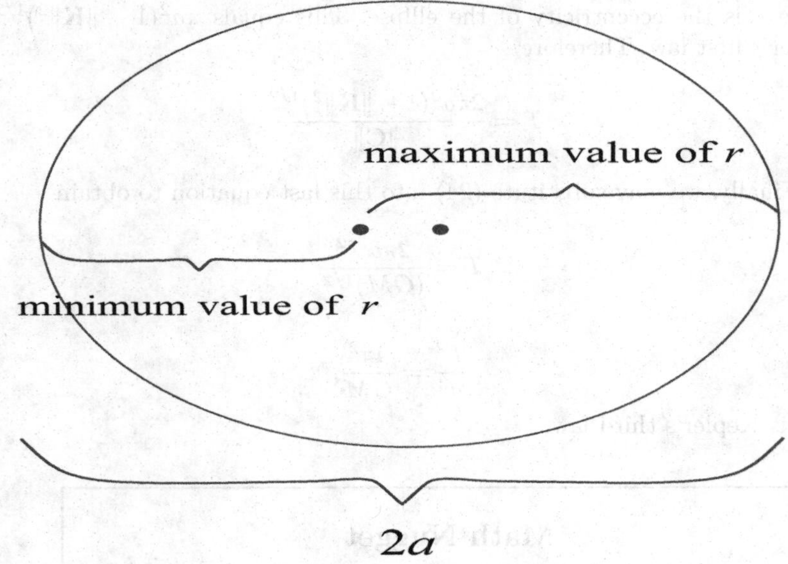

FIGURE 5.12
Analysis of the major axis.

We conclude that

$$\|\mathbf{C}\| = \left(aGM(1 - \|\mathbf{K}\|^2)\right)^{1/2}. \tag{5.9}$$

Now recall from our proof of the second law that

$$\frac{dA}{dt} = \frac{1}{2}\|\mathbf{C}\|.$$

Then, by antidifferentiating, we find that

$$A(t) = \frac{1}{2}\|\mathbf{C}\|t.$$

(There is no constant term since $A(0) = 0$.) Let $\mathcal{A}$ denote the total area inside the elliptical orbit and T the time it takes to sweep out one orbit. Then

$$\mathcal{A} = A(T) = \frac{1}{2}\|\mathbf{C}\|T.$$

Solving for T we obtain

$$T = \frac{2\mathcal{A}}{\|\mathbf{C}\|}.$$

But the area inside an ellipse with major axis $2a$ and minor axis $2b$ is

$$\mathcal{A} = \pi ab = \pi a^2(1 - e^2)^{1/2},$$

where e is the eccentricity of the ellipse. This equals $\pi a^2(1 - \|\mathbf{K}\|^2)^{1/2}$ by Kepler's first law. Therefore

$$T = \frac{2\pi a^2(1 - \|\mathbf{K}\|^2)^{1/2}}{\|\mathbf{C}\|}.$$

Finally, we may substitute (24) into this last equation to obtain

$$T = \frac{2\pi a^{3/2}}{(GM)^{1/2}}$$

or

$$\frac{T^2}{a^3} = \frac{4\pi^2}{GM}.$$

This is Kepler's third law.

Math Nugget

Johannes Kepler and the Motion of the Planets

Johannes Kepler (1571–1630) is said to have been an energetic and affectionate man. He married his first wife in part for her money, and soon realized the error of his ways. When she died, he decided to apply scientific methods in the selection of a second wife: he carefully analyzed and compared the virtues and defects of several ladies before selecting his second partner in matrimony. That marriage too was an unhappy one.

Kepler's scientific career also had its ups and downs. His attempt at collaboration with his hero Tycho Brahe fell victim to the incompatibility of their strong personalities. In his position as Royal Astronomer in Prague, a post which he inherited from Tycho Brahe, he was often unpaid.

It appears that Kepler's personal frustration, his terrific energy, and his scientific genius found a focus in questions about planetary motion. Kepler formulated his three laws by studying many years worth of data about the motion of the planets that had been gathered by Tycho Brahe. It is amazing that he could stare at hundreds of pages of numerical data and come up with the three elegant laws that we have discussed here.

Kepler could have simplified his task considerably by using the tables of logarithms that John Napier (1550-1617) and his assistants were developing at the time. But Kepler could not understand Napier's mathematical justifications for his tables, so he refused to use them.

Later, Newton conceived the idea that Kepler's laws could be *derived*, using calculus, from his inverse square law of gravitational attraction. In fact it seems clear that this problem is one of the main reasons that Newton developed the calculus. Newton's idea was a fantastic insight: that physical laws could be derived from a set of physical axioms was a new technique in the philosophy of science. On top of this, considerable technical proficiency was required to actually carry out the program. Newton's derivation of Kepler's laws, presented here in a modernized and streamlined form, is a model for the way that mathematical physics is done today.

EXAMPLE 5.2.1 The planet Uranus describes an elliptical orbit about the sun. It is known that the semimajor axis of this orbit has length 2870×10^6 kilometers. The gravitational constant is $G = 6.637 \times 10^{-8}$ $\text{cm}^3/(g \cdot \text{sec}^2)$. Finally, the mass of the sun is 2×10^{33} grams. Determine the period of the orbit of Uranus.

Solution: Refer to the explicit formulation of Kepler's third law that we proved above. We have

$$\frac{T^2}{a^3} = \frac{4\pi^2}{GM}.$$

We must be careful to use consistent units. The gravitational constant G is given in terms of grams, centimeters, and seconds. The mass of the sun is in grams. We convert the semimajor axis to centimeters: $a = 2870 \times 10^{11}$ cm. $= 2.87 \times 10^{14}$ cm. Then we calculate that

$$\begin{aligned} T &= \left(\frac{4\pi^2}{GM} \cdot a^3\right)^{1/2} \\ &= \left(\frac{4\pi^2}{(6.637 \times 10^{-8})(2 \times 10^{33})} \cdot (2.87 \times 10^{14})^3\right)^{1/2} \\ &\approx [70.308 \times 10^{17}]^{1/2}\text{sec.} \\ &= 26.516 \times 10^8 \text{ sec.} \end{aligned}$$

Notice how the units mesh perfectly so that our answer is in seconds. There are 3.16×10^7 seconds in an Earth year. We divide by this number to find that the time of one orbit of Uranus is

$$T \approx 83.9 \text{ Earth years}. \qquad \square$$

Exercises

1. It is common to take the "mean distance" of the planet from the sun to be the length of the semimajor axis of the ellipse. That is because this number is the average of the least distance to the sun and the greatest distance to the sun. Now you can answer these questions:
   (a) Mercury's "year" is 88 Earth days. What is Mercury's mean distance from the sun?
   (b) The mean distance of the planet Saturn from the sun is 9.54 astronomical units.[2] What is Saturn's period of revolution about the sun?
2. Show that the speed v of a planet at any point of its orbit is given by

$$v^2 = k\left(\frac{2}{r} - \frac{1}{a}\right).$$

3. Suppose that the Earth explodes into fragments, which fly off at the same speed in different directions into orbits of their own. Use Kepler's third law and the result of Exercise 2 to show that all fragments that do not fall into the sun or escape from the solar system will eventually reunite later at the same point where they began to diverge (i.e., where the explosion took place).
4. Kepler's first law may be written as

$$r = \frac{h^2/k}{1 + e\cos\theta}.$$

Prove this assertion. Kepler's second law may be written as

$$r^2\frac{d\theta}{dt} = h.$$

Prove this assertion too. Let $\mathbf{F}$ be the central attractive force that the sun exerts on the planet and F its magnitude. Now verify these statements:

(a) $F_\theta = 0$ (c) $\dfrac{d^2r}{dt^2} = \dfrac{ke\cos\theta}{r^2}$

(b) $\dfrac{dr}{dt} = \dfrac{ke}{h}\sin\theta$ (d) $F_r = -\dfrac{mk}{r^2} = -G\dfrac{Mm}{r^2}$

[2]Here one astronomical unit is the Earth's mean distance from the sun, which is 93, 000, 000 miles or 150, 000, 000 kilometers.

Use these facts to prove that a planet of mass m is attracted toward the origin (the center of the sun) with a force whose magnitude is inversely proportional to the square of r. (Newton's discovery of this fact caused him to formulate his law of universal gravitation and to investigate its consequences.)

5. It is common to take $h = \|C\|$ and $k = GM$. Kepler's third law may then be formulated as

$$T^2 = \frac{4\pi^2 a^2 b^2}{h^2} = \left(\frac{4\pi^2}{k}\right) a^3. \qquad (*)$$

Prove this formula. In working with Kepler's third law, it is customary to measure T in Earth years and a in astronomical units (see Exercise 1, the footnote, for the definition of this term). With these convenient units of measure, $(*)$ takes the simpler form $T^2 = a^3$. What is the period of revolution T of a planet whose mean distance from the sun is

(a) twice that of the Earth?

(b) three times that of the Earth?

(c) 25 times that of the Earth?

6. Use your symbol manipulation software, such as `Maple` or `Mathematica`, to calculate the orbit of a planet having mass m about a "sun" of mass M, assuming that the planet is given an initial velocity of v_0.

Historical Note

Euler

Leonhard Euler (1707–1783), a Swiss by birth, was one of the foremost mathematicians of all time. He was also arguably the most prolific author of all time in any field. The publication of Euler's complete works was begun in 1911 and the end is still not in sight. The works were originally planned for 72 volumes, but new manuscripts have been discovered and the work continues.

Euler's interests were vast, and ranged over all parts of mathematics and science. He wrote effortlessly and fluently. When he was stricken with blindness at the age of 59, he enlisted the help of assistants to record his thoughts. Aided by his powerful memory and fertile imagination, Euler's output actually increased.

Euler was a native of Basel, Switzerland and a student of the noted mathematician Johann Bernoulli (mentioned elsewhere in this text). His working life was spent as a member of the Academies of Science at Berlin and St. Petersburg. He was a man of broad culture, well-versed in the classical languages and literatures (he knew the *Aeneid* by heart), physiology, medicine, botany, geography, and the entire body of physical science.

Euler had 13 children. Even so, his personal life was uneventful and placid. It is said that he died while dandling a grandchild on his knee. He never taught, but his influence on the teaching of mathematics has been considerable.

Euler wrote three great treatises: *Introductio in Analysin Infinitorum* (1748), *Institutiones Calculi Differentialis* (1755), and *Institutiones Calculi Integralis* (1768–1794). These works both assessed and codified the works of all Euler's predecessors, and they contained many of his own ideas as well. It has been said that all elementary and advanced calculus textbooks since 1748 are either copies of Euler or copies of copies of Euler.

Among many other important advances, Euler's work extended and perfected plane and solid analytic geometry, introduced the analytic approach to trigonometry, and was responsible for the modern treatment of the functions $\ln x$ and e^x. He created a consistent theory of logarithms of negative and imaginary numbers, and discovered that $\ln z$ has infinitely many values. Euler's work established the use of the symbols e, π, and i (for $\sqrt{-1}$). Euler linked these three important numbers together with the remarkable formula

$$e^{\pi i} = -1 .$$

Euler was also the one who established the notation $\sin x$ and $\cos x$, the use of $f(x)$ for an arbitrary function, and the use of $\sum$ to denote a series.

The distinction between pure and applied mathematics did not exist in Euler's day. For him, the entire physical universe was grist for his mill. The foundations of classical mechanics were laid by Isaac Newton, but Euler was the principal architect of the subject. In his treatise of 1736 he was the first to explicitly introduce the concept of a mass-point or particle, and he was also the first to study the acceleration of a particle moving along any curve and to use the notion of a vector in connection with velocity and acceleration. Euler's discoveries were so pervasive that many of them continue to be used without any explicit credit to Euler. Among his named contributions are Euler's equations of motion for the rotation of a rigid body, Euler's hydrodynamic equation for the flow of an ideal incompressible fluid, Euler's law for the bending of elastic beams, and Euler's critical load in the theory of the buckling of columns.

Few mathematicians have had the fluency, the clarity of thought, and the profound influence of Leonhard Euler. His ideas are an integral part of modern mathematics.

Anatomy of an Application

Bessel Functions and the Vibrating Membrane

Bessel functions arise typically in the solution of Bessel's differential equation

$$x^2y'' + xy' + (x^2 - p^2)y = 0 .$$

They are among the most important special functions of mathematical physics. In the present discussion we shall explore the use of these functions to describe Euler's analysis of a vibrating circular membrane. The approach is similar to

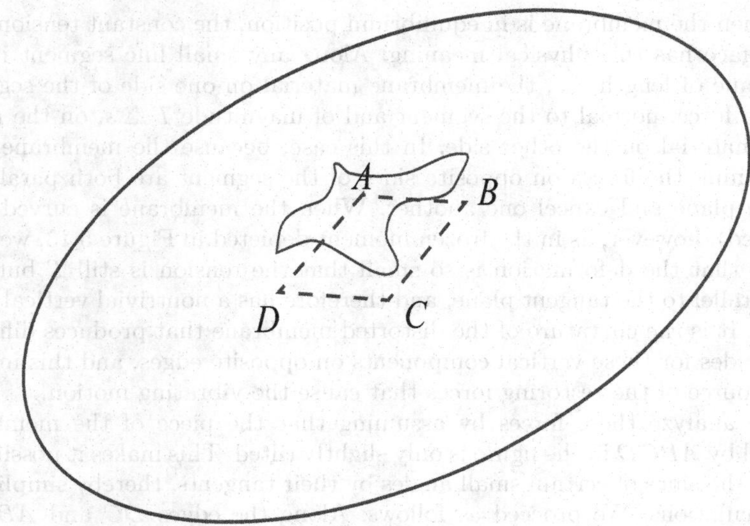

FIGURE 5.13
A segment of the vibrating membrane.

that for the vibrating string, which is treated elsewhere in the present book (Section 11.2). Here we shall anticipate some of the ideas that you will see later.

We shall be considering a uniform thin sheet of flexible material (polyester, for instance). The sheet will be pulled taut into a state of uniform tension and clamped along a given closed curve (a circle, perhaps) in the x-y plane. When the membrane is displaced slightly from its equilibrium position and then released, the restoring forces created by the deformation cause it to vibrate. For instance, this is how a drum works. To simplify the mathematics, we shall consider only *small oscillations* of a *freely vibrating* membrane.

We shall assume that the displacement of the membrane is so small that each point of the surface is moved only in the z direction (i.e., perpendicular to the plane of the membrane). The displacement is given by a function $z = z(x, y, t)$, where t is time. We shall consider a small, rectangular piece of the membrane with dimensions $\triangle x$ and $\triangle y$. The corners of this rectangle are the points (x, y), $(x + \triangle x, y)$, $(x, y + \triangle y)$, and $(x + \triangle x, y + \triangle y)$. This rectangle, and the portion of the displaced membrane that lies above it, are depicted in Figure 5.13.

If m is the constant mass per unit area of the membrane, then the mass of the rectangular piece is $m \triangle x \triangle y$. Newton's second law of motion then tells us that

$$F = m \triangle x \triangle y \frac{\partial^2 z}{\partial t^2} \tag{5.10}$$

is the force acting on the membrane in the z-direction.

When the membrane is in equilibrium position, the constant tension T in the surface has this physical meaning: Along any small line segment in the membrane of length $\triangle s$, the membrane material on one side of the segment exerts a force, normal to the segment and of magnitude $T \triangle s$, on the membrane material on the other side. In this case, because the membrane is in equilibrium, the forces on opposite sides of the segment are both parallel to the x-y plane and cancel one another. When the membrane is curved (i.e., displaced), however, as in the frozen moment depicted in Figure 5.13, we shall assume that the deformation is so small that the tension is still T but now acts parallel to the tangent plane, and therefore has a nontrivial vertical component. It is the curvature of the distorted membrane that produces different magnitudes for these vertical components on opposite edges, and this in turn is the source of the restoring forces that cause the vibrating motion.

We analyze these forces by assuming that the piece of the membrane denoted by $ABCD$ in the figure is only slightly tilted. This makes it possible to replace the sines of certain small angles by their tangents, thereby simplifying the calculations. We proceed as follows: Along the edges DC and AB, the forces are perpendicular to the x-axis and almost parallel to the y-axis, with small z-components approximately equal to

$$T \triangle x \left(\frac{\partial z}{\partial y}\right)_{y+\triangle y} \qquad \text{and} \qquad -T \triangle x \left(\frac{\partial z}{\partial y}\right)_{y} .$$

Hence their sum is approximately equal to

$$T \triangle x \left\{ \left(\frac{\partial z}{\partial y}\right)_{y+\triangle y} - \left(\frac{\partial z}{\partial y}\right)_{y} \right\} .$$

The subscripts on these partial derivatives indicate their values at the points $(x, y + \triangle y)$ and (x, y).

Performing the same type of analysis on the edges BC and AD, we find that the total force in the z-direction—coming from all four edges—is

$$T \triangle y \left\{ \left(\frac{\partial z}{\partial x}\right)_{x+\triangle x} - \left(\frac{\partial z}{\partial x}\right)_{x} \right\} + T \triangle x \left\{ \left(\frac{\partial z}{\partial y}\right)_{y+\triangle y} - \left(\frac{\partial z}{\partial y}\right)_{y} \right\} .$$

As a result, equation (10) can be rewritten as

$$T \frac{(\partial z/\partial x)_{x+\triangle x} - (\partial z/\partial x)_x}{\triangle x} + T \frac{(\partial z/\partial y)_{y+\triangle y} - (\partial z/\partial y)_y}{\triangle y} = m \frac{\partial^2 z}{\partial t^2} .$$

If we now set $a^2 = T/m$ and let $\triangle x \to 0$, $\triangle y \to 0$, then we find that

$$a^2 \left(\frac{\partial^2 z}{\partial x^2} + \frac{\partial^2 z}{\partial y^2} \right) = \frac{\partial^2 z}{\partial t^2} ; \tag{5.11}$$

this is the *two-dimensional wave equation.* We note that this is a *partial differential equation*: it involves functions of three variables, and of course partial derivatives.

Now we shall consider the displacement of a circular membrane. So our study will be the model for a drum. Of course we shall use polar coordinates, with the origin at the center of the drum. The wave equation now has the form

$$a^2\left(\frac{\partial^2 z}{\partial r^2} + \frac{1}{r}\frac{\partial z}{\partial r} + \frac{1}{r^2}\frac{\partial^2 z}{\partial \theta^2}\right) = \frac{\partial^2 z}{\partial t^2} \, . \tag{5.12}$$

Here, naturally, $z = z(r, \theta, t)$ is a function of the polar coordinates r, θ and of time t. We assume, without loss of generality, that the membrane has radius 1. Thus it is clamped to its plane of equilibrium along the circle $r = 1$ in the polar coordinates. Thus our boundary condition is

$$z(1, \theta, t) = 0 \, , \tag{5.13}$$

because the height of the membrane at the edge of the disc-shaped displaced region is 0. The problem, then, is to find a solution of (12) that satisfies the boundary condition (13) together with certain initial conditions that we shall not consider at the moment.

We shall apply the method of separation of variables. Thus we seek a solution of the form

$$z(r, \theta, t) = u(r)v(\theta)w(t) \, . \tag{5.14}$$

We substitute (14) into (12) and perform a little algebra to obtain

$$\frac{u''(r)}{u(r)} + \frac{1}{r}\frac{u'(r)}{u(r)} + \frac{1}{r^2}\frac{v''(\theta)}{v(\theta)} = \frac{1}{a^2}\frac{w''(t)}{w(t)} \, . \tag{5.15}$$

Now our analysis follows familiar lines: Since the left-hand side of (15) depends on r and θ only and the right-hand side depends on t only, we conclude that both sides are equal to some constant K. In order for the membrane to vibrate, $w(t)$ must be periodic. Thus the constant λ must be negative (if it is positive, then the solutions of $w'' - Ka^2w = 0$ will be real exponentials and hence *not* periodic). We thus equate both sides of (15) with $K = -\lambda^2$ for some $\lambda > 0$ and obtain the two ordinary differential equations

$$w''(t) + \lambda^2 a^2 w(t) = 0 \tag{5.16}$$

and

$$\frac{u''(r)}{u(r)} + \frac{1}{r}\frac{u'(r)}{u(r)} + \frac{1}{r^2}\frac{v''(\theta)}{v(\theta)} = -\lambda^2 \, . \tag{5.17}$$

Now (16) is easy to solve, and its general solution is

$$w(t) = c_1 \cos \lambda a t + c_2 \sin \lambda a t \, . \tag{5.18}$$

We can rewrite (17) as

$$r^2\frac{u''(r)}{u(r)} + r\frac{u'(r)}{u(r)} + \lambda^2 r^2 = -\frac{v''(\theta)}{v(\theta)} \, . \tag{5.19}$$

Notice that in equation (19) we have a function of r only on the left and a function of θ only on the right. So, as usual, both sides must be equal to a constant L. Now we know, by the physical nature of our problem, that v must be 2π-periodic. Looking at the right-hand side of (19) then tells us that $L = n^2$ for $n \in \{0, 1, 2, \dots\}$.

With these thoughts in mind, equation (19) splits into

$$v''(\theta) + n^2 v(\theta) = 0 \tag{5.20}$$

and

$$r^2 u''(r) + r u'(r) + (\lambda^2 r^2 - n^2) u(r) = 0 . \tag{5.21}$$

Of course equation (20) has, as its general solution,

$$v(\theta) = d_1 \cos n\theta + d_2 \sin n\theta .$$

(Note that this solution is not valid when $n = 0$. But, when $n = 0$ the equation has no nontrivial periodic solutions.) Also observe that equation (21) is a slight variant of Bessel's equation (in fact a change of variables of the form $r = w^b$, $u = v \cdot w^c$, for appropriately chosen b and c, will transform the Bessel's equation given at the beginning of this discussion to equation (21)). It turns out that, according to physical considerations, the solution that we want of equation (21) is

$$u(r) = k \cdot J_n(r) .$$

Here k is a physical constant and J_n is the nth Bessel function, discussed in detail in Chapters 6 and 13. Note for now that the Bessel functions are transcendental functions, best described with a power series.

Let us conclude this discussion by considering the boundary condition (13) for our problem. It can now be written simply as $u(1) = 0$ or

$$J_n(\lambda) = 0 .$$

Thus the permissible values of λ are the positive zeros of the Bessel function J_n (see also the discussion in Chapter 13 below). It is known that there are infinitely many such zeros, and they have been studied intensively over the years. The reference [WAT] is a great source of information about Bessel functions.

Problems for Review and Discovery

A. Drill Exercises

1. Find the general solution of each of these differential equations.

(a) $y'' - 3y' + y = 0$
(b) $y'' + y' + y = 0$
(c) $y'' + 6y' + 9y = 0$
(d) $y'' - y' + 6y = 0$
(e) $y'' - 2y' - 5y = x$
(f) $y'' + y = e^x$
(g) $y'' + y' + y = \sin x$
(h) $y'' - y = e^{3x}$

2. Solve each of the following initial value problems.
(a) $y'' + 9y = 0\,, \quad y(0) = 1\,,\ y'(0) = 2$
(b) $y'' - y' + 4y = x\,, \quad y(1) = 2\,,\ y'(1) = 1$
(c) $y'' + 2y' + 5y = e^x\,, \quad y(0) = -1\,,\ y'(0) = 1$
(d) $y'' + 3y' + 4y = \sin x\,, \quad y(\pi/2) = 1\,,\ y'(\pi/2) = -1$
(e) $y'' + y = e^{-x}\,, \quad y(2) = 0\,,\ y'(2) = -2$
(f) $y'' - y = \cos x\,, \quad y(0) = 3\,,\ y'(0) = 2$
(g) $y'' = \tan x\,, \quad y(1) = 1\,,\ y'(1) = -1$
(h) $y'' - 2y' = \ln x\,, \quad y(1) = e\,,\ y'(1) = 1/e$

3. Solve each of these differential equations.
(a) $y'' + 3y' + 2y = 2x - 1$
(b) $y'' - 3y' + 2y = e^{-x}$
(c) $y'' - y' - 2y = \cos x$
(d) $y'' + 2y' - y = xe^x \sin x$
(e) $y'' + 9y = \sec 2x$
(f) $y'' + 4y' + 4y = x \ln x$
(g) $x^2y'' + 3xy' + y = 2/x$
(h) $y'' + 4y = \tan^2 x$

4. Use the given solution of the differential equation to find the general solution.
(a) $y'' - y = 3e^{2x}\,, \quad y_1(x) = e^{2x}$
(b) $y'' + y = -8\sin 3x\,, \quad y_1(x) = \sin 3x$
(c) $y'' + y' + y = x^2 + 2x + 2\,, \quad y_1(x) = x^2$
(d) $y'' + y' = \frac{x-1}{x^2}\,, \quad y_1(x) = \ln x$

B. Challenge Exercises

1. Consider the differential equation $y'' + 4y = 0$. Convert it to a system of first-order, linear ordinary differential equations by setting $v = y'$. Hence we have

$$\begin{aligned} y' &= v \\ v' &= -4y \end{aligned}$$

Find solutions $y(x), v(x)$ for this system. If we think of x as a parameter, then the map

$$x \longmapsto (y(x), v(x))$$

describes a curve in the plane. Draw this curve for a variety of different sets of initial conditions. What sort of curve is it?

2. Explain why $y_1(x) = \sin x$ and $y_2(x) = 2x$ cannot both be solutions of the same ordinary differential equation

$$y'' = F(x, y, y')$$

for a smooth F.

3. Show that the Euler equation

$$x^2y'' - 2xy' + 2y = 0$$

with initial conditions

$$y(0) = 0\ ,\ y'(0) = 0$$

has infinitely many solutions. Why does this surprising result not contradict the general ideas that we learned in this chapter?

4. Does the differential equation

$$y'' + 9y = -3\cos 2x$$

have any periodic solutions? Say exactly what you mean by "periodic" as you explain your answer.

C. Problems for Discussion and Exploration

1. Show that the ordinary differential equation $y' + y = \cos x$ has a unique periodic solution.

2. Find the regions where the solution of the initial value problem

$$y'' = -3y\,, \quad y(0) = -1$$

is concave down. In what regions is it concave up? What do these properties tell us about the solution? Do you need to actually solve the differential equation in order to answer this question?

3. Consider solutions of the differential equation

$$\frac{d^2y}{dx^2} - c\frac{dy}{dx} + y = 0$$

for a constant c. Describe how the behavior of this solution changes as c varies.

4. Endeavor to find an approximate solution to the differential equation

$$\frac{d^2y}{dx^2} + \sin y = 0$$

by guessing that the solution is a polynomial. Think of this polynomial as the Taylor polynomial of the actual solution. Can you say anything about how accurately your polynomial solution approximates the true solution?

6

Power Series Solutions and Special Functions

- Power series basics
- Convergence of power series
- Series solutions of first-order equations
- Series solutions of second-order equations
- Ordinary points
- Regular singular points
- Frobenius method
- The hypergeometric equation

6.1 Introduction and Review of Power Series

It is useful to classify the functions that we know, or will soon know, in an informal way. The *polynomials* are functions of the form

$$a_0 + a_1 x + a_2 x^2 + \cdots + a_{k-1} x^{k-1} + a_k x^k \,,$$

where $a_0, a_1, \ldots, a_k$ are constants. This is a polynomial of degree k. A *rational function* is a quotient of polynomials. For example,

$$r(x) = \frac{3x^3 - x + 4}{x^2 + 5x + 1}$$

is a rational function.

A *transcendental function* is one that is not a polynomial or a rational function. The *elementary transcendental functions* are the ones that we encounter in calculus class: sine, cosine, logarithm, exponential, and their inverses and combinations using arithmetic/algebraic operations.

The *higher transcendental functions* are ones that are defined using power series (although they often arise by way of integrals or asymptotic expansions

DOI: 10.1201/9781003214526-6

or other means). These often are discovered as solutions of differential equations. These functions are a bit difficult to understand, just because they are not given by elementary formulas. But they are frequently very important because they come from fundamental problems of mathematical physics. As an example, solutions of Bessel's equation, which we saw at the end of the last chapter, are called Bessel functions and are studied intensively (see [WAT]).

Higher transcendental functions are frequently termed *special functions.* These important functions were studied extensively in the eighteenth and nineteenth centuries—by Gauss, Euler, Abel, Jacobi, Weierstrass, Riemann, Hermite, Poincaré, and other leading mathematicians of the day. Although many of the functions that they studied were quite recondite, and are no longer of much interest today, others (such as the Riemann zeta function, the gamma function, and elliptic functions) are still intensively studied.

In the present chapter we shall learn to solve differential equations using the method of power series, and we shall have a very brief introduction to how special functions arise from this process. It is a long chapter, with many new ideas. But there are many rewards along the way.

6.1.1 Review of Power Series

We begin our study with a quick review of the key ideas from the theory of power series.

I. A series of the form

$$\sum_{j=0}^{\infty} a_j x^j = a_0 + a_1 x + a_2 x^2 + \cdots \tag{6.1}$$

is called a *power series in* x. Slightly more general is the series

$$\sum_{j=0}^{\infty} a_j (x-a)^j ,$$

which is a *power series in* $x - a$ (or *expanded about the point* a).

II. The series (1) is said to *converge* at a point x if the limit

$$\lim_{k\to\infty} \sum_{j=0}^{k} a_j x^j$$

exists. The value of the limit is called the *sum* of the series. (This is just the familiar idea of defining the value of a series to be the limit of its partial sums.)

Obviously (1) converges when $x = 0$, since all terms but the first (or zeroeth) will then be equal to 0. The following three examples illustrate, in an informal way, what the convergence properties might be at other values of x.

(a) The series

$$\sum_{j=0}^{\infty} j!x^j = 1 + x + 2!x^2 + 3!x^3 + \cdots$$

diverges[1] at every $x \neq 0$. This can be seen by using the ratio test from the theory of series. It of course converges at $x = 0$.

(b) The series

$$\sum_{j=0}^{\infty} \frac{x^j}{j!} = 1 + x + \frac{x^2}{2!} + \frac{x^3}{3!} + \cdots$$

converges at every value of x, including $x = 0$. This can be seen by applying the ratio test from the theory of series.

(c) The series

$$\sum_{j=0}^{\infty} x^j = 1 + x + x^2 + x^3 + \cdots$$

converges when $|x| < 1$ and diverges when $|x| \geq 1$.

These three examples are special instances of a general phenomenon that governs the convergence behavior of power series. There will *always* be a number R, $0 \leq R \leq \infty$ such that the series converges for $|x| < R$ and diverges for $|x| > R$. In the first example, $R = 0$; in the second example, $R = +\infty$; in the third example, $R = 1$. We call R the *radius of convergence* of the power series. The interval $(-R, R)$ is called the *interval of convergence*. In practice, we check convergence at the endpoints of the interval of convergence by hand in each example. We add those points to the interval of convergence as appropriate. The next three examples will illustrate how we calculate R in practice.

EXAMPLE 6.1.1 Calculate the radius of convergence and interval of convergence of the series

$$\sum_{j=0}^{\infty} \frac{x^j}{j^2} .$$

Solution: We apply the ratio test:

$$\lim_{j\to\infty} \left| \frac{x^{j+1}/(j+1)^2}{x^j/j^2} \right| = \left| \lim_{j\to\infty} \frac{j^2}{(j+1)^2} \cdot x \right| = |x| .$$

[1]Here we use the notation $n! = n \cdot (n-1) \cdot (n-2) \cdots 3 \cdot 2 \cdot 1$. This is called the *factorial notation*. By convention, $1! = 1$ and $0! = 1$.

We know that the series will converge when this limit is less than 1, or $|x| < 1$. Likewise, it diverges when $|x| > 1$. Thus the radius of convergence is $R = 1$.

In practice, one has to check the *endpoints* of the interval of convergence by hand for each case. In this example, we see immediately that the series converges at ± 1. Thus we may say that the interval of convergence is $[-1, 1]$. □

EXAMPLE 6.1.2 Calculate the radius of convergence and interval of convergence of the series

$$\sum_{j=0}^{\infty} \frac{x^j}{j}\,.$$

Solution: We apply the ratio test:

$$\lim_{j\to\infty} \left| \frac{x^{j+1}/(j+1)}{x^j/j} \right| = \left| \lim_{j\to\infty} \frac{j}{j+1} \cdot x \right| = |x|\,.$$

We know that the series will converge when this limit is less than 1, or $|x| < 1$. Likewise, it diverges when $|x| > 1$. Thus the radius of convergence is $R = 1$.

In this example, we see immediately that the series converges at -1 (by the alternating series test) and diverges at $+1$ (since this gives the harmonic series). Thus we may say that the interval of convergence is $[-1, 1)$. □

EXAMPLE 6.1.3 Calculate the radius of convergence and interval of convergence of the series

$$\sum_{j=0}^{\infty} \frac{x^j}{j^j}\,.$$

Solution: We use the root test:

$$\lim_{j\to\infty} \left| \frac{x^j}{j^j} \right|^{1/j} = \lim_{j\to\infty} \left| \frac{x}{j} \right| = 0\,.$$

Of course $0 < 1$, regardless of the value of x. So the series converges for all x. The radius of convergence is $+\infty$ and the interval of convergence is $(-\infty, +\infty)$. There is no need to check the endpoints of the interval of convergence, because there are none. □

III. Suppose that our power series (1) converges for $|x| < R$ with $R > 0$. Denote its sum by $f(x)$, so

$$f(x) = \sum_{j=0}^{\infty} a_j x^j = a_0 + a_1 x + a_2 x^2 + \cdots\,.$$

Thus the power series defines a *function*, and we may consider differentiating it. In fact the function f is continuous and has derivatives of all orders. We may calculate the derivatives by differentiating the series termwise:

$$\begin{aligned} f'(x) &= \sum_{j=1}^{\infty} j a_j x^{j-1} = a_1 + 2a_2 x + 3a_3 x^2 + \cdots , \\ f''(x) &= \sum_{j=2}^{\infty} j(j-1)x^{j-2} = 2a_2 + 3 \cdot 2a_3 x + \cdots , \end{aligned}$$

and so forth. Each of these series converges on the same interval $|x| < R$.

Observe that, if we evaluate the formula for f at $x = 0$, we find that

$$a_0 = f(0) .$$

If instead we evaluate the formula for f' at $x = 0$, then we obtain the useful fact that

$$a_1 = \frac{f'(0)}{1!} .$$

If we evaluate the formula for f'' at $x = 0$, then we obtain the analogous fact that

$$a_2 = \frac{f^{(2)}(0)}{2!} .$$

In general, we can derive (by successive differentiations) the formula

$$a_j = \frac{f^{(j)}(0)}{j!} , \tag{6.2}$$

which gives us an explicit way to determine the coefficients of the power series expansion of a function. It follows from these considerations that a power series is identically equal to 0 if and only if each of its coefficients is 0.

We may also note that a power series may be integrated termwise. If

$$f(x) = \sum_{j=0}^{\infty} a_j x^j = a_0 + a_1 x + a_2 x^2 + \cdots ,$$

then

$$\int f(x)\, dx = \sum_{j=0}^{\infty} a_j \frac{x^{j+1}}{j+1} = a_0 x + a_1 \frac{x^2}{2} + a_2 \frac{x^3}{3} + \cdots + C .$$

If

$$f(x) = \sum_{j=0}^{\infty} a_j x^j = a_0 + a_1 x + a_2 x^2 + \cdots$$

and

$$g(x) = \sum_{j=0}^{\infty} b_j x^j = b_0 + b_1 x + b_2 x^2 + \cdots$$

for $|x| < R$, then these functions may be added or subtracted by adding or subtracting the series termwise:

$$f(x) \pm g(x) = \sum_{j=0}^{\infty} (a_j \pm b_j)x^j = (a_0 \pm b_0) + (a_1 \pm b_1)x + (a_2 \pm b_2)x^2 + \cdots .$$

Also f and g may be multiplied as if they were polynomials, to wit

$$f(x) \cdot g(x) = \sum_{j=0}^{\infty} c_n x^n ,$$

where

$$c_n = a_0 b_n + a_1 b_{n-1} + \cdots + a_n b_0 .$$

We shall say more about operations on power series below.

Finally, we note that if two different power series converge to the same function, then (2) tells us that the two series are precisely the same (i.e., have the same coefficients). In particular, if $f(x) \equiv 0$ for $|x| < R$ then all the coefficients of the power series expansion for f are equal to 0.

IV. Suppose that f is a function that has derivatives of all orders on $|x| < R$. We may calculate the coefficients

$$a_j = \frac{f^{(j)}(0)}{j!}$$

and then write the (formal) series

$$\sum_{j=0}^{\infty} a_j x^j . \tag{6.3}$$

It is then natural to ask whether the series (3) *converges to* f. When the function f is sine or cosine or logarithm or the exponential then the answer is "yes." But these are very special functions. Actually, the answer to our question is generically "no." Most infinitely differentiable functions do *not* have power series expansion that converges back to the function. In fact most have power series that do not converge at all; but even in the unlikely circumstance that the series does converge, it will *not* converge to the original f.

This circumstance may seem rather strange, but it explains why mathematicians spent so many years trying to understand power series. The functions that *do* have convergent power series are called *real analytic* and they are very particular functions with remarkable properties. Even though the subject of real analytic functions is more than 300 years old, the first and only book written on the subject is [KRP1].

We do have a way of coming to grips with the unfortunate state of affairs that has just been described, and that is the theory of *Taylor expansions.* For a function with sufficiently many derivatives, here is what is actually true:

$$f(x) = \sum_{j=0}^{n} \frac{f^{(j)}(0)}{j!} x^j + R_n(x), \tag{6.4}$$

where the remainder term $R_n(x)$ is given by

$$R_n(x) = \frac{f^{(n+1)}(\xi)}{(n+1)!} x^{n+1}$$

for some number ξ between 0 and x. The power series converges to f precisely when the partial sums in (4) converge, and that happens precisely when the remainder term goes to zero. What is important for you to understand is that, generically, the remainder term *does not go to zero.* But formula (4) is still valid.

We can use formula (4) to obtain the familiar power series expansions for several important functions:

$$\begin{aligned} e^x &= \sum_{j=0}^{\infty} \frac{x^j}{j!} = 1 + x + \frac{x^2}{2!} + \frac{x^3}{3!} + \cdots \\ \sin x &= \sum_{j=0}^{\infty} (-1)^j \frac{x^{2j+1}}{(2j+1)!} = x - \frac{x^3}{3!} + \frac{x^5}{5!} - + \cdots \\ \cos x &= \sum_{j=0}^{\infty} (-1)^j \frac{x^{2j}}{(2j)!} = 1 - \frac{x^2}{2!} + \frac{x^4}{4!} - + \cdots \end{aligned}$$

Of course there are many others, including the logarithm and the other trigonometric functions. Just for practice, let us verify that the first of these formulas is actually valid.

First,

$$\frac{d^j}{dx^j} e^x = e^x \qquad \text{for every } j.$$

Thus

$$a_j = \frac{[d^j/dx^j]e^x\big|_{x=0}}{j!} = \frac{1}{j!}.$$

This confirms that the formal power series for e^x is just what we assert it to be. To check that it converges back to e^x, we must look at the remainder term, which is

$$R_n(x) = \frac{f^{(n+1)}(\xi)}{(n+1)!} x^{n+1} = \frac{e^{\xi} \cdot x^{n+1}}{(n+1)!}.$$

Of course, for x fixed, we have that $|\xi| < |x|$; also $n \to \infty$ implies that $(n+1)! \to \infty$ much faster than x^{n+1}. So the remainder term goes to zero and the series converges to e^x.

V. We conclude by summarizing some properties of real analytic functions.

1. Polynomials and the functions e^x, $\sin x$, $\cos x$ are all real analytic at all points.

2. If f and g are real analytic at x_0 then $f \pm g$, $f \cdot g$, and f/g (provided $g(x_0) \neq 0$) are real analytic at x_0.

3. If f is real analytic at x_0 and if f^{-1} is a continuous inverse on an interval containing $f(x_0)$ and $f'(x_0) \neq 0$, then f^{-1} is real analytic at $f(x_0)$.

4. If g is real analytic at x_0 and f is real analytic at $g(x_0)$, then $f \circ g$ is real analytic at x_0.

5. A function defined by the sum of a power series is real analytic at all interior points of its interval of convergence.

VI. It is worth recording that all the basic properties of *real* power series that we have discussed above are still valid for *complex* power series. Such a series has the form

$$\sum_{j=0}^{\infty} c_j z^j \qquad \text{or} \qquad \sum_{j=0}^{\infty} c_j (z-a)^j ,$$

where the coefficients c_j are real or complex numbers and z is a complex variable. The series has a radius of convergence R, and the domain of convergence is now a disc $D(0, R)$ or $D(a, R)$ rather than an interval. In practice, a complex analytic function has radius of convergence about a point a that is equal to the distance to the nearest singularity from a. See [KNO] or [GRK] or [KRP1] for further discussion of these ideas.

Exercises

1. Use the ratio test (for example) to verify that $R = 0$, $R = \infty$, and $R = 1$ for the series **(a)**, **(b)**, **(c)** that are discussed in the text.
2. If $p \neq 0$ and p is not a positive integer then show that the series

$$\sum_{j=1}^{\infty} \frac{p(p-1)(p-2)\cdots(p-j+1)}{j!} \cdot x^j$$

converges for $|x| < 1$ and diverges for $|x| > 1$.
3. Verify that $R = +\infty$ for the power series expansions of sine and cosine.
4. Use Taylor's formula to confirm the validity of the power series expansions for e^x, $\sin x$, and $\cos x$.

5. When we first encounter geometric series we learn that

$$1 + x + x^2 + \cdots + x^n = \frac{1 - x^{n+1}}{1 - x}$$

provided $x \neq 1$. Indeed, let S be the expression on the left and consider $(1 - x) \cdot S$. Use this formula to show that the expansions

$$\frac{1}{1 - x} = 1 + x + x^2 + x^3 + \cdots$$

and

$$\frac{1}{1 + x} = 1 - x + x^2 - x^3 + \cdots$$

are valid for $|x| < 1$. Use these formulas to show that

$$\ln(1 + x) = x - \frac{x^2}{2} + \frac{x^3}{3} - \frac{x^4}{4} + - \cdots$$

and

$$\arctan x = x - \frac{x^3}{3} + \frac{x^5}{5} - \frac{x^7}{7} + - \cdots$$

for $|x| < 1$.

6. Use the first expansion given in Exercise 5 to find the power series for $1/(1 - x)^2$

(a) by squaring;

(b) by differentiating.

7. **(a)** Show that the series

$$y = 1 - \frac{x^2}{2!} + \frac{x^4}{4!} - \frac{x^6}{6!} + - \cdots$$

satisfies $y'' = -y$.

(b) Show that the series

$$y = 1 - \frac{x^2}{2^2} + \frac{x^4}{2^2 \cdot 4^2} - \frac{x^6}{2^2 \cdot 4^2 \cdot 6^2} + - \cdots$$

converges for all x. Verify that it defines a solution of the equation

$$xy'' + y' + xy = 0 .$$

This function is the *Bessel function of order 0.* It will be encountered later in other contexts.

6.2 Series Solutions of First-Order Differential Equations

Now we get our feet wet and use power series to solve first-order linear equations. This will turn out to be misleadingly straightforward to do, but it will show us the basic moves.

At first we consider the equation

$$y' = y\,.$$

Of course we know that the solution to this equation is $y = C \cdot e^x$, but let us pretend that we do not know this. We proceed by *guessing* that the equation has a solution given by a power series, and we proceed to solve for the coefficients of that power series.

So our guess is a solution of the form

$$y = a_0 + a_1 x + a_2 x^2 + a_3 x^3 + \cdots\,.$$

Then

$$y' = a_1 + 2a_2 x + 3a_3 x^2 + \cdots$$

and we may substitute these two expressions into the differential equation. Thus

$$a_1 + 2a_2 x + 3a_3 x^2 + \cdots = a_0 + a_1 x + a_2 x^2 + \cdots\,.$$

Now the powers of x must match up (i.e., the coefficients must be equal). We conclude that

$$\begin{aligned} a_1 &= a_0 \\ 2a_2 &= a_1 \\ 3a_3 &= a_2 \end{aligned}$$

and so forth. Let us take a_0 to be an unknown constant C. Then we see that

$$\begin{aligned} a_1 &= C \\ a_2 &= \frac{C}{2} \\ a_3 &= \frac{C}{3 \cdot 2} \\ &\text{etc.} \end{aligned}$$

In general,

$$a_n = \frac{C}{n!}\,.$$

In summary, our power series solution of the original differential equation is

$$y = \sum_{j=0}^{\infty} \frac{C}{j!} x^j = C \cdot \sum_{j=0}^{\infty} \frac{x^j}{j!} = C \cdot e^x\,.$$

Thus we have a new way, using power series, of discovering the general solution of the differential equation $y' = y$.

The next example illustrates the point that, by running our logic a bit differently, we can *use* a differential equation to derive the power series expansion for a given function.

EXAMPLE 6.2.1 Let p be an arbitrary real constant. Use a differential equation to derive the power series expansion for the function

$$y = (1 + x)^p .$$

Solution: Of course the given y is a solution of the initial value problem

$$(1 + x) \cdot y' = py \ , \quad y(0) = 1 .$$

We assume that the equation has a power series solution

$$y = \sum_{j=0}^{\infty} a_j x^j = a_0 + a_1 x + a_2 x^2 + \cdots$$

with positive radius of convergence R. Then

$$y' = \sum_{j=1}^{\infty} j \cdot a_j x^{j-1} = a_1 + 2a_2 x + 3a_3 x^2 + \cdots$$

$$xy' = \sum_{j=1}^{\infty} j \cdot a_j x^j = a_1 x + 2a_2 x^2 + 3a_3 x^3 + \cdots$$

$$py = \sum_{j=0}^{\infty} pa_j x^j = pa_0 + pa_1 x + pa_2 x^2 + \cdots .$$

We rewrite the differential equation as

$$y' + xy' = py .$$

Now we see that the equation tells us that the sum of the first two of our series equals the third. Thus

$$\sum_{j=1}^{\infty} ja_j x^{j-1} + \sum_{j=1}^{\infty} ja_j x^j = \sum_{j=0}^{\infty} pa_j x^j .$$

We immediately see two interesting anomalies: the powers of x on the left-hand side do not match up, so the two series cannot be immediately added. Also the summations do not all begin in the same place. We address these two concerns as follows.

First, we can change the index of summation in the first sum on the left to obtain

$$\sum_{j=0}^{\infty} (j + 1)a_{j+1} x^j + \sum_{j=1}^{\infty} ja_j x^j = \sum_{j=0}^{\infty} pa_j x^j .$$

Write out the first few terms of the sum we have changed, and the original sum, to see that they are just the same.

Now every one of our series has x^j in it, but they begin at different places. So we break off the extra terms as follows:

$$\sum_{j=1}^{\infty}(j+1)a_{j+1}x^j + \sum_{j=1}^{\infty} ja_jx^j - \sum_{j=1}^{\infty} pa_jx^j = -a_1x^0 + pa_0x^0 . \qquad (6.5)$$

Notice that all we have done is to break off the zeroth terms of the first and third series, and put them on the right.

The three series on the left-hand side of (5) are begging to be put together: they have the same form, they all involve the same powers of x, and they all begin at the same index. Let us do so:

$$\sum_{j=1}^{\infty}[(j+1)a_{j+1} + ja_j - pa_j]x^j = -a_1 + pa_0 .$$

Now the powers of x that appear on the left are 1, 2, ..., and there are none of these on the right. We conclude that each of the coefficients on the left is zero; by the same reasoning, the coefficient $(-a_1 + pa_0)$ on the right (i.e., the constant term) equals zero. Here we are using the uniqueness property of the power series representation. This gives us infinitely many equations.

So we have the equations[2]

$$\begin{aligned} -a_1 + pa_0 &= 0 \\ (j+1)a_{j+1} + (j-p)a_j &= 0 \quad \text{for } j \geq 1 . \end{aligned}$$

Our initial condition tells us that $a_0 = 1$. Then our first equation implies that $a_1 = p$. The next equation, with $j = 1$, says that

$$2a_2 + (1-p)a_1 = 0 .$$

Hence $a_2 = (p-1)a_1/2 = p(p-1)/2$. Continuing, we take $j = 2$ in the second equation to get

$$3a_3 + (2-p)a_2 = 0$$

so $a_3 = (p-2)a_2/3 = p(p-1)(p-2)/(3 \cdot 2)$.

We may continue in this manner to obtain that

$$a_j = \frac{p(p-1)(p-2)\cdots(p-j+1)}{j!} .$$

Thus the power series expansion for our solution y is

$$\begin{aligned} y &= 1 + px + \frac{p(p-1)}{2!}x^2 + \frac{p(p-1)(p-2)}{3!}x^3 + \cdots \\ &\quad + \frac{p(p-1)(p-2)\cdots(p-j+1)}{j!}x^j + \cdots . \end{aligned}$$

[2]A set of equations like this is called a *recursion*. It expresses later indexed a_j's in terms of earlier indexed a_j's.

Since we knew in advance that the solution of our initial value problem was

$$y = (1+x)^p$$

(and this function is certainly analytic at 0), we find that we have derived Isaac Newton's general binomial theorem (or binomial series):

$$(1+x)^p = 1 + px + \frac{p(p-1)}{2!}x^2 + \frac{p(p-1)(p-2)}{3!}x^3 + \cdots + \frac{p(p-1)(p-2)\cdots(p-j+1)}{j!}x^j + \cdots \qquad \square$$

Exercises

1. For each of the following differential equations, find a power series solution of the form $\sum_j a_j x^j$. Endeavor to recognize this solution as the series expansion of a familiar function. Now solve the equation directly, using a method from the earlier part of the book, to confirm your series solution.
   (a) $y' = 2xy$
   (b) $y' + y = 1$
   (c) $y' - y = 2$
   (d) $y' + y = 0$
   (e) $y' - y = 0$
   (f) $y' - y = x^2$
2. For each of the following differential equations, find a power series solution of the form $\sum_j a_j x^j$. Then solve the equation directly. Compare your two answers, and explain any discrepancies that may arise.
   (a) $xy' = y$
   (b) $x^2 y' = y$
   (c) $y' - (1/x)y = x^2$
   (d) $y' + (1/x)y = x$
3. Express the function $\arcsin x$ in the form of a power series $\sum_j a_j x^j$ by solving the differential equation $y' = (1-x^2)^{-1/2}$ in two different ways (*Hint*: Remember the binomial series.) Use this result to obtain the formula

$$\frac{\pi}{6} = \frac{1}{2} + \frac{1}{2}\cdot\frac{1}{3\cdot 2^3} + \frac{1\cdot 3}{2\cdot 4}\cdot\frac{1}{5\cdot 2^5} + \frac{1\cdot 3\cdot 5}{2\cdot 4\cdot 6}\cdot\frac{1}{7\cdot 2^7} + \cdots .$$

4. The differential equations considered in the text and in the preceding exercises are all linear. By contrast, the equation

$$y' = 1 + y^2 \qquad (*)$$

is nonlinear. One may verify directly that $y(x) = \tan x$ is the particular solution for which $y(0) = 0$. Show that

$$\tan x = x + \frac{1}{3}x^3 + \frac{2}{15}x^5 + \cdots$$

by assuming a solution for equation $(*)$ in the form of a power series $\sum_j a_j x^j$ and then finding the coefficients a_j in two ways:

(a) by the method of the examples in the text (note particularly how the nonlinearity of the equation complicates the formulas);

(b) by differentiating equation $(*)$ repeatedly to obtain

$$y'' = 2yy', \quad y''' = 2yy'' + 2(y')^2, \quad \text{etc.}$$

and using the formula $a_j = f^{(j)}(0)/j!$.

5. Solve the equation

$$y' = x - y, \quad y(0) = 0$$

by each of the methods suggested in the last exercise. What familiar function does the resulting series represent? Verify your conclusion by solving the equation directly as a first-order linear equation.

6. Use your symbol manipulation software, such as `Maple` or `Mathematica`, to write a routine that will find the coefficients of the power series solution for a given first-order ordinary differential equation.

6.3 Second-Order Linear Equations: Ordinary Points

We have invested considerable effort in studying equations of the form

$$y'' + p \cdot y' + q \cdot y = 0\,. \tag{6.6}$$

Here p and q could be constants or, more generally, functions. In some sense, our investigations thus far have been misleading; for we have only considered particular equations in which a closed-form solution can be found. These cases are really the exception rather than the rule. For most such equations, there is no "formula" for the solution. Power series then give us some extra flexibility. Now we may seek a power series solution; that solution is valid, and may be calculated and used in applications, even though it may not be expressed in a compact formula.

A number of the differential equations that arise in mathematical physics—Bessel's equation, Lagrange's equation, and many others—in fact fit the description that we have just presented. So it is worthwhile to develop techniques for studying (6). In the present section we shall concentrate on finding a power series solution to equation (6)—written in standard form—expanded about a point x_0, where x_0 has the property that p and q have convergent power series expansions about x_0. In this circumstance we call x_0

an *ordinary point* of the differential equation. Later sections will treat the situation where either p or q (or both) has a singularity at x_0.

We begin our study with a familiar equation, just to see the basic steps, and how the solution will play out.[3] Namely, we endeavor to solve

$$y'' + y = 0$$

by power series methods. As usual, we guess a solution of the form

$$y = \sum_{j=0}^{\infty} a_j x^j = a_0 + a_1 x + a_2 x^2 + \cdots .$$

Of course it follows that

$$y' = \sum_{j=1}^{\infty} j a_j x^{j-1} = a_1 + 2a_2 x + 3a_3 x^2 + \cdots$$

and

$$y'' = \sum_{j=2}^{\infty} j(j-1) a_j x^{j-2} = 2 \cdot 1 \cdot a_2 + 3 \cdot 2 \cdot a_3 x + 4 \cdot 3 \cdot a_4 x^2 \cdots .$$

Plugging the first and third of these into the differential equation gives

$$\sum_{j=2}^{\infty} j(j-1) a_j x^{j-2} + \sum_{j=0}^{\infty} a_j x^j = 0 .$$

As in the last example of Section 6.2, we find that the series has x raised to different powers, and that the summation begins in different places. We follow the standard procedure for repairing these matters.

First, we change the index of summation in the second series. So

$$\sum_{j=2}^{\infty} j(j-1) a_j x^{j-2} + \sum_{j=2}^{\infty} a_{j-2} x^{j-2} = 0 .$$

We invite the reader to verify that the new second series is just the same as the original second series (merely write out a few terms of each to check). We are fortunate in that both series now begin at the same index. So we may add them together to obtain

$$\sum_{j=2}^{\infty} [j(j-1) a_j + a_{j-2}] x^{j-2} = 0 .$$

[3] Of course this is an equation that we know how to solve by other means. Now we are learning a new solution technique.

The only way that such a power series can be identically zero is if each of the coefficients is zero. So we obtain the recursion equations

$$j(j-1)a_j + a_{j-2} = 0, \qquad j = 2, 3, 4, \ldots .$$

Then $j = 2$ gives us

$$a_2 = -\frac{a_0}{2 \cdot 1}.$$

It will be convenient to take a_0 to be an arbitrary constant A, so that

$$a_2 = -\frac{A}{2 \cdot 1}.$$

The recursion for $j = 4$ says that

$$a_4 = -\frac{a_2}{4 \cdot 3} = \frac{A}{4 \cdot 3 \cdot 2 \cdot 1}.$$

Continuing in this manner, we find that

$$\begin{aligned} a_{2j} &= (-1)^j \cdot \frac{A}{2j \cdot (2j-1) \cdot (2j-2) \cdots 3 \cdot 2 \cdot 1} \\ &= (-1)^j \cdot \frac{A}{(2j)!}, \quad j = 1, 2, \ldots . \end{aligned}$$

Thus we have complete information about the coefficients with even index.

Now let us consider the odd indices. Look at the recursion for $j = 3$. This is

$$a_3 = -\frac{a_1}{3 \cdot 2}.$$

It is convenient to take a_1 to be an arbitrary constant B. Thus

$$a_3 = -\frac{B}{3 \cdot 2 \cdot 1}.$$

Continuing with $j = 5$, we find that

$$a_5 = -\frac{a_3}{5 \cdot 4} = \frac{B}{5 \cdot 4 \cdot 3 \cdot 2 \cdot 1}.$$

In general,

$$a_{2j+1} = (-1)^j \frac{B}{(2j+1)!}, \quad j = 1, 2, \ldots .$$

In summary, then, the general solution of our differential equation is given by

$$y = A \cdot \left(\sum_{j=0}^{\infty} (-1)^j \cdot \frac{1}{(2j)!} x^{2j} \right) + B \cdot \left(\sum_{j=0}^{\infty} (-1)^j \frac{1}{(2j+1)!} x^{2j+1} \right).$$

Of course we recognize the first power series as the cosine function and the second as the sine function. So we have rediscovered that the general solution of $y'' + y = 0$ is

$$y = A \cdot \cos x + B \cdot \sin x .$$

EXAMPLE 6.3.1 Use the method of power series to solve the differential equation

$$(1-x^2)y'' - 2xy' + p(p+1)y = 0\,. \tag{6.7}$$

Here p is an arbitrary real constant. This is called *Legendre's equation.*

Solution: First we write the equation in standard form:

$$y'' - \frac{2x}{1-x^2}y' + \frac{p(p+1)}{1-x^2}y = 0\,.$$

Observe that, near $x = 0$, division by 0 is avoided and the coefficients p and q are real analytic. So 0 is an ordinary point.

We therefore guess a solution of the form

$$y = \sum_{j=0}^{\infty} a_j x^j = a_0 + a_1 x + a_2 x^2 + \cdots$$

and calculate

$$y' = \sum_{j=1}^{\infty} j a_j x^{j-1} = a_1 + 2a_2 x + 3a_3 x^2 + \cdots$$

and

$$y'' = \sum_{j=2}^{\infty} j(j-1) a_j x^{j-2} = 2a_2 + 3\cdot 2\cdot a_3 x + \cdots\,.$$

It is most convenient to treat the differential equation in the form (7). We calculate

$$-x^2 y'' = -\sum_{j=2}^{\infty} j(j-1)a_j x^j$$

and

$$-2xy' = -\sum_{j=1}^{\infty} 2ja_j x^j\,.$$

Substituting into the differential equation now yields

$$\sum_{j=2}^{\infty} j(j-1)a_j x^{j-2} - \sum_{j=2}^{\infty} j(j-1)a_j x^j - \sum_{j=1}^{\infty} 2ja_j x^j + \sum_{j=0}^{\infty} p(p+1)a_j x^j = 0\,.$$

We adjust the index of summation in the first sum so that it contains x^j rather than x^{j-2} and we break off spare terms and collect them together after the infinite sums. The result is

$$\sum_{j=2}^{\infty}(j+2)(j+1)a_{j+2}x^j - \sum_{j=2}^{\infty}j(j-1)a_jx^j$$
$$-\sum_{j=2}^{\infty}2ja_jx^j \quad + \sum_{j=2}^{\infty}p(p+1)a_jx^j$$
$$+\left[2a_2 + 6a_3x - 2a_1x + p(p+1)a_0 + p(p+1)a_1x\right] = 0\,.$$

In other words,

$$\sum_{j=2}^{\infty}\Big((j+2)(j+1)a_{j+2} - j(j-1)a_j - 2ja_j + p(p+1)a_j\Big)x^j$$
$$+\left[2a_2 + 6a_3x - 2a_1x + p(p+1)a_0 + p(p+1)a_1x\right] = 0\,.$$

Finally, we rewrite this as

$$\sum_{j=2}^{\infty}\Big((j+2)(j+1)a_{j+2} - j(j-1)a_j - 2ja_j + p(p+1)a_j\Big)x^j$$
$$+\Big(2a_2 + p(p+1)a_0\Big) + \Big(6a_3 - 2a_1 + p(p+1)a_1\Big)x = 0\,.$$

As a result,

$$(j+2)(j+1)a_{j+2} - j(j-1)a_j - 2ja_j + p(p+1)a_j = 0 \qquad \text{for } j = 2, 3, \ldots$$

together with

$$2a_2 + p(p+1)a_0 = 0$$

and

$$6a_3 - 2a_1 + p(p+1)a_1 = 0\,.$$

We have arrived at the recursion relation

$$a_2 = -\frac{(p+1)p}{2\cdot 1}\cdot a_0$$

$$a_3 = -\frac{(p+2)(p-1)}{3\cdot 2\cdot 1}\cdot a_1$$

$$a_{j+2} = -\frac{(p+j+1)(p-j)}{(j+2)(j+1)}\cdot a_j \qquad \text{for } j = 2, 3, \ldots\,. \tag{6.8}$$

We recognize a familiar pattern: The coefficients a_0 and a_1 are unspecified, so we set $a_0 = A$ and $a_1 = B$. Then we may proceed to solve for the rest of the coefficients. Now

$$a_2 = -\frac{(p+1)p}{2!} \cdot A$$

$$a_3 = -\frac{(p+2)(p-1)}{3!} \cdot B$$

$$a_4 = -\frac{(p+3)(p-2)}{4 \cdot 3} a_2 = \frac{(p+3)(p+1)p(p-2)}{4!} \cdot A$$

$$a_5 = -\frac{(p+4)(p-3)}{5 \cdot 4} a_3 = \frac{(p+4)(p+2)(p-1)(p-3)}{5!} \cdot B$$

$$a_6 = -\frac{(p+5)(p-4)}{6 \cdot 5} a_4 = -\frac{(p+5)(p+3)(p+1)p(p-2)(p-4)}{6!} \cdot A$$

$$a_7 = -\frac{(p+6)(p-5)}{7 \cdot 6} a_5 = -\frac{(p+6)(p+4)(p+2)(p-1)(p-3)(p-5)}{7!} \cdot B \,,$$

and so forth. Putting these coefficient values into our supposed power series solution, we find that the general solution of our differential equation is

$$\begin{aligned} y \;=\; & A\biggl(1 - \frac{(p+1)p}{2!}x^2 + \frac{(p+3)(p+1)p(p-2)}{4!}x^4 \\ & \quad - \frac{(p+5)(p+3)(p+1)p(p-2)(p-4)}{6!}x^6 + - \cdots \biggr) \\ & + B\biggl(x - \frac{(p+2)(p-1)}{3!}x^3 + \frac{(p+4)(p+2)(p-1)(p-3)}{5!}x^5 \\ & \quad - \frac{(p+6)(p+4)(p+2)(p-1)(p-3)(p-5)}{7!}x^7 + - \cdots \biggr). \end{aligned}$$

We assure the reader that, when p is not an integer, then these are *not* familiar elementary transcendental functions. These are what we call *Legendre functions.* In the special circumstance that p is a positive even integer, the first function (that which is multiplied by A) terminates as a polynomial. In the special circumstance that p is a positive odd integer, the second function (that which is multiplied by B) terminates as a polynomial. These are called *Legendre polynomials* P_n, and they play an important role in mathematical physics, representation theory, and interpolation theory. We shall encounter the Legendre polynomials later in the chapter. □

It is actually possible, without much effort, to check the radius of convergence of the functions we discovered as solutions in the last example. In fact we use the recursion relation (8) to see that

$$\left| \frac{a_{2j+2}x^{2j+2}}{a_{2j}x^{2j}} \right| = \left| -\frac{(p-2j)(p+2j+1)}{(2j+1)(2j+2)} \right| \cdot |x|^2 \to |x|^2$$

as $j \to \infty$. Thus the series expansion of the first Legendre function converges when $|x| < 1$, so the radius of convergence is 1. A similar calculation shows that the radius of convergence for the second Legendre function is 1.

We now enunciate a general result about the power series solution of an ordinary differential equation at an ordinary point. The proof is omitted.

Theorem 6.3.2 *Let x_0 be an ordinary point of the differential equation*

$$y'' + p \cdot y' + q \cdot y = 0\,, \tag{6.9}$$

and let α and β be arbitrary real constants. Then there exists a unique real analytic function $y = y(x)$ that has a power series expansion about x_0 and so that

(a) *The function y solves the differential equation (9).*

(b) *The function y satisfies the initial conditions $y(x_0) = \alpha$, $y'(x_0) = \beta$.*

If the functions p and q have power series expansions about x_0 with radius of convergence R then so does y.

We conclude with this remark. The examples that we have worked in detail resulted in solutions with *two-term* (or *binary*) recursion formulas: a_2 was expressed in terms of a_0, and a_3 was expressed in terms of a_1, etc. In general, the recursion formulas that arise in solving an ordinary differential equation at an ordinary point may result in more complicated recursion relations.

Exercises

1. In each of the following problems, verify that 0 is an ordinary point. Then find the power series solution of the given equation.

(a) $y'' + xy' + y = 0$

(b) $y'' - y' + xy = 0$

(c) $y'' + 2xy' - y = x$

(d) $y'' + y' - x^2y = 1$

(e) $(1 + x^2)y'' + xy' + y = 0$

(f) $y'' + (1 + x)y' - y = 0$

2. Find the general solution of

$$(1 + x^2)y'' + 2xy' - 2y = 0$$

in terms of power series in x. Can you express this solution by means of elementary functions?

3. Consider the equation $y'' + xy' + y = 0$.

(a) Find its general solution $y = \sum_j a_j x^j$ in the form $y = c_1 y_1(x) + c_2 y_2(x)$, where y_1, y_2 are power series.

(b) Use the ratio test to check that the two series y_1 and y_2 from part **(a)** converge for all x (as Theorem 4.3.2 actually asserts).

(c) Show that one of these two solutions, say y_1, is the series expansion of $e^{-x^2/2}$ and use this fact to find a second independent solution by the method discussed in Section 2.4. Convince yourself that this second solution is the function y_2 found in part **(a)**.

4. Verify that the equation $y'' + y' - xy = 0$ has a three-term recursion formula and find its series solutions y_1 and y_2 such that

(a) $y_1(0) = 1, \quad y_1'(0) = 0$

(b) $y_2(0) = 0, \quad y_2'(0) = 1$

Theorem 4.3.2 guarantees that both series converge at every $x \in \mathbf{R}$. Notice how difficult it would be to verify this assertion by working directly with the series themselves.

5. The equation $y'' + (p + 1/2 - x^2/4)y = 0$, where p is a constant, certainly has a series solution of the form $y = \sum_j a_j x^j$.

(a) Show that the coefficients a_j are related by the three-term recursion formula

$$(n+1)(n+2)a_{n+2} + \left(p + \frac{1}{2}\right)a_n - \frac{1}{4}a_{n-2} = 0\,.$$

(b) If the dependent variable is changed from y to w by means of $y = we^{-x^2/4}$, then show that the equation is transformed into $w'' - xw' + pw = 0$.

(c) Verify that the equation in **(b)** has a two-term recursion formula and find its general solution.

6. Solutions of *Airy's equation* $y'' + xy = 0$ are called *Airy functions*, and have applications to the theory of diffraction.

(a) Apply the theorems of the last chapter to verify that every nontrivial Airy function has infinitely many positive zeros and at most one negative zero.

(b) Find the Airy functions in the form of power series, and verify directly that these series converge for all x.

(c) Use the results of **(b)** to write down the general solution of $y'' - xy = 0$ without calculation.

7. *Chebyshev's equation* is

$$(1 - x^2)y'' - xy' + p^2 y = 0\,,$$

where p is constant.

(a) Find two linearly independent series solutions valid for $|x| < 1$.

(b) Show that if $p = n$ where n is an integer ≥ 0, then there is a polynomial solution of degree n. When these polynomials are multiplied by suitable constants, then they are called the *Chebyshev polynomials.* We shall see this topic again later in the book.

8. *Hermite's equation* is

$$y'' - 2xy' + 2py = 0\,,$$

where p is a constant.

(a) Show that its general solution is $y(x) = a_1 y_1(x) + a_2 y_2(x)$, where

$$\begin{aligned} y_1(x) &= 1 - \frac{2p}{2!}x^2 + \frac{2^2 p(p-2)}{4!}x^4 \\ &\quad - \frac{2^3 p(p-2)(p-4)}{6!}x^6 + - \cdots \end{aligned}$$

and

$$\begin{aligned} y_2(x) &= x - \frac{2(p-1)}{3!}x^3 + \frac{2^2(p-1)(p-3)}{5!}x^5 \\ &\quad - \frac{2^3(p-1)(p-3)(p-5)}{7!}x^7 + - \cdots . \end{aligned}$$

By Theorem 4.3.2, both series converge for all x. Verify this assertion directly.

(b) If p is a nonnegative integer, then one of these series terminates and is thus a polynomial—y_1 if p is even and y_2 if p is odd—while the other remains an infinite series. Verify that for $p = 0, 1, 2, 3, 4, 5$, these polynomials are $1, x, 1 - 2x^2, x - 2x^3/3, 1 - 4x^2 + 4x^4/3, x - 4x^3/3 + 4x^5/15$.

(c) It is clear that the only polynomial solutions of Hermite's equation are constant multiples of the polynomials described in part **(b)**. Those constant multiples, which have the additional property that the terms containing the highest powers of x are of the form $2^n x^n$, are denoted by $H_n(x)$ and called the *Hermite polynomials.* Verify that $H_0(x) = 1$, $H_1(x) = 2x$, $H_2(x) = 4x^2 - 2$, $H_3(x) = 8x^3 - 12x$, $H_4(x) = 16x^4 - 48x^2 + 12$, and $H_5(x) = 32x^5 - 160x^3 + 120x$.

(d) Verify that the polynomials listed in **(c)** are given by the general formula

$$H_n(x) = (-1)^n e^{x^2} \frac{d^n}{dx^n} e^{-x^2}\,.$$

9. Use your symbol manipulation software, such as `Maple` or `Mathematica`, to write a routine that will find the coefficients of the power series solution, expanded about an ordinary point, for a given second-order ordinary differential equation.

6.4 Regular Singular Points

Consider a differential equation in the standard form

$$y'' + p \cdot y' + q \cdot y = 0\,.$$

Let us consider a solution about a point x_0. If either of the coefficient functions p or q fails to be analytic at x_0, then x_0 is not an ordinary point for the differential equation and the methods of the previous section do not apply. In this circumstance we call x_0 a *singular point*.

There is some temptation to simply avoid the singular points. But often these are the points that are of the greatest physical interest. We must learn techniques for analyzing singular points. A simple example begins to suggest what the behavior near a singular point might entail. Consider the differential equation

$$y'' + \frac{2}{x}y' - \frac{2}{x^2}y = 0\,. \tag{6.10}$$

Obviously the point $x_0 = 0$ is a singular point for this equation. One may verify directly that the functions $y_1(x) = x$ and $y_2(x) = x^{-2}$ are solutions of this differential equation. Thus the general solution is

$$y = Ax + Bx^{-2}\,. \tag{6.11}$$

If we are interested only in solutions that are bounded near the origin, then we must take $B = 0$ and the solutions will have the form $y = Ax$. Most likely, the important physical behavior will take place when $B \neq 0$; we want to consider (11) in full generality.

The solution of ordinary differential equations near singular points of arbitrary type is extremely difficult. Many equations are intractable. Fortunately, many equations that arise in practice, or from physical applications, are of a particularly tame sort. We say that a singular point x_0 for the differential equation

$$y'' + p \cdot y' + q \cdot y = 0$$

is a *regular singular point* if

$$(x - x_0) \cdot p(x) \qquad \text{and} \qquad (x - x_0)^2 q(x)$$

are analytic at x_0. As an example, equation (10) has a regular singular point at 0 because

$$x \cdot p(x) = x \cdot \frac{2}{x} = 2$$

is analytic at 0 and

$$x^2 \cdot \left(-\frac{2}{x^2}\right) = -2$$

is analytic at 0.

Let us now consider some important differential equations from mathematical physics, just to illustrate how regular singular points arise in practice. Recall Legendre's equation

$$y'' - \frac{2x}{1 - x^2}y' + \frac{p(p+1)}{1 - x^2}y = 0\,.$$

We see immediately that ± 1 are singular points. The point $x_0 = 1$ is a regular singular point because

$$(x-1)\cdot p(x) = (x-1)\cdot\left(-\frac{2x}{1-x^2}\right) = \frac{2x}{x+1}$$

and

$$(x-1)^2 q(x) = (x-1)^2\cdot\left(\frac{p(p+1)}{1-x^2}\right) = -\frac{(x-1)p(p+1)}{x+1}$$

are both real analytic at $x = 1$ (namely, we *avoid* division by 0). A similar calculation shows that $x_0 = -1$ is a regular singular point.

As a second example, consider *Bessel's equation of order p*, where $p \geq 0$ is a constant:

$$x^2 y'' + xy' + (x^2 - p^2)y = 0\,.$$

Written in the form

$$y'' + \frac{1}{x}y' + \frac{x^2 - p^2}{x^2}y = 0\,,$$

the equation is seen to have a singular point at $x_0 = 0$. But it is regular because

$$x\cdot p(x) = 1 \quad \text{and} \quad x^2\cdot q(x) = x^2 - p^2$$

are both real analytic at 0.

Let us assume for the rest of this discussion that the regular singular point is at $x_0 = 0$.

The key idea in solving a differential equation at a regular singular point is to guess a solution of the form

$$y = y(x) = x^m\cdot\left(a_0 + a_1 x + a_2 x^2 + \cdots\right). \tag{6.12}$$

We see that we have modified the guess used in the last section by adding a factor of x^m in front. Here the exponent m can be positive or negative or zero—and m *need not be an integer*. In practice—and this is conceptually important to avoid confusion—we assume that we have factored out the greatest possible power of x; thus the coefficient a_0 will always be nonzero.

We call a series of the type (12) a *Frobenius series*. We now solve the differential equation at a regular singular point just as we did in the last section, except that now our recursion relations will be more complicated—as they will involve both the coefficients a_j and also the new exponent m. The method is best understood by examining an example.

EXAMPLE 6.4.1 Use the method of Frobenius series to solve the differential equation

$$2x^2 y'' + x(2x+1)y' - y = 0 \tag{6.13}$$

about the regular singular point 0.

Solution: Writing the equation in the standard form

$$y'' + \frac{x(2x+1)}{2x^2}y' - \frac{1}{2x^2}y = 0,$$

we readily check that

$$x \cdot \left(\frac{x(2x+1)}{2x^2}\right) = \frac{2x+1}{2} \quad \text{and} \quad x^2 \cdot \left(-\frac{1}{2x^2}\right) = -\frac{1}{2}$$

are both real analytic at $x_0 = 0$. So $x_0 = 0$ is a regular singular point for this differential equation.

We guess a solution of the form

$$y = x^m \cdot \sum_{j=0}^{\infty} a_j x^j = \sum_{j=0}^{\infty} a_j x^{m+j}$$

and therefore calculate that

$$y' = \sum_{j=0}^{\infty} (m+j) a_j x^{m+j-1}$$

and

$$y'' = \sum_{j=0}^{\infty} (m+j)(m+j-1) a_j x^{m+j-2}.$$

Notice that we do not begin the series for y' at 1 and we do not begin the series for y'' at 2, just because m may not be an integer so none of the summands may vanish (as in the case of an ordinary point).

Plugging these calculations into the differential equation, written in the form (13), yields

$$2\sum_{j=0}^{\infty}(m+j)(m+j-1)a_j x^{m+j}$$

$$+2\sum_{j=0}^{\infty}(m+j)a_j x^{m+j+1} + \sum_{j=0}^{\infty}(m+j)a_j x^{m+j} - \sum_{j=0}^{\infty} a_j x^{m+j} = 0.$$

We make the usual adjustments in the indices so that all powers of x are x^{m+j}, and break off the odd terms to push over to the right. We obtain

$$2\sum_{j=1}^{\infty}(m+j)(m+j-1)a_j x^{m+j} + 2\sum_{j=1}^{\infty}(m+j-1)a_{j-1} x^{m+j}$$

$$+\sum_{j=1}^{\infty}(m+j)a_j x^{m+j} - \sum_{j=1}^{\infty} a_j x^{m+j} + 2m(m-1)a_0 x^m + m a_0 x^m - a_0 x^m = 0.$$

The result is

$$\Big(2(m+j)(m+j-1)a_j+2(m+j-1)a_{j-1}+(m+j)a_j-a_j\Big) = 0 \quad \text{for } j = 1, 2, 3, \ldots \tag{6.14}$$

together with

$$[2m(m-1) + m - 1]a_0 = 0\,.$$

It is clearly not to our advantage to let $a_0 = 0$. Thus

$$2m(m-1) + m - 1 = 0\,.$$

This is the *indicial equation.*

The roots of this quadratic indicial equation are $m = -1/2, 1$. We put each of these values into (14) and solve the two resulting recursion relations.

Now (14) says that

$$(2m^2 + 2j^2 + 4mj - j - m - 1)a_j = (-2m - 2j + 2)a_{j-1}\,.$$

For $m = -1/2$ this is

$$a_j = \frac{3-2j}{-3j+2j^2}a_{j-1}\,.$$

We set $a_0 = A$ so that

$$a_1 = -a_0 = -A\,, \quad a_2 = -\frac{1}{2}a_1 = \frac{1}{2}a_0 = \frac{1}{2}A, \text{ etc.}$$

For $m = 1$ we have

$$a_j = \frac{-2j}{3j+2j^2}a_{j-1}\,.$$

We set $a_0 = B$ so that

$$a_1 = -\frac{2}{5}a_0 = -\frac{2}{5}B\,, \quad a_2 = -\frac{4}{14}a_1 = \frac{4}{35}B\,, \text{ etc.}$$

Thus we have found two linearly independent solutions. When $m = -1/2$ we obtain the solution

$$Ax^{-1/2}\cdot\big(1 - x + \frac{1}{2}x^2 - + \cdots\big)\,.$$

And when $m = 1$ we obtain the solution

$$Bx\cdot\big(1 - \frac{2}{5}x + \frac{4}{35}x^2 - + \cdots\big)\,.$$

The general solution of our differential equation is then

$$y = Ax^{-1/2}\cdot\big(1 - x + \frac{1}{2}x^2 - + \cdots\big) + Bx\cdot\big(1 - \frac{2}{5}x + \frac{4}{35}x^2 - + \cdots\big)\,. \quad \square$$

There are some circumstances (such as when the indicial equation has a repeated root) that the method we have presented will not yield two linearly independent solutions. We explore these circumstances in the next section.

Exercises

1. For each of the following differential equations, locate and classify its singular points on the x-axis.

(a) $x^3(x-1)y'' - 2(x-1)y' + 3xy = 0$
(b) $x^2(x^2-1)y'' - x(1-x)y' + 2y = 0$
(c) $x^2y'' + (2-x)y' = 0$
(d) $(3x+1)xy'' - (x+1)y' + 2y = 0$

2. Determine the nature of the point $x = 0$ (i.e., what type of singular point it is) for each of the following differential equations.

(a) $y'' + (\sin x)y = 0$
(b) $xy'' + (\sin x)y = 0$
(c) $x^2y'' + (\sin x)y = 0$
(d) $x^3y'' + (\sin x)y = 0$
(e) $x^4y'' + (\sin x)y = 0$

3. Find the indicial equation and its roots (for a series solution in powers of x) for each of the following differential equations.

(a) $x^3y'' + (\cos 2x - 1)y' + 2xy = 0$
(b) $4x^2y'' + (2x^4 - 5x)y' + (3x^2 + 2)y = 0$
(c) $x^2y'' + 3xy' + 4xy = 0$
(d) $x^3y'' - 4x^2y' + 3xy = 0$

4. For each of the following differential equations, verify that the origin is a regular singular point and calculate two independent Frobenius series solutions:

(a) $4x^2y'' + 3y' + y = 0$
(b) $2xy'' + (3-x)y' - y = 0$
(c) $2xy'' + (x+1)y' + 3y = 0$
(d) $2x^2y'' + xy' - (x+1)y = 0$

5. When $p = 0$, Bessel's equation becomes

$$x^2y'' + xy' + x^2y = 0 .$$

Show that its indicial equation has only one root, and use the method of this section to deduce that

$$y(x) = \sum_{j=0}^{\infty} \frac{(-1)^j}{2^{2j}(j!)^2} x^{2j}$$

is the corresponding Frobenius series solution.

6. Consider the differential equation

$$y'' + \frac{1}{x^2}y' - \frac{1}{x^3}y = 0 .$$

(a) Show that $x = 0$ is an irregular singular point.
(b) Use the fact that $y_1 = x$ is a solution to find a second independent solution y_2 by the method discussed earlier in the book.

(c) Show that the second solution y_2 that we found in part **(b)** cannot be expressed as a Frobenius series.

7. Consider the differential equation

$$y'' + \frac{p}{x^b}y' + \frac{q}{x^c}y = 0\,,$$

where p and q are nonzero real numbers and b and c are positive integers. It is clear that, if $b > 1$ or $c > 2$, then $x = 0$ is an irregular singular point.

(a) If $b = 2$ and $c = 3$, then show that there is only one possible value of m (from the indicial equation) for which there might exist a Frobenius series solution.

(b) Show similarly that m satisfies a quadratic equation—and hence we can hope for two Frobenius series solutions, corresponding to the roots of this equation—if and only if $b = 1$ and $c \leq 2$. Observe that these are exactly the conditions that characterize $x = 0$ as a "weak" or regular singular point as opposed to a "strong" or irregular singular point.

8. The differential equation

$$x^2y'' + (3x - 1)y' + y = 0 \qquad (\star)$$

has $x = 0$ as an irregular singular point. If

$$\begin{aligned} y &= x^m(a_0 + a_1x + a_2x^2 + \cdots) \\ &= a_0x^m + a_1x^{m+1} + a_2x^{m+2} + \cdots \end{aligned}$$

is inserted into $(\star)$, then show that $m = 0$ and the corresponding Frobenius series "solution" is the power series

$$y = \sum_{j=0}^{\infty} j!x^j\,,$$

which converges only at $x = 0$. This demonstrates that even when a Frobenius series formally satisfies such an equation, it is not necessarily a valid solution.

6.5 More on Regular Singular Points

We now look at the Frobenius series solution of

$$y'' + p \cdot y' + q \cdot y = 0$$

at a regular singular point from a theoretical point of view.

Assuming that 0 is regular singular, we may suppose that

$$x \cdot p(x) = \sum_{j=0}^{\infty} p_jx^j \qquad \text{and} \qquad x^2 \cdot q(x) = \sum_{j=0}^{\infty} q_jx^j\,,$$

valid on a nontrivial interval $(-R, R)$. We guess a solution of the form

$$y = x^m \sum_{j=0}^{\infty} a_j x^j = \sum_{j=0}^{\infty} a_j x^{j+m}$$

and calculate

$$y' = \sum_{j=0}^{\infty} (j+m) a_j x^{j+m-1}$$

and

$$y'' = \sum_{j=0}^{\infty} (j+m)(j+m-1) a_j x^{j+m-2} .$$

Then

$$\begin{aligned} p(x) \cdot y' &= \frac{1}{x} \left(\sum_{j=0}^{\infty} p_j x^j \right) \left(\sum_{j=0}^{\infty} a_j (m+j) x^{m+j-1} \right) \\ &= x^{m-2} \left(\sum_{j=0}^{\infty} p_j x^j \right) \left(\sum_{j=0}^{\infty} a_j (m+j) x^j \right) \\ &= x^{m-2} \sum_{j=0}^{\infty} \left(\sum_{k=0}^{j} p_{j-k} a_k (m+k) \right) x^j , \end{aligned}$$

where we have used the formula (Section 6.1) for the product of two power series. Now, breaking off the summands corresponding to $k = j$, we find that this last is equal to

$$x^{m-2} \sum_{j=0}^{\infty} \left(\sum_{k=0}^{j-1} p_{j-k} a_k (m+k) + p_0 a_j (m+j) \right) x^j .$$

Similarly,

$$\begin{aligned} q(x) \cdot y &= \frac{1}{x^2} \left(\sum_{j=0}^{\infty} q_j x^j \right) \left(\sum_{j=0}^{\infty} a_j x^{m+j} \right) \\ &= x^{m-2} \left(\sum_{j=0}^{\infty} q_j x^j \right) \left(\sum_{j=0}^{\infty} a_j x^j \right) \\ &= x^{m-2} \sum_{j=0}^{\infty} \left(\sum_{k=0}^{j} q_{j-k} a_k \right) x^j \\ &= x^{m-2} \sum_{j=0}^{\infty} \left(\sum_{k=0}^{j-1} q_{j-k} a_k + q_0 a_j \right) x^j . \end{aligned}$$

We put the series expressions for y'', $p \cdot y'$, and $q \cdot y$ into the differential equation, and cancel the common factor of x^{m-2}. The result is

$$\sum_{j=0}^{\infty}\Bigg\{a_j[(m+j)(m+j-1)+(m+j)p_0+q_0] \\ + \sum_{k=0}^{j-1} a_k[(m+k)p_{j-k}+q_{j-k}]\Bigg\}x^j = 0\,.$$

Now of course each coefficient of x^j must be 0, so we obtain the following recursion relations:

$$a_j[(m+j)(m+j-1)+(m+j)p_0+q_0] \\ + \sum_{k=0}^{j-1} a_k[(m+k)p_{j-k}+q_{j-k}] = 0$$

for $j = 0, 1, 2, \ldots$. (Incidentally, this illustrates a point we made earlier, in Section 6.3: That recursion relations need not be binary.)

It is convenient, and helps us to emphasize the indicial equation, to write $f(m) = m(m-1) + mp_0 + q_0$. Then the recursion relation for $j = 0$ is

$$a_0 f(m) = 0$$

(because the sum in the recursion is vacuous when $j = 0$).

The successive recursion relations are

$$a_1 f(m+1) + a_0(mp_1 + q_1) = 0$$

$$a_2 f(m+2) + a_0(mp_2 + q_2) + a_1[(m+1)p_1 + q_1] = 0$$

$$\ldots$$

$$a_j f(m+j) + a_0(mp_j + q_j) + \cdots + a_{j-1}[(m+j-1)p_1 + q_1] = 0$$

etc.

The first recursion formula tells us, since $a_0 \neq 0$, that

$$\boxed{f(m) = m(m-1) + mp_0 + q_0 = 0\,.}$$

This is of course the indicial equation. The roots m_1 and m_2 of this equation are called the *exponents* of the differential equation at the regular singular point. In practice you will use this formula to solve for the exponents m when you are applying the Frobenius method to solve a differential equation with a regular singular point.

If the roots m_1 and m_2 are distinct and *do not differ by an integer*, then our procedures will produce two linearly independent solutions for the differential equation. If $m_1 \leq m_2$ differs by an integer, say $m_2 = m_1 + j$ for some integer $j \geq 1$, then the recursion procedure breaks down because the coefficient of a_j in the jth recursion relation for m_1 will be 0—so that we cannot solve for a_j. The case $m_1 = m_2$ also leads to difficulties—because then our methods only generate one solution. *When m_1 and m_2 differ by a positive integer, then it is only necessary to do the analysis (leading to a solution of the differential equation) for the exponent m_1. The exponent m_2 is not guaranteed to lead to anything new.* The exponent m_2 *could* lead to something new, as the next example shows. But we cannot depend on it.

We conclude this section by **(i)** enunciating a theorem that summarizes our positive results and **(ii)** providing an example that illustrates how to handle the degenerate cases just described.

Theorem 6.5.1 *Suppose that $x_0 = 0$ is a regular singular point for the differential equation*

$$y'' + p \cdot y' + q \cdot y = 0\,.$$

Assume that the power series for $x \cdot p(x)$ and $x^2 \cdot q(x)$ have radius of convergence $R > 0$. Suppose that the indicial equation $f(m) = m(m-1) + mp_0 + q_0 = 0$ has roots m_1, m_2 with $m_1 \leq m_2$. Then the differential equation has at least one solution of the form

$$y_1 = x^{m_1} \sum_{j=0}^{\infty} a_j x^j$$

on the interval $(-R, R)$.

In case $m_2 - m_1$ is not zero or a positive integer, then the differential equation has a second independent solution

$$y_2 = x^{m_2} \sum_{j=0}^{\infty} b_j x^j$$

on the interval $(-R, R)$.

EXAMPLE 6.5.2 Find two independent Frobenius series solutions of

$$xy'' + 2y' + xy = 0\,.$$

Solution: Notice that the constant term of the coefficient of y' is 2 and the constant term of the coefficient of y is 0. Thus the indicial equation is

$$m(m-1)+2m+0=0\,.$$

The exponents for the regular singular point 0 are then $0, -1$.

Corresponding to $m_1 = 0$, we guess a solution of the form

$$y=\sum_{j=0}^{\infty} a_j x^j$$

which entails

$$y'=\sum_{j=1}^{\infty} j a_j x^{j-1}$$

and

$$y''=\sum_{j=2}^{\infty} j(j-1) a_j x^{j-2}\,.$$

Putting these expressions into the differential equation yields

$$x\sum_{j=2}^{\infty} j(j-1)a_j x^{j-2}+2\sum_{j=1}^{\infty} j a_j x^{j-1}+x\sum_{j=0}^{\infty} a_j x^j=0\,.$$

We adjust the indices so that all powers are x^j, and break off the lower indices so that all sums begin at $j = 1$. The result is

$$\sum_{j=1}^{\infty}\Big(a_{j+1}[j(j+1)+2(j+1)]+a_{j-1}\Big)x^j+2a_1=0\,.$$

We read off the recursion relations

$$a_1=0$$

and

$$a_{j+1}=-\frac{a_{j-1}}{(j+2)(j+1)}$$

for $j \geq 1$. Thus all the coefficients with odd index are 0 and we calculate that

$$a_2=-\frac{a_0}{3\cdot 2}\,,\quad a_4=\frac{a_0}{5\cdot 4\cdot 3\cdot 2}\,,\quad a_6=-\frac{a_0}{7\cdot 6\cdot 5\cdot 4\cdot 3\cdot 2}\,,$$

etc. Thus we have found one Frobenius solution

$$\begin{aligned} y_1 &= a_0\left(1-\frac{1}{3!}x^2+\frac{1}{5!}x^4-\frac{1}{7!}x^6-+\cdots\right) \\ &= a_0\cdot\frac{1}{x}\cdot\left(x-\frac{x^3}{3!}+\frac{x^5}{5!}-+\cdots\right) \\ &= a_0\cdot\frac{1}{x}\cdot\sin x\,. \end{aligned}$$

Corresponding to $m_2 = -1$ we guess a solution of the form

$$y = x^{-1} \cdot \sum_{j=0}^{\infty} b_j x^j = \sum_{j=0}^{\infty} b_j x^{j-1} .$$

Thus we calculate that

$$y' = \sum_{j=0}^{\infty} (j-1) b_j x^{j-2}$$

and

$$y'' = \sum_{j=0}^{\infty} (j-1)(j-2) b_j x^{j-3} .$$

Putting these calculations into the differential equation gives

$$\sum_{j=0}^{\infty} (j-1)(j-2) b_j x^{j-2} + \sum_{j=0}^{\infty} 2(j-1) b_j x^{j-2} + \sum_{j=0}^{\infty} b_j x^j = 0 .$$

We adjust the indices so that all powers that occur are x^{j-2} and break off extra terms so that all sums begin at $j = 2$. The result is

$$\sum_{j=2}^{\infty} [(j-1)(j-2) b_j + 2(j-1) b_j + b_{j-2}] x^{j-2}$$
$$+(-1)(-2) b_0 x^{-2} + (0)(-1) b_1 x^{-1} + 2(-1) b_0 x^{-2} + 2(0) b_1 x^{-1} = 0 .$$

The four terms at the end cancel out. Thus our recursion relation is

$$b_j = -\frac{b_{j-2}}{j(j-1)}$$

for $j \geq 2$.

It is now easy to calculate that

$$b_2 = -\frac{b_0}{2 \cdot 1} , \; b_4 = \frac{b_0}{4 \cdot 3 \cdot 2 \cdot 1} , \; b_6 = -\frac{b_0}{6 \cdot 5 \cdot 4 \cdot 3 \cdot 2 \cdot 1} , \text{ etc.}$$

Also

$$b_3 = -\frac{b_1}{3 \cdot 2 \cdot 1} , \; b_5 = \frac{b_1}{5 \cdot 4 \cdot 3 \cdot 2 \cdot 1} , \; b_7 = -\frac{b_1}{7 \cdot 6 \cdot 5 \cdot 4 \cdot 3 \cdot 2 \cdot 1} , \text{ etc.}$$

This gives the solution

$$\begin{aligned} y &= b_0 \cdot \frac{1}{x} \cdot \left(1 - \frac{1}{2!} x^2 + \frac{1}{4!} x^4 - + \cdots \right) + b_1 \cdot \frac{1}{x} \left(x - \frac{1}{3!} x^3 + \frac{1}{5!} x^5 - + \cdots \right) \\ &= b_0 \cdot \frac{1}{x} \cdot \cos x + b_1 \cdot \frac{1}{x} \cdot \sin x . \end{aligned}$$

Now we already discovered the solution $(1/x) \sin x$ in our first calculation with $m_1 = 0$. Our second calculation, with $m_2 = -1$, reproduces that solution and discovers the new, linearly independent solution $(1/x) \cos x$. □

EXAMPLE 6.5.3 The equation

$$4x^2y'' - 8x^2y' + (4x^2 + 1)y = 0$$

has only one Frobenius series solution. Find the general solution.

Solution: The indicial equation is

$$m(m-1) + m \cdot 0 + \frac{1}{4} = 0\,.$$

The roots are $m_1 = 1/2, m_2 = 1/2$. This is a repeated indicial root, and it will lead to complications.

We guess a solution of the form

$$y = x^{1/2}\sum_{j=0}^{\infty} a_j x^j = \sum_{j=0}^{\infty} a_j x^{j+1/2}\,.$$

Thus

$$y' = \sum_{j=0}^{\infty}(j+1/2)a_j x^{j-1/2}$$

and

$$y'' = \sum_{j=0}^{\infty}(j+1/2)(j-1/2)a_j x^{j-3/2}\,.$$

Putting these calculations into the differential equation yields

$$4x^2 \cdot \sum_{j=0}^{\infty}(j+1/2)(j-1/2)a_j x^{j-3/2}$$
$$-8x^2 \cdot \sum_{j=0}^{\infty}(j+1/2)a_j x^{j-1/2} + (4x^2+1)\cdot\sum_{j=0}^{\infty} a_j x^{j+1/2} = 0\,.$$

We may rewrite this as

$$\sum_{j=0}^{\infty}(2j+1)(2j-1)a_j x^{j+1/2}$$
$$-\sum_{j=0}^{\infty}(8j+4)a_j x^{j+3/2} + \sum_{j=0}^{\infty}4a_j x^{j+5/2} + \sum_{j=0}^{\infty}a_j x^{j+1/2} = 0\,.$$

We adjust the indices so that all powers of x are $x^{j+1/2}$ and put extra terms on the right so that all sums begin at $j = 2$. The result is

$$\sum_{j=2}^{\infty}\left[(2j+1)(2j-1)a_j - (8j-4)a_{j-1} + 4a_{j-2} + a_j\right]x^{j+1/2}$$
$$= \quad -1\cdot(-1)a_0x^{1/2} - 3\cdot 1\cdot a_1x^{3/2} + 4a_0x^{3/2} - a_0x^{1/2} - a_1x^{3/2}$$

or

$$\sum_{j=2}^{\infty}[4j^2 a_j - (8j-4)a_{j-1} + 4a_{j-2}]x^{j+1/2} = x^{3/2}(4a_0 - 4a_1).$$

We thus discover the recursion relations

$$a_1 = a_0$$

$$a_j = \frac{(2j-1)a_{j-1} - a_{j-2}}{j^2}.$$

Therefore

$$a_2 = \frac{a_0}{2!}, \; a_3 = \frac{a_0}{3!}, \; a_4 = \frac{a_0}{4!}, \; a_5 = \frac{a_0}{5!}, \text{ etc.}$$

We thus have found the Frobenius series solution to our differential equation given by

$$y_1(x) = x^{1/2} \cdot \left(1 + \frac{x}{1!} + \frac{x^2}{2!} + \frac{x^3}{3!} + \frac{x^4}{4!} + \cdots\right) = x^{1/2} \cdot e^x.$$

But now we are stuck because $m = 1/2$ is a repeated root of the indicial equation. There is no other. We can rescue the situation by thinking back to how we handled matters for second-order linear equations with constant coefficients. In that circumstance, if the associated polynomial had distinct roots r_1, r_2 then the differential equation had solutions $y_1 = e^{r_1 x}$ and $y_2 = e^{r_2 x}$. But if the associated polynomial had repeated roots r and r, then solutions of the differential equation were given by $y_1 = e^{rx}$ and $y_2 = x \cdot e^{rx}$. Reasoning by analogy, and considering that x (or its powers) is a *logarithm* of e^x, we might hope that when m is a repeated root of the indicial equation, and $y_1 = y_1(x)$ is a solution corresponding to the root m, then $y_2(x) = \ln x \cdot y_1(x)$ is a second solution. We in fact invite the reader to verify that

$$y_2(x) = \ln x \cdot x^{1/2} \cdot e^x$$

is a second, linearly independent solution of the differential equation. So its general solution is

$$y = Ax^{1/2}e^x + B\ln x \cdot x^{1/2}e^x. \qquad \square$$

Exercises

1. The equation
$$x^2y'' - 3xy' + (4x+4)y = 0$$
has only one Frobenius series solution. Find it.
2. The equation
$$4x^2y'' - 8x^2y' + (4x^2+1)y = 0$$
has only one Frobenius series solution. Find the general solution.

3. Find two independent Frobenius series solutions of each of the following equations.

(a) $xy'' + 2y' + xy = 0$ **(c)** $xy'' - y' + 4x^3y = 0$
(b) $x^2y'' - x^2y' + (x^2 - 2)y = 0$

4. Verify that the point 1 is a regular singular point for the equation

$$(x-1)^2y'' - 3(x-1)y' + 2y = 0\,.$$

Now use Frobenius's method to solve the equation.

5. Verify that the point -1 is a regular singular point for the equation

$$3(x+1)^2y'' - (x+1)y' - y = 0\,.$$

Now use Frobenius's method to solve the equation.

6. Bessel's equation of order $p = 1$ is

$$x^2y'' + xy' + (x^2 - 1)y = 0\,.$$

We see that $x_0 = 0$ is a regular singular point. Show that $m_1 - m_2 = 2$ and that the equation has only one Frobenius series solution. Then find it.

7. Bessel's equation of order $p = 1/2$ is

$$x^2y'' + xy' + \left(x^2 - \frac{1}{4}\right)y = 0\,.$$

Show that $m_1 - m_2 = 1$, but that the equation still has two independent Frobenius series solutions. Then find them.

8. The only Frobenius series solution of Bessel's equation of order $p = 0$ is given in Exercise 5 of Section 4.4. By taking this to be solution y_1, and substituting the formula

$$y_2 = y_1 \ln x + x^{m_2} \sum_{j=0}^{\infty} c_j x^j$$

into the differential equation, obtain the second independent solution given by

$$y_2(x) = y_1(x) \cdot \ln x + \sum_{j=1}^{\infty} \frac{(-1)^{j+1}}{2^{2j}(j!)^2}\left(1 + \frac{1}{2} + \cdots + \frac{1}{j}\right)x^{2j}\,.$$

Historical Note

Gauss

Often called the "prince of mathematicians," Carl Friedrich Gauss (1777–1855) left a legacy of mathematical genius that exerts considerable influence even today.

Gauss was born in the city of Brunswick in northern Germany. He showed an early aptitude with arithmetic and numbers. He was finding errors in his father's bookkeeping at the age of 3, and his facility with calculations soon became well known. He came to the attention of the Duke of Brunswick, who supported the young man's further education.

Gauss attended the Caroline College in Brunswick (1792–1795), where he completed his study of the classical languages and explored the works of Newton, Euler, and Lagrange. Early in this period he discovered the prime number theorem—legend has it by staring for hours at tables of primes. Gauss did not prove the theorem (it was finally proved in 1896 by Hadamard and de la Vallee Poussin). He also, at this time, invented the method of least squares for minimizing errors—a technique that is still widely used today. He also conceived the Gaussian (or normal) law of distribution in the theory of probability.

At the university, Gauss was at first attracted by philology and put off by the mathematics courses. But at the age of 18, he made a remarkable discovery—of which regular polygons can be constructed by ruler and compass—and that set his future for certain. During these years Gauss was flooded, indeed nearly overwhelmed, by mathematical ideas. In 1795, just as an instance, Gauss discovered the fundamental law of quadratic reciprocity. It took a year of concerted effort for him to prove it. It is the core of his celebrated treatise *Disquisitiones Arithmeticae*, published in 1801. That book is arguably the cornerstone of modern number theory, as it contains the fundamental theorem of arithmetic as well as foundational ideas on congruences, forms, residues, and the theory of cyclotomic equations. The hallmark of Gauss's *Disquisitiones* is a strict adherence to rigor (unlike much of the mathematics of Gauss's day) and a chilling formality and lack of motivation.

Gauss's work touched all parts of mathematics—not just number theory. He discovered what is now known as the Cauchy integral formula, developed the intrinsic geometry of surfaces, discovered the mean-value property for harmonic functions, proved a version of Stokes's theorem, developed the theory of elliptic functions, and he anticipated Bolyai and Lobachevsky's ideas on non-Euclidean geometry. With regard to the latter—which was really an earth-shaking breakthrough—Gauss said that he did not publish his ideas because nobody (i.e., none of the mathematicians of the time) would appreciate or understand them.

Beginning in the 1830s, Gauss was increasingly occupied by physics. He had already a real *coup* in helping astronomers to locate the planet Ceres using strictly mathematical reasoning. Now he turned his attention to conservation of energy, the calculus of variations, optics, geomagnetism, and potential theory.

Carl Friedrich Gauss was an extraordinarily powerful and imaginative mathematician who made fundamental contributions to all parts of the subject. He had a long and productive scientific career. When, one day, a messenger came to him with the news that his wife was dying, Gauss said, "Tell her to wait a bit until I am done with this theorem." Such is the life of a master of mathematics.

Historical Note

Abel

Niels Henrik Abel (1802–1829) was one of the foremost mathematicians of the nineteenth century, and perhaps the greatest genius ever produced by the Scandinavian countries. Along with his contemporaries Gauss and Cauchy, Abel helped to lay the foundations for the modern mathematical method.

Abel's genius was recognized when he was still young. In spite of grinding poverty, he managed to attend the University of Oslo. When only 21 years old, Abel produced a proof that the fifth degree polynomial cannot be solved by an elementary formula (involving only arithmetic operations and radicals). Recall that the quadratic equation can be solved by the quadratic formula, and cubic and quartic equations can be solved by similar but more complicated formulas. This was an age-old problem, and Abel's solution was a personal triumph. He published the proof in a small pamphlet at his own expense. This was typical of the poor luck and lack of recognition that plagued Abel's short life.

Abel desired to spend time on the Continent and commune with the great mathematicians of the day. He finally got a government fellowship. His first year was spent in Berlin, where he became a friend and colleague of August Leopold Crelle. He helped Crelle to found the famous *Journal für die Reine und Angewandte Mathematik*, now the oldest extant mathematics journal.

There are many parts of modern mathematics that bear Abel's name. These include Abel's integral equation, Abelian integrals and functions, Abelian groups, Abel's series, Abel's partial summation formula, Abel's limit theorem, and Abel summability. The basic theory of elliptic functions is due to Abel (the reader may recall that these played a decisive role in Andrew Wiles's solution of Fermat's Last Problem).

Like Riemann (discussed elsewhere in this book), Abel lived in penury. He never held a proper academic position (although, shortly before Abel's death, Crelle finally secured for him a professorship in Berlin). The young genius contracted tuberculosis at the age of 26 and died soon thereafter.

Crelle eulogized Abel in his *Journal* with these words:

> All of Abel's works carry the imprint of an ingenuity and force of thought which is amazing. One may say that he was able to penetrate all obstacles down to the very foundation of the problem, with a force which appeared irresistible ... He distinguished himself equally by the purity and nobility

of his character and by a rare modesty which made his person cherished to the same unusual degree as was his genius.

It is difficult to imagine what Abel might have accomplished had he lived a normal lifespan and had an academic position with adequate financial support. His was one of the great minds of mathematics.

Today one may see a statue of Abel in the garden of the Norwegian King's palace—he is depicted stamping out the serpents of ignorance. Also the Abel Prize, one of the most distinguished of mathematical encomia, is named for this great scientist.

Anatomy of an Application

Steady-State Temperature in a Ball

Let us show how to reduce the analysis of the heat equation in a three-dimensional example to the study of Legendre's equation. Imagine a solid ball of radius 1. We shall work in spherical coordinates (r, θ, ϕ) on this ball. We hypothesize that the surface temperature at the boundary is held at $g(\theta) = T_0 \sin^4 \theta$. Of course the steady-state temperature T will satisfy Laplace's equation

$$\nabla^2 T = \left(\frac{\partial^2}{\partial x^2} + \frac{\partial^2}{\partial y^2} + \frac{\partial^2}{\partial z^2} \right) T = 0 .$$

The Laplace operator is rotationally invariant and the boundary condition does not depend on the azimuthal angle ϕ; hence, the solution will not depend on ϕ either. One may calculate that Laplace's equation thus takes the form

$$\frac{1}{r^2} \frac{\partial}{\partial r} \left(r^2 \frac{\partial T}{\partial r} \right) + \frac{1}{r^2 \sin \theta} \left(\sin \theta \frac{\partial T}{\partial \theta} \right) = 0 . \tag{6.15}$$

We seek a solution by the method of separation of variables. Thus we set $T(r, \theta) = A(r) \cdot B(\theta)$. Substituting into (15) gives

$$B \cdot \frac{d}{dr} \left(r^2 \frac{dA}{dr} \right) + \frac{A}{\sin \theta} \cdot \frac{d}{d\theta} \left(\sin \theta \frac{dB}{d\theta} \right) = 0 .$$

Thus we find that

$$\frac{1}{A} \frac{d}{dr} \left(r^2 \frac{dA}{dr} \right) = - \frac{1}{B \sin \theta} \frac{d}{d\theta} \left(\sin \theta \frac{dB}{d\theta} \right) .$$

The left-hand side depends only on r and the right-hand side depends only on θ. We conclude that both sides are equal to a constant. Looking ahead to the use of the Legendre equation, we are going to suppose that the constant has the form $c = n(n+1)$ for n a nonnegative integer. This rather surprising hypothesis will be justified later. Also refer back to the discussion in Example 6.3.1.

Now we have this ordinary differential equation for B:

$$\frac{1}{\sin\theta}\frac{d}{d\theta}\left(\sin\theta\frac{dB}{d\theta}\right) + n(n+1) = 0\,.$$

We make the change of variable

$$\nu = \cos\theta, \quad y(\nu) = B(\theta)\,.$$

With the standard identities

$$\sin^2\theta = 1 - \nu^2 \quad \text{and} \quad \frac{d}{d\nu} = \frac{d\theta}{d\nu}\frac{d}{d\theta} = -\frac{1}{\sin\theta}\frac{d}{d\theta}\,,$$

we find our differential equation converted to

$$\frac{d}{d\nu}\left((1-\nu^2)\frac{dy}{d\nu}\right) + n(n+1)y = 0\,.$$

This is equivalent to

$$(1-\nu^2)y'' - 2\nu y' + n(n+1)y = 0\,. \qquad (6.16)$$

This is Legendre's equation.

Observe that $\nu = \pm 1$ corresponds to $\theta = 0, \pi$, i.e., the poles of the sphere. A physically meaningful solution will certainly be finite at these points. We conclude that the solution of our differential equation (16) is the Legendre polynomial $y(\nu) = P_n(\nu)$. Therefore the solution to our original problem is $B_n(\theta) = P_n(\cos\theta)$.

Our next task is to solve for $A(r)$. The differential equation (resulting from our separation of variables) is

$$\frac{d}{dr}\left(r^2\frac{dA}{dr}\right) = n(n+1)A\,.$$

One can use the method of power series to solve this equation. Or one can take a shortcut and simply guess that the solution will be a power of r. Using the latter method, we find that

$$A_n(r) = c_n r^n + d_n r^{-1-n}\,.$$

Again, physical considerations can guide our thinking. We know that the temperature must be finite at the center of the sphere. Thus d_n must equal 0.

Hence $A_n(r) = c_n r^n$. Putting this information together with our solution in θ, we find the solution of Laplace's equation to be

$$T = c_n r^n P_n(\cos\theta)\,.$$

Here, of course, c_n is an arbitrary real constant.

Now we invoke the familiar idea of taking a linear combination of the solutions we have found to produce more solutions. We write our general solution as

$$T = \sum_{n=0}^{\infty} c_n r^n P_n(\cos\theta)\,.$$

Recall that we specified the initial temperature on the sphere (the boundary of the ball) to be $T = T_0 \sin^4\theta$ when $r = 1$. Thus we know that

$$T_0 \sin^4\theta = f(\theta) = \sum_{n=0}^{\infty} c_n P_n(\cos\theta)\,.$$

It is then possible to use the theory of Fourier–Legendre expansions to solve for the c_n. Since we have not developed that theory in the present book, we shall not carry out these calculations. We merely record the fact that the solution turns out to be

$$T = T_0\left(\frac{8}{15}P_0(\cos\theta) - \frac{16}{21}r^2 P_2(\cos\theta) + \frac{8}{35}r^4 P_4(\cos\theta)\right).$$

Problems for Review and Discovery

A. Drill Exercises

1. Use the method of power series to find a solution of each of these differential equations.

(a) $y'' + 2xy = x^2$
(b) $y'' - xy' + y = x$
(c) $y'' + y' + y = x^3 - x$
(d) $2y'' + xy' + y = 0$
(e) $(4 + x^2)y'' - y' + y = 0$
(f) $(x^2 + 1)y'' - xy' + y = 0$
(g) $y'' - (x + 1)y' - xy = 0$
(h) $(x - 1)y'' + (x + 1)y' + y = 0$

2. For each of the following differential equations, verify that the origin is a regular singular point and calculate two independent Frobenius series solutions.

(a) $(x^2+1)x^2y'' - xy' + (2+x)y = 0$
(b) $x^2y'' + xy' + (1+x)y = 0$
(c) $xy'' - 4y' + xy = 0$
(d) $4x^2y'' + 4x^2y' + 2y = 0$
(e) $2xy'' + (1-x)y' + y = 0$
(f) $xy'' - (x-1)y' + 2y = 0$
(g) $x^2y'' + x(1-x)y' + y = 0$
(h) $xy'' + (x+1)y' + y = 0$

3. In each of these problems, use the method of Frobenius to find the first four nonzero terms in the series solution about $x = 0$ for a solution to the given differential equation.

(a) $x^3y''' + 2x^2y'' + (x+x^2)y' + xy = 0$
(b) $x^3y''' + x^2y'' - 3xy' + (x-1)y = 0$
(c) $x^3y''' - 2x^2y'' + (x^2+2x)y' - xy = 0$
(d) $x^3y''' + (2x^3 - x^2)y'' - xy' + y = 0$

B. Challenge Problems

1. For some applications it is useful to have a series expansion about the point at infinity. In order to study such an expansion, we institute the change of variables $z = 1/x$ (of course we must remember to use the chain rule to transform the derivative as well) and expand about $z = 0$. In each of the following problems, use this idea to verify that ∞ is a regular singular point for the given differential equation by checking that $z = 0$ is a regular singular point for the transformed equation. Then find the first four nonzero terms in the series expansion about ∞ of a solution to the original differential equation.

(a) $x^3y'' + x^2y' + y = 0$
(b) $9(x-2)^2(x-3)y'' + 6x(x-2)y' + 16y = 0$
(c) $(1-x^2)y'' - 2xy' + p(p+1)y = 0$ (Legendre's equation)
(d) $x^2y'' + xy' + (x^2 - p^2)y = 0$ (Bessel's equation)

2. *Laguerre's equation* is

$$xy'' + (1-x)y' + py = 0\,,$$

where p is a constant. Show that the only solutions that are bounded near the origin are series of the form

$$1 + \sum_{n=1}^{\infty} \frac{-p(-p+1)\cdots(-p+n-1)}{(n!)^2} x^n\,.$$

This is the series representation of a *confluent hypergeometric function*, and is often denoted by the symbol $F(-p, 1, x)$. In case $p \geq 0$ is an integer, then show that this solution is in fact a polynomial. These solutions are called *Laguerre polynomials*, and they play an important role in the study of the quantum mechanics of the hydrogen atom.

3. The ordinary differential equation

$$x^4 \frac{d^2y}{dx^2} + \lambda^2 y = 0\,, \quad x > 0$$

is the mathematical model for the buckling of a column in the shape of a truncated cone. The positive constant λ depends on the rigidity of the column, the moment of inertia at the top of the column, and the load. Use the substitution $x = 1/z$ to reduce this differential equation to the form

$$\frac{d^2y}{dz^2} + \frac{2}{z}\frac{dy}{dz} + \lambda^2 y = 0\,.$$

Find the first five terms in the series expansion about the origin of a solution to this new equation. Convert it back to an expansion for the solution of the original equation.

C. Problems for Discussion and Exploration

1. Consider a nonlinear ordinary differential equation such as

$$[\sin y]y'' + e^y y' - y^2 = 0\,.$$

Why would it be neither efficient nor useful to guess a power series solution for this equation?

2. A celebrated theorem of Cauchy and Kowalewska guarantees that a nonsingular ordinary differential equation with real analytic coefficients will have a real analytic solution. What will happen if you seek a real analytic (i.e., a power series) solution to a differential equation that does *not* have real analytic coefficients?

3. Show that if y is a solution of *Bessel's equation*

$$x^2 y'' + xy' + (x^2 - p^2)y = 0$$

of order p, then $u(x) = x^{-c} y(ax^b)$ is a solution of

$$x^2 u'' + (2c+1)xu' + [a^2 b^2 x^{2b} + (c^2 - p^2 b^2)u = 0\,.$$

Use this result to show that the general solution of *Airy's equation* $y'' - xy = 0$ is

$$y = |x|^{1/2}\left(AJ_{1/3}\left(\frac{2|x|^{3/2}}{3}\right) + BJ_{-1/3}\left(\frac{2|x|^{3/2}}{3}\right)\right).$$

Here J_n is the *Bessel function* defined by

$$J_n(x) = \left(\frac{x}{2}\right)^n \sum_{k=0}^{\infty} \frac{(-1)^k}{k!(k+n)!}\left(\frac{x}{2}\right)^{2k}$$

as long as $n \geq 0$ is an integer. In case n is replaced by p not an integer, then we replace $(k+n)!$ by $\Gamma(k+p+1)$, where

$$\Gamma(z) \equiv \int_0^\infty x^{z-1} e^{-x}\, dx\,.$$

7

Fourier Series: Basic Concepts

- The idea of Fourier series
- Calculating a Fourier series
- Convergence of Fourier series
- Odd and even functions
- Fourier series on arbitrary intervals
- Orthogonality

7.1 Fourier Coefficients

Trigonometric and Fourier series constitute one of the oldest parts of analysis. They arose, for instance, in classical studies of the heat and wave equations. Today they play a central role in the study of sound, heat conduction, electromagnetic waves, mechanical vibrations, signal processing, and image analysis and compression. Whereas power series (see Chapter 4) can only be used to represent very special functions (most functions, even smooth ones, do *not* have convergent power series), Fourier series can be used to represent very broad classes of functions.

For us, a trigonometric series is one of the form

$$f(x) = \frac{1}{2}a_0 + \sum_{n=1}^{\infty}\Big(a_n \cos nx + b_n \sin nx\Big). \tag{7.1}$$

We shall be concerned with two main questions:

1. Given a function f, how do we calculate the coefficients a_n, b_n?

2. Once the series for f has been calculated, can we determine that it converges, and that it converges to f?

We begin our study with some classical calculations that were first performed by Euler (1707–1783). It is convenient to assume that our function f is defined on the interval $[-\pi, \pi] = \{x \in \mathbf{R} : -\pi \le x \le \pi\}$. We shall temporarily make the important assumption that the *trigonometric series* (1) *for f converges uniformly.* While this turns out to be true for a large class of functions (continuously differentiable functions, for example), for now this is merely a convenience so that our calculations are justified.

DOI: 10.1201/9781003214526-7

We apply the integral to both sides of (1). The result is

$$\begin{aligned}\int_{-\pi}^{\pi} f(x)\,dx &= \int_{-\pi}^{\pi}\left(\frac{1}{2}a_0 + \sum_{n=1}^{\infty}\left(a_n\cos nx + b_n\sin nx\right)\right)dx \\ &= \int_{-\pi}^{\pi}\frac{1}{2}a_0\,dx + \sum_{n=1}^{\infty}\int_{-\pi}^{\pi} a_n\cos nx\,dx + \sum_{n=1}^{\infty}\int_{-\pi}^{\pi} b_n\sin nx\,dx\,.\end{aligned}$$

The change in order of summation and integration is justified by the uniform convergence of the series (see [KRA2, page 202, ff.]).

Now each of $\cos nx$ and $\sin nx$ integrates to 0. The result is that

$$a_0 = \frac{1}{\pi}\int_{-\pi}^{\pi} f(x)\,dx\,.$$

In effect, then, a_0 is (twice) the *average* of f over the interval $[-\pi, \pi]$.

To calculate a_j, we multiply the formula (1) by $\cos jx$ and then integrate as before. The result is

$$\begin{aligned}\int_{-\pi}^{\pi} f(x)\cos jx\,dx &= \int_{-\pi}^{\pi}\left\{\frac{1}{2}a_0 + \sum_{n=1}^{\infty}\left(a_n\cos nx + b_n\sin nx\right)\right\}\cos jx\,dx \\ &= \int_{-\pi}^{\pi}\frac{1}{2}a_0\cos jx\,dx + \sum_{n=1}^{\infty}\int_{-\pi}^{\pi} a_n\cos nx\cos jx\,dx \\ &\quad + \sum_{n=1}^{\infty}\int_{-\pi}^{\pi} b_n\sin nx\cos jx\,dx\,. \qquad (2)\end{aligned}$$

Now the first integral on the right vanishes, as we have already noted. Further recall that

$$\cos nx\cos jx = \frac{1}{2}\left(\cos(n+j)x + \cos(n-j)x\right)$$

and

$$\sin nx\cos jx = \frac{1}{2}\left(\sin(n+j)x + \sin(n-j)x\right).$$

It follows immediately that

$$\int_{-\pi}^{\pi}\cos nx\cos jx\,dx = 0 \qquad \text{when } n \neq j$$

and

$$\int_{-\pi}^{\pi}\sin nx\cos jx\,dx = 0 \qquad \text{for all } n, j\,.$$

Thus our formula (2) reduces to

$$\int_{-\pi}^{\pi} f(x)\cos jx\,dx = \int_{-\pi}^{\pi} a_j\cos jx\cos jx\,dx\,.$$

We may use our formula above for the product of cosines to integrate the right-hand side. The result is

$$\int_{-\pi}^{\pi} f(x)\cos jx\,dx = a_j\cdot\pi$$

or

$$a_j = \frac{1}{\pi}\int_{-\pi}^{\pi} f(x)\cos jx\,dx\,.$$

A similar calculation shows that

$$b_j = \frac{1}{\pi}\int_{-\pi}^{\pi} f(x)\sin jx\,dx\,.$$

In summary, we now have formulas for calculating all the a_j's and b_j's.

EXAMPLE 7.1.1 Find the Fourier series of the function

$$f(x) = x, \qquad -\pi \le x \le \pi\,.$$

Solution: Of course

$$a_0 = \frac{1}{\pi}\int_{-\pi}^{\pi} x\,dx = \frac{1}{\pi}\cdot\frac{x^2}{2}\bigg|_{-\pi}^{\pi} = 0\,.$$

For $j \ge 1$, we calculate a_j as follows:

$$\begin{aligned} a_j &= \frac{1}{\pi}\int_{-\pi}^{\pi} x\cos jx\,dx \\ &\overset{\text{(parts)}}{=} \frac{1}{\pi}\left(x\frac{\sin jx}{j}\bigg|_{-\pi}^{\pi} - \int_{-\pi}^{\pi}\frac{\sin jx}{j}\,dx\right) \\ &= \frac{1}{\pi}\left\{0 - \left(-\frac{\cos jx}{j^2}\bigg|_{-\pi}^{\pi}\right)\right\} \\ &= 0\,. \end{aligned}$$

Similarly, we calculate the b_j:

$$\begin{aligned} b_j &= \frac{1}{\pi}\int_{-\pi}^{\pi} x\sin jx\,dx \\ &\overset{\text{(parts)}}{=} \frac{1}{\pi}\left(x\frac{-\cos jx}{j}\bigg|_{-\pi}^{\pi} - \int_{-\pi}^{\pi}\frac{-\cos jx}{j}\,dx\right) \\ &= \frac{1}{\pi}\left\{-\frac{2\pi\cos j\pi}{j} - \left(-\frac{\sin jx}{j^2}\bigg|_{-\pi}^{\pi}\right)\right\} \\ &= \frac{2\cdot(-1)^{j+1}}{j}\,. \end{aligned}$$

Now that all the coefficients have been calculated, we may summarize the result as

$$x = f(x) = 2\left(\sin x - \frac{\sin 2x}{2} + \frac{\sin 3x}{3} - + \cdots\right)\,. \qquad \square$$

It is sometimes convenient, in the study of Fourier series, to think of our functions as defined on the entire real line. We extend a function that is initially given on the interval $[-\pi,\pi]$ to the entire line using the idea of *periodicity.* The sine function and cosine function are periodic in the sense that $\sin(x+2\pi) = \sin x$ and

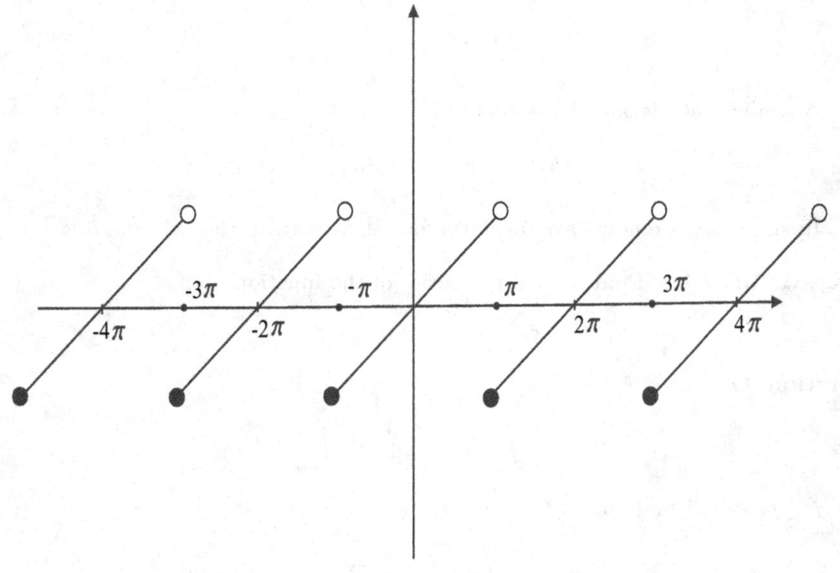

FIGURE 7.1
Periodic extension of $f(x) = x$.

$\cos(x + 2\pi) = \cos x$. We say that sine and cosine are *periodic with period* 2π. Thus it is natural, if we are given a function f on $[-\pi, \pi)$, to define $f(x + 2\pi) = f(x)$, $f(x + 2 \cdot 2\pi) = f(x)$, $f(x - 2\pi) = f(x)$, etc.[1]

Figure 7.1 exhibits the periodic extension of the function $f(x) = x$ on $[-\pi, \pi)$ to the real line.

Figure 7.2 shows the first four summands of the Fourier series for $f(x) = x$. The finest dashes show the curve $y = 2\sin x$, the next finest is $-\sin 2x$, the next is $(2/3)\sin 3x$, and the coarsest is $-(1/2)\sin 4x$.

Figure 7.3 shows the sum of the first four terms of the Fourier series and also of the first six terms, as compared to $f(x) = x$. Figure 7.4 shows the sum of the first eight terms of the Fourier series and also of the first ten terms, as compared to $f(x) = x$.

EXAMPLE 7.1.2 Calculate the Fourier series of the function

$$g(x) = \begin{cases} 0 & \text{if} \quad -\pi \leq x < 0 \\ \pi & \text{if} \quad 0 \leq x \leq \pi \,. \end{cases}$$

Solution: Following our formulas, we calculate

$$a_0 = \frac{1}{\pi} \int_{-\pi}^{\pi} g(x)\, dx = \frac{1}{\pi} \int_{-\pi}^{0} 0\, dx + \frac{1}{\pi} \int_{0}^{\pi} \pi\, dx = \pi \,.$$

[1]Notice that we take the original function f to be defined on $[-\pi, \pi)$ rather than $[-\pi, \pi]$ to avoid any ambiguity at the endpoints.

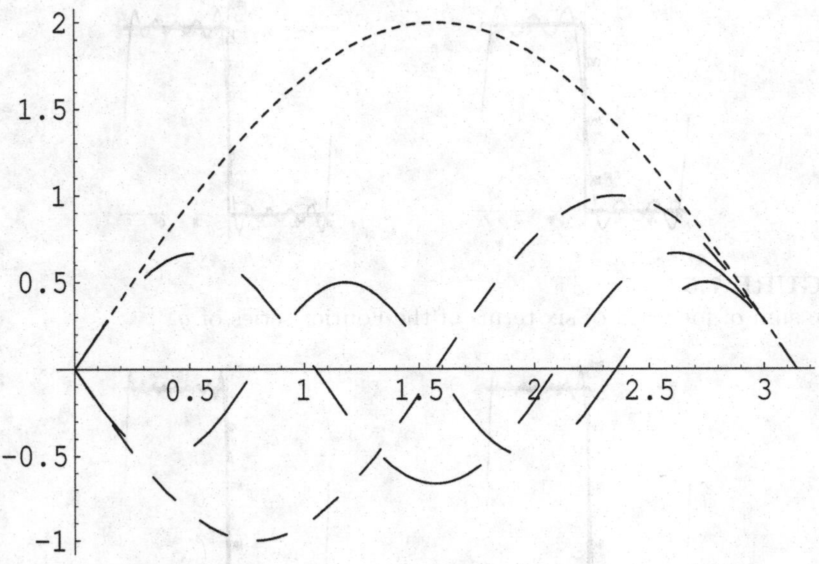

FIGURE 7.2
The first four summands for $f(x) = x$.

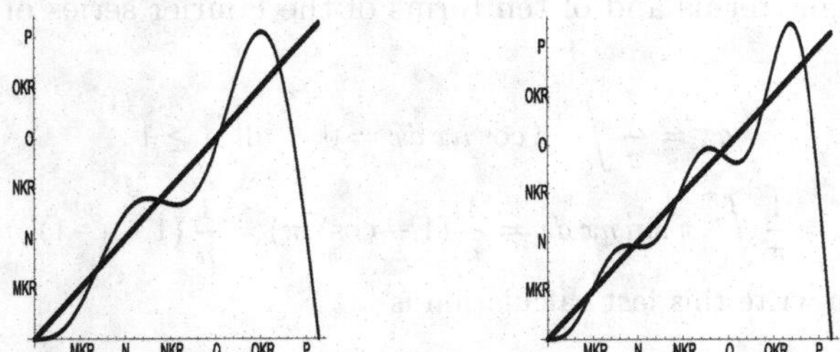

FIGURE 7.3
The sum of four terms and of six terms of the Fourier series of $f(x) = x$.

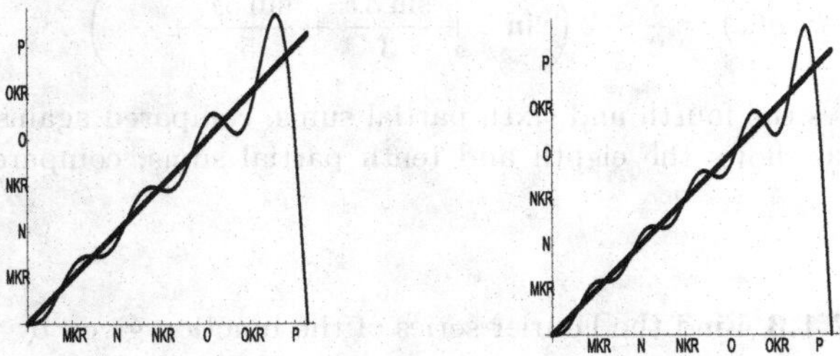

FIGURE 7.4
The sum of eight terms and of ten terms of the Fourier series of $f(x) = x$.

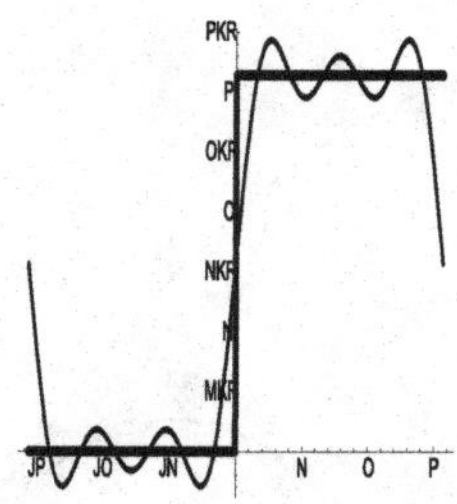

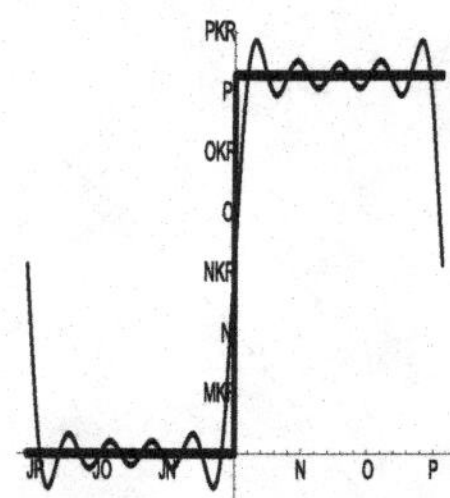

FIGURE 7.5
The sum of four and of six terms of the Fourier series of g.

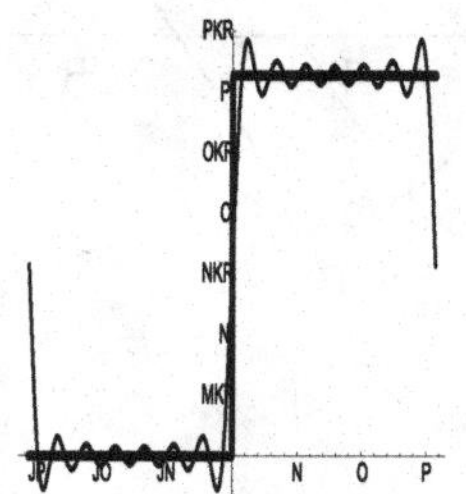

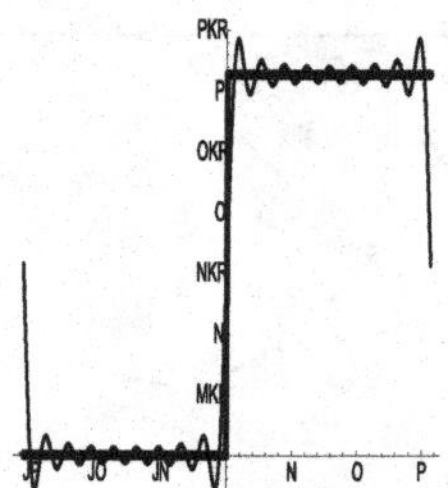

FIGURE 7.6
The sum of eight terms and of ten terms of the Fourier series of g.

$$a_n = \frac{1}{\pi}\int_0^\pi \pi \cos nx\, dx = 0, \quad \text{all } n \geq 1\,.$$

$$b_n = \frac{1}{\pi}\int_0^\pi \pi \sin nx\, dx = \frac{1}{n}(1 - \cos n\pi) = \frac{1}{n}(1 - (-1)^n)\,.$$

Another way to write this last calculation is

$$b_{2n} = 0, b_{2n-1} = \frac{2}{2n-1}\,.$$

In sum, the Fourier expansion for g is

$$g(x) = \frac{\pi}{2} + 2\left(\sin x + \frac{\sin 3x}{3} + \frac{\sin 5x}{5} + \cdots\right)\,.$$

Figure 7.5 shows the fourth and sixth partial sums, compared against the function $g(x)$. Figure 7.6 shows the eighth and tenth partial sums, compared against the function $g(x)$. □

EXAMPLE 7.1.3 Find the Fourier series of the function given by

$$h(x) = \begin{cases} -\frac{\pi}{2} & \text{if} \quad -\pi \leq x < 0 \\ \frac{\pi}{2} & \text{if} \quad 0 \leq x \leq \pi\,. \end{cases}$$

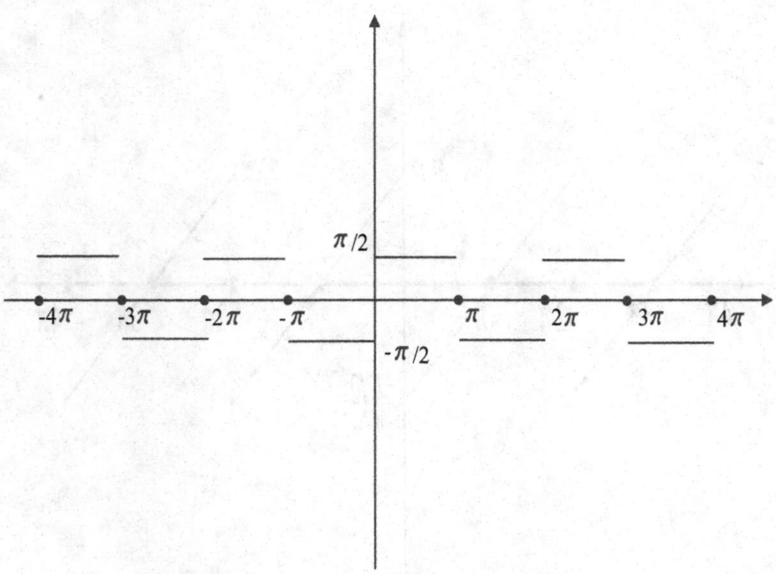

FIGURE 7.7
Graph of the function h in Example 7.1.3.

Solution: This is the same function as in the last example, with $\pi/2$ subtracted. Thus the Fourier series may be obtained by subtracting $\pi/2$ from the Fourier series that we obtained in that example. The result is

$$f(x) = 2\left(\sin x + \frac{\sin 3x}{3} + \frac{\sin 5x}{5} + \cdots\right).$$

The graph of this function, suitably periodized, is shown in Figure 7.7. □

EXAMPLE 7.1.4 Calculate the Fourier series of the function

$$k(x) = \begin{cases} -\frac{\pi}{2} - \frac{x}{2} & \text{if} \quad -\pi \le x < 0 \\ \frac{\pi}{2} - \frac{x}{2} & \text{if} \quad 0 \le x \le \pi. \end{cases}$$

Solution: This function is simply the function from Example 7.1.3 minus half the function from Example 7.1.1. In other words, $k(x) = h(x) - [1/2]f(x)$. Thus we may obtain the requested Fourier series by subtracting half the series from Example 7.1.1

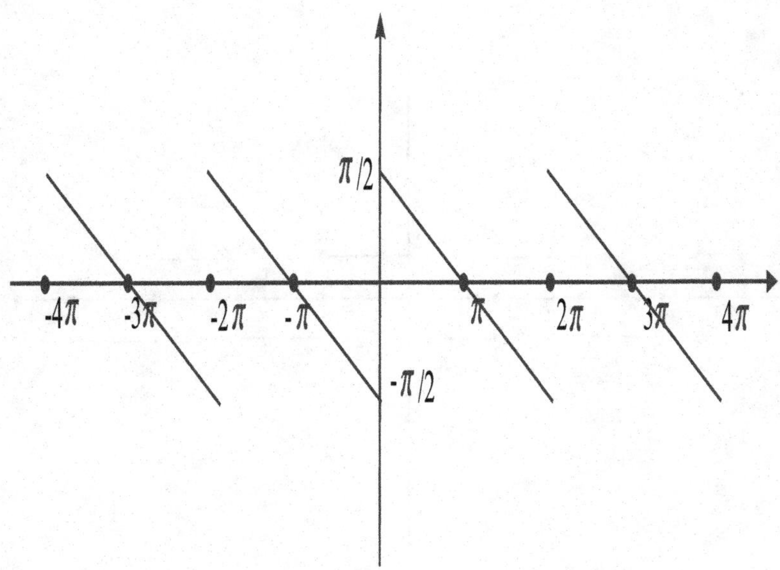

FIGURE 7.8
The sawtooth wave.

from the series in Example 7.1.3. The result is

$$\begin{aligned} f(x) &= 2\left(\sin x + \frac{\sin 3x}{3} + \frac{\sin 5x}{5} + \cdots\right) \\ &\quad - \left(\sin x - \frac{\sin 2x}{2} + \frac{\sin 3x}{3} - + \cdots\right) \\ &= \sin x + \frac{\sin 2x}{2} + \frac{\sin 3x}{3} + \cdots \\ &= \sum_{n=1}^{\infty} \frac{\sin nx}{n} . \end{aligned}$$

The graph of this series is the sawtooth wave shown in Figure 7.8. □

Exercises

1. Find the Fourier series of the function

$$f(x) = \begin{cases} \pi & \text{if} \quad -\pi \leq x \leq \dfrac{\pi}{2} \\ 0 & \text{if} \quad \dfrac{\pi}{2} < x \leq \pi . \end{cases}$$

2. Find the Fourier series for the function

$$f(x) = \begin{cases} 0 & \text{if} \quad -\pi \le x < 0 \\ 1 & \text{if} \quad 0 \le x \le \frac{\pi}{2} \\ 0 & \text{if} \quad \dfrac{\pi}{2} < x \le \pi\,. \end{cases}$$

3. Find the Fourier series of the function

$$f(x) = \begin{cases} 0 & \text{if} \quad -\pi \le x < 0 \\ \sin x & \text{if} \quad 0 \le x \le \pi\,. \end{cases}$$

4. Solve Exercise 3 with $\sin x$ replaced by $\cos x$.

5. Find the Fourier series for each of these functions. Pay special attention to the reasoning used to establish your conclusions; consider alternative lines of thought.

(a) $f(x) = \pi, \quad -\pi \le x \le \pi$

(b) $f(x) = \sin x, \quad -\pi \le x \le \pi$

(c) $f(x) = \cos x, \quad -\pi \le x \le \pi$

(d) $f(x) = \pi + \sin x + \cos x, \quad -\pi \le x \le \pi$

Solve Exercises 6 and 7 by using the methods of Examples 6.1.3 and 6.1.4, without actually calculating the Fourier coefficients.

6. Find the Fourier series for the function given by

(a)

$$f(x) = \begin{cases} -a & \text{if} \quad -\pi \le x < 0 \\ a & \text{if} \quad 0 \le x \le \pi \end{cases}$$

for a, a positive real number.

(b)

$$f(x) = \begin{cases} -1 & \text{if} \quad -\pi \le x < 0 \\ 1 & \text{if} \quad 0 \le x \le \pi \end{cases}$$

(c)

$$f(x) = \begin{cases} -\frac{\pi}{4} & \text{if} \quad -\pi \le x < 0 \\ \frac{\pi}{4} & \text{if} \quad 0 \le x \le \pi \end{cases}$$

(d)

$$f(x) = \begin{cases} -1 & \text{if} \quad -\pi \le x < 0 \\ 2 & \text{if} \quad 0 \le x \le \pi \end{cases}$$

(e)

$$f(x) = \begin{cases} 1 & \text{if} \quad -\pi \le x < 0 \\ 2 & \text{if} \quad 0 \le x \le \pi \end{cases}$$

7. Obtain the Fourier series for the function in Exercise 2 from the results of Exercise 1. (*Hint:* Begin by forming the difference of π and the function in Example 6.1.2.)

8. Without using the theory of Fourier series at all, show graphically that the sawtooth wave of Figure 7.1 can be represented as the sum of a sawtooth wave of period π and a square wave of period 2π.

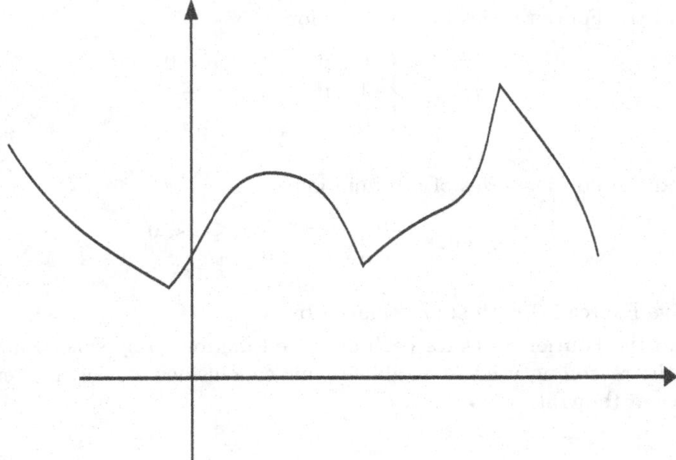

FIGURE 7.9
A piecewise smooth function.

7.2 Some Remarks about Convergence

The study of convergence of Fourier series is both deep and subtle. It would take us far afield to consider this matter in any detail. In the present section we shall very briefly describe a few of the basic results, but we shall not prove them. See [KRA3] for a more thoroughgoing discussion of these matters.

Our basic pointwise convergence result for Fourier series, which finds its genesis in work of Dirichlet (1805–1859), is this:

Definition 7.2.1 Let f be a function on $[-\pi, \pi]$. We say that f is *piecewise smooth* f is continuous and if the graph of f consists of finitely many continuously differentiable curves, and furthermore that the one-sided derivatives exist at each of the endpoints $\{p_1, \ldots, p_k\}$ of the definition of the curves, in the sense that

$$\lim_{h\to 0^+} \frac{f(p_j + h) - f(p_j)}{h} \qquad \text{and} \qquad \lim_{h\to 0^-} \frac{f(p_j + h) - f(p_j)}{h}$$

exist. Further, we require that f' extend continuously to $[p_j, p_{j+1}]$ for each $j = 1, \ldots, k-1$. See Figure 7.9.

> **Theorem 7.2.2** *Let f be a function on $[-\pi, \pi]$ which is piecewise smooth and overall continuous. Then the Fourier series of f converges at each point c of $[-\pi, \pi]$ to $f(c)$.*

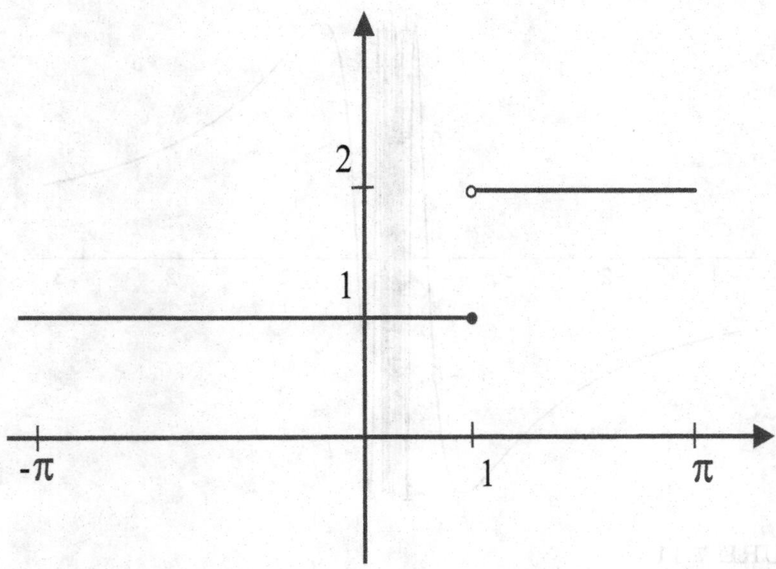

FIGURE 7.10
A simple discontinuity.

Let f be a function on the interval $[-\pi, \pi]$. We say that f has a *simple discontinuity* (or a *discontinuity of the first kind*) at the point $c \in (-\pi, \pi)$ if the limits $\lim_{x\to c^-} f(x)$ and $\lim_{x\to c^+} f(x)$ exist and

$$\lim_{x\to c^-} f(x) \neq \lim_{x\to c^+} f(x)\,.$$

The reader should understand that a simple discontinuity is in contradistinction to the other kind of discontinuity. That is to say, f has a *discontinuity of the second kind* at c if either $\lim_{x\to c^-} f(x)$ or $\lim_{x\to c^+} f(x)$ does not exist.

EXAMPLE 7.2.3 The function

$$f(x) = \begin{cases} 1 & \text{if} \quad -\pi \leq x \leq 1 \\ 2 & \text{if} \quad 1 < x \leq \pi \end{cases}$$

has a simple discontinuity at $x = 1$. It is continuous at all other points of the interval $[-\pi, \pi]$. See Figure 7.10.

The function

$$g(x) = \begin{cases} \sin \frac{1}{x} & \text{if} \quad x \neq 0 \\ 0 & \text{if} \quad x = 0 \end{cases}$$

has a discontinuity of the second kind at the origin. See Figure 7.11. ∎

Our next result about convergence is a bit more technical to state, but it is important in practice, and has historically been very influential. It is due to L. Fejér.

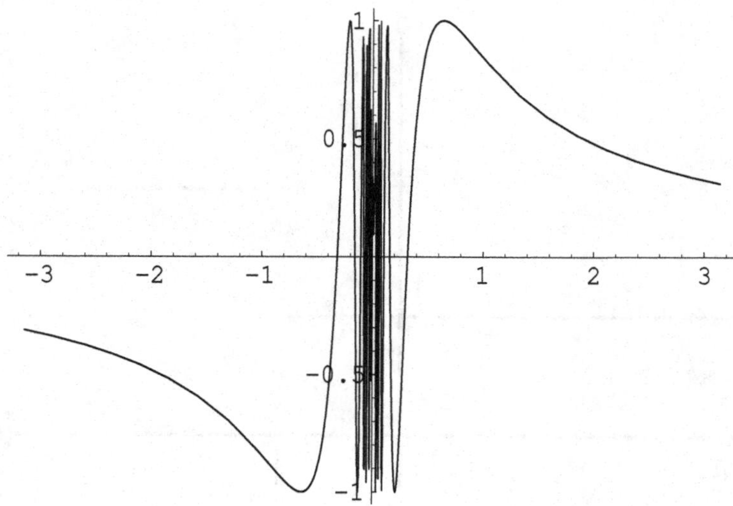

FIGURE 7.11
A discontinuity of the second kind.

Definition 7.2.4 Let f be a function and let

$$\frac{1}{2}a_0 + \sum_{n=1}^{\infty}\left(a_n \cos nx + b_n \sin nx\right)$$

be its Fourier series. The Nth *partial sum* of this series is

$$S_N(f)(x) = \frac{1}{2}a_0 + \sum_{n=1}^{N}\left(a_n \cos nx + b_n \sin nx\right).$$

The *Cesàro mean* of the series is

$$\sigma_N(f)(x) = \frac{1}{N+1}\sum_{j=0}^{N} S_j(f)(x).$$

In other words, the Cesàro means are simply the averages of the partial sums.

Theorem 7.2.5 *Let f be a continuous function on the interval $[-\pi, \pi]$. Then the Cesàro means $\sigma_N(f)$ of the Fourier series for f converge uniformly to f.*

It is worth noting explicitly that if the Fourier series of a function f converges at a point x_0, then the Cesàro means of the series also converge at x_0—and to the very same limit.

A useful companion result is this:

Theorem 7.2.6 (Fejér) *Let f be a piecewise continuous function on $[-\pi, \pi]$—meaning that the graph of f consists of finitely many continuous curves. Let p be the endpoint of one of those curves, and assume that $\lim_{x\to p-} f(x) \equiv f(p^-)$ and $\lim_{x\to p^+} f(x) \equiv f(p^+)$ exist. Then the Cesàro means of the Fourier series of f at p converges to $[f(p^-) + f(p^+)]/2$.*

In fact, with a few more hypotheses, we may make the result even sharper. Recall that a function f is *monotone increasing* if $x_1 \leq x_2$ implies $f(x_1) \leq f(x_2)$. The function is *monotone decreasing* if $x_1 \leq x_2$ implies $f(x_1) \geq f(x_2)$. If the function is either monotone increasing or monotone decreasing, then we just call it *monotone*. Now we have this result of Dirichlet:

Theorem 7.2.7 (Dirichlet) *Let f be a function on $[-\pi, \pi]$ which is piecewise continuous. Assume that each piece of f is monotone. Then the Fourier series of f converges at each point of continuity c of f in $[-\pi, \pi]$ to $f(c)$. At other points x it converges to $[f(x^-) + f(x^+)]/2$.*

The hypotheses in this theorem are commonly referred to as the *Dirichlet conditions.*

By linearity, we may extend this last result to functions that are piecewise the difference of two monotone functions. Such functions are said to be of *bounded variation*, and exceed the scope of the present book. See [KRA2] for a detailed discussion. The book [TIT] discusses convergence of the Fourier series of such functions.

Exercises

1. In Exercises 1, 2, 3, 4, and 6 of the last section, sketch the graph of the sum of each Fourier series on the interval $-5\pi \leq x \leq 5\pi$. Now do the same on the interval $-2\pi \leq x \leq 8\pi$.

2. Find the Fourier series for the periodic function defined by

$$f(x) = \begin{cases} -\pi & \text{if} \quad -\pi \leq x < 0 \\ x & \text{if} \quad 0 \leq x < \pi \end{cases}$$

Sketch the graph of the sum of this series on the interval $-5\pi \leq x \leq 5\pi$

and find what numerical sums are implied by the convergence behavior at the points of discontinuity $x = 0$ and $x = \pi$.

3. **(a)** Show that the Fourier series for the periodic function

$$f(x) = \begin{cases} 0 & \text{if} \quad -\pi \le x < 0 \\ x^2 & \text{if} \quad 0 \le x < \pi \end{cases}$$

is

$$\begin{aligned} f(x) &= \frac{\pi^2}{6} + 2\sum_{j=1}^{\infty}(-1)^j \frac{\cos jx}{j^2} \\ &\quad + \pi\sum_{j=1}^{\infty}(-1)^{j+1}\frac{\sin jx}{j} - \frac{4}{\pi}\sum_{j=1}^{\infty}\frac{\sin(2j-1)x}{(2j-1)^3} . \end{aligned}$$

(b) Sketch the graph of the sum of this series on the interval $-5\pi \le x \le 5\pi$.

(c) Use the series in part **(a)** with $x = 0$ and $x = \pi$ to obtain the two sums

$$1 - \frac{1}{2^2} + \frac{1}{3^2} - \frac{1}{4^2} + - \cdots = \frac{\pi^2}{12}$$

and

$$1 + \frac{1}{2^2} + \frac{1}{3^2} + \frac{1}{4^2} + \cdots = \frac{\pi^2}{6} .$$

(d) Derive the second sum in **(c)** from the first. *Hint:* Add $2\sum_j(1/[2j])^2$ to both sides.

4. What can you say about the convergence of the Fourier series of the function

$$f(x) = \begin{cases} -1 & \text{if} \quad x < 0 \\ 0 & \text{if} \quad x = 0 \\ 1 & \text{if} \quad x > 0 \end{cases}$$

at the origin?

5. **(a)** Find the Fourier series for the periodic function defined by $f(x) = e^x$, $-\pi \le x \le \pi$. *Hint:* Recall that $\cosh x = (e^x + e^{-x})/2$.

(b) Sketch the graph of the sum of this series on the interval $-5\pi \le x \le 5\pi$.

(c) Use the series in **(a)** to establish the sums

$$\sum_{j=1}^{\infty}\frac{1}{j^2+1} = \frac{1}{2}\left(\frac{\pi}{\tanh \pi} - 1\right)$$

and

$$\sum_{j=1}^{\infty}\frac{(-1)^j}{j^2+1} = \frac{1}{2}\left(\frac{\pi}{\sinh \pi} - 1\right) .$$

6. It is usually most convenient to study classes of functions that form linear spaces, that is, that are closed under the operations of addition and scalar multiplication. Unfortunately, this linearity condition does not hold for the class of functions defined on the interval $[-\pi, \pi]$ by the Dirichlet conditions. Verify this statement by examining the functions

$$f(x) = \begin{cases} x^2 \sin \frac{1}{x} + 2x & \text{if} \quad x \neq 0 \\ 0 & \text{if} \quad x = 0 \end{cases}$$

and

$$g(x) = -2x\,.$$

7. If f is defined on the interval $[-\pi, \pi]$ and satisfies the Dirichlet conditions there, then prove that $f(x^-) \equiv \lim_{\substack{t \to x \\ t < x}} f(t)$ and $f(x^+) \equiv \lim_{\substack{t \to x \\ t > x}} f(t)$ exist at every interior point, and also that $f(x^+)$ exists at the left endpoint and $f(x^-)$ exists at the right endpoint. *Hint:* Each interior point of discontinuity is isolated from other such points, in the sense that the function is continuous at all nearby points. Also, on each side of such a point and near enough to it, the function does not oscillate; it is therefore increasing or decreasing.

7.3 Even and Odd Functions: Cosine and Sine Series

A function f is said to be *even* if $f(-x) = f(x)$. A function g is said to be *odd* if $g(-x) = -g(x)$.

EXAMPLE 7.3.1 The function $f(x) = \cos x$ is even because $\cos(-x) = \cos x$. The function $g(x) = \sin x$ is odd because $\sin(-x) = -\sin x$. ∎

The graph of an even function is symmetric about the y-axis. The graph of an odd function is skew-symmetric about the y-axis. Refer to Figure 7.12.

If f is even on the interval $[-a, a]$, then

$$\int_{-a}^{a} f(x)\, dx = 2 \int_{0}^{a} f(x)\, dx \tag{7.2}$$

and if f is odd on the interval $[-a, a]$ then

$$\int_{-a}^{a} f(x)\, dx = 0\,. \tag{7.3}$$

Finally, we have the following parity relations

$$(\text{even}) \cdot (\text{even}) = (\text{even}) \qquad (\text{even}) \cdot (\text{odd}) = (\text{odd})$$

$$(\text{odd}) \cdot (\text{odd}) = (\text{even})\,.$$

Now suppose that f is an even function on the interval $[-\pi, \pi]$. Then $f(x) \cdot \sin nx$ is odd, and therefore

$$b_n = \frac{1}{\pi} \int_{-\pi}^{\pi} f(x) \sin nx\, dx = 0\,.$$

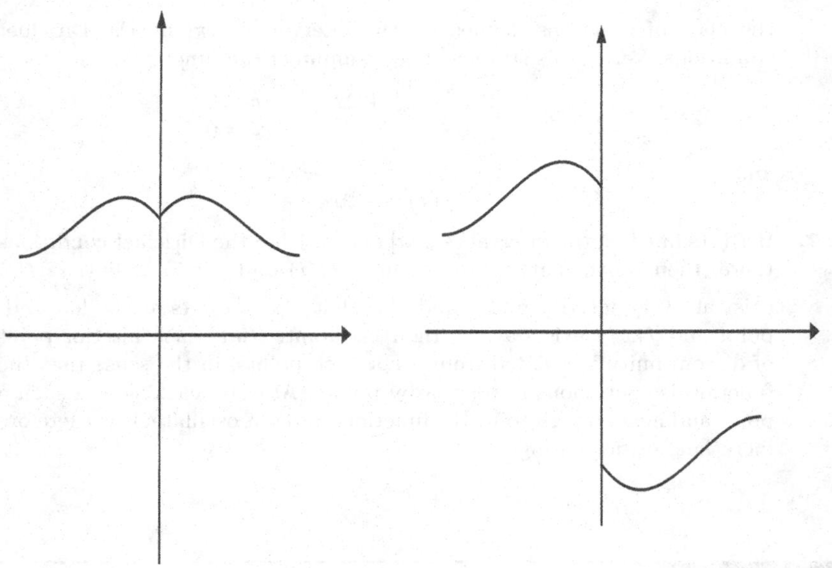

FIGURE 7.12
An even and an odd function.

For the cosine coefficients, we have

$$a_n = \frac{1}{\pi}\int_{-\pi}^{\pi} f(x)\cos nx\,dx = \frac{2}{\pi}\int_{0}^{\pi} f(x)\cos nx\,dx\,.$$

Thus the Fourier series for an even function contains only cosine terms.

By the same token, suppose now that f is an odd function on the interval $[-\pi,\pi]$. Then $f(x)\cdot\cos nx$ is an odd function, and therefore

$$a_n = \frac{1}{\pi}\int_{-\pi}^{\pi} f(x)\cos nx\,dx = 0\,.$$

For the sine coefficients, we have

$$b_n = \frac{1}{\pi}\int_{-\pi}^{\pi} f(x)\sin nx\,dx = \frac{2}{\pi}\int_{0}^{\pi} f(x)\sin nx\,dx\,.$$

Thus the Fourier series for an odd function contains only sine terms.

EXAMPLE 7.3.2 Examine the Fourier series of the function $f(x) = x$ from the point of view of even/odd.

Solution: The function is odd, so the Fourier series must be a sine series. We calculated in Example 7.1.1 that the Fourier series is in fact

$$x = f(x) = 2\left(\sin x - \frac{\sin 2x}{2} + \frac{\sin 3x}{3} - + \cdots\right). \tag{7.4}$$

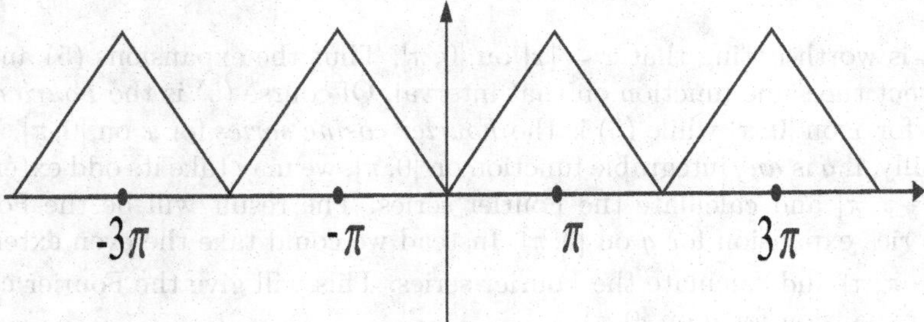

FIGURE 7.13
Periodic extension of $f(x) = |x|$.

The expansion is valid on $(-\pi, \pi)$, but not at the endpoints (since the series of course sums to 0 at $-\pi$ and π). □

EXAMPLE 7.3.3 Examine the Fourier series of the function $f(x) = |x|$ from the point of view of even/odd.

Solution: The function is even, so the Fourier series must be a cosine series. In fact we see that

$$a_0 = \frac{1}{\pi} \int_{-\pi}^{\pi} |x|\, dx = \frac{2}{\pi} \int_0^{\pi} x\, dx = \pi\,.$$

Also, for $n \geq 1$,

$$a_n = \frac{2}{\pi} \int_0^{\pi} |x| \cos nx\, dx = \frac{2}{\pi} \int_0^{\pi} x \cos nx\, dx\,.$$

An integration by parts gives that

$$a_n = \frac{2}{\pi n^2}(\cos n\pi - 1) = \frac{2}{\pi n^2}[(-1)^n - 1]\,.$$

As a result,

$$a_{2j} = 0 \quad \text{and} \quad a_{2j-1} = -\frac{4}{\pi(2j-1)^2}\,.$$

In conclusion,

$$|x| = \frac{\pi}{2} - \frac{4}{\pi}\left(\cos x + \frac{\cos 3x}{3^2} + \frac{\cos 5x}{5^2} + \cdots\right). \tag{7.5}$$

The periodic extension of the original function $f(x) = |x|$ on $[-\pi, \pi]$ is depicted in Figure 7.13. By Theorem 7.2.7 (see also Theorem 7.2.2), the series converges to f at every point of $[-\pi, \pi]$. □

It is worth noting that $x = |x|$ on $[0, \pi]$. Thus the expansions (5) and (6) represent the same function on that interval. Of course (5) is the *Fourier sine series* for x on $[0, \pi]$ while (6) is the *Fourier cosine series* for x on $[0, \pi]$. More generally, if g is *any* integrable function on $[0, \pi]$, we may take its odd extension $\widetilde{g}$ to $[-\pi, \pi]$ and calculate the Fourier series. The result will be the Fourier sine series expansion for g on $[0, \pi]$. Instead we could take the even extension $\widetilde{\widetilde{g}}$ to $[-\pi, \pi]$ and calculate the Fourier series. This will give the Fourier cosine series expansion for g on $[0, \pi]$.

Let us look at some concrete examples.

EXAMPLE 7.3.4 Find the Fourier sine series and the Fourier cosine series expansions for the function $f(x) = \cos x$ on the interval $[0, \pi]$.

Solution: The odd extension of $\cos x$ is

$$\widetilde{f}(x) = \begin{cases} \cos x & \text{if} \quad 0 \le x \le \pi \\ -\cos x & \text{if} \quad -\pi \le x < 0 \, . \end{cases}$$

Draw the graph of $\widetilde{f}$ so that you understand that this is indeed the odd extentsion.

Of course the Fourier series expansion of the odd extension $\widetilde{f}$ contains only sine terms. Its coefficients will be

$$b_n = \frac{2}{\pi} \int_0^\pi \cos x \sin nx \, dx = \begin{cases} 0 & \text{if} \quad n = 1 \\ \frac{2n}{\pi}\left(\frac{1+(-1)^n}{n^2-1}\right) & \text{if} \quad n > 1 \, . \end{cases}$$

As a result,

$$b_{2j-1} = 0 \quad \text{and} \quad b_{2j} = \frac{8j}{\pi(4j^2 - 1)} \, .$$

The sine series for f is therefore

$$\cos x = f(x) = \frac{8}{\pi} \sum_{j=1}^{\infty} \frac{j \sin 2jx}{4j^2 - 1} \, , \quad 0 < x < \pi \, .$$

To obtain the cosine series for f, we consider the even extension $\widetilde{\widetilde{f}}$. Of course all the b_n will vanish. Also

$$a_n = \frac{2}{\pi} \int_0^\pi \cos x \cos nx \, dx = \begin{cases} 1 & \text{if} \quad n = 1 \\ 0 & \text{if} \quad n > 1 \, . \end{cases}$$

We therefore see, not surprisingly, that the Fourier cosine series for cosine on $[0, \pi]$ is the single summand $\cos x$. □

Exercises

1. Determine whether each of the following functions is even, odd, or neither.

$$x^5 \sin x,\ x^2 \sin 2x,\ e^x,\ (\sin x)^3,\ \sin x^2,\ \cos(x + x^3),$$

$$x + x^2 + x^3,\ \ln \frac{1+x}{1-x}.$$

2. Show that any function f defined on a symmetrically placed interval can be written as the sum of an even function and an odd function. *Hint:* $f(x) = \frac{1}{2}[f(x) + f(-x)] + \frac{1}{2}[f(x) - f(-x)]$.
3. Prove properties (3) and (4) analytically, by dividing the integral and making a suitable change of variables.
4. Show that the sine series of the constant function $f(x) = \pi/4$ is

$$\frac{\pi}{4} = \sin x + \frac{\sin 3x}{3} + \frac{\sin 5x}{5} + \cdots$$

for $0 < x < \pi$. What sum is obtained by setting $x = \pi/2$? What is the cosine series of this function?

5. Find the Fourier series for the function of period 2π defined by $f(x) = \cos x/2$, $-\pi \le x \le \pi$. Sketch the graph of the sum of this series on the interval $-5\pi \le x \le 5\pi$.
6. Find the sine and the cosine series for $f(x) = \sin x$.
7. Find the Fourier series for the 2π-periodic function defined on its fundamental period $[-\pi, \pi]$ by

$$f(x) = \begin{cases} x + \frac{\pi}{2} & \text{if} \quad -\pi \le x < 0 \\ -x + \frac{\pi}{2} & \text{if} \quad 0 \le x \le \pi \end{cases}$$

(a) by computing the Fourier coefficients directly;

(b) using the formula

$$|x| = \frac{\pi}{2} - \frac{4}{\pi}\left(\cos x + \frac{\cos 3x}{3^2} + \frac{\cos 5x}{5^2} + \cdots\right)$$

from the text.

Sketch the graph of the sum of this series (a triangular wave) on the interval $-5\pi \le x \le 5\pi$.

8. For the function $f(x) = \pi - x$, find

(a) its Fourier series on the interval $-\pi < x < \pi$;

(b) its cosine series on the interval $0 \le x \le \pi$;

(c) its sine series on the interval $0 < x \le \pi$.

Sketch the graph of the sum of each of these series on the interval $[-5\pi, 5\pi]$.

9. Let

$$f(x) = \begin{cases} x & \text{if} \quad 0 \le x \le \pi/2 \\ \pi - x & \text{if} \quad \pi/2 < x \le \pi . \end{cases}$$

Show that the cosine series for this function is

$$f(x) = \frac{\pi}{4} - \frac{2}{\pi} \sum_{j=1}^{\infty} \frac{\cos 2(2j-1)x}{(2j-1)^2} .$$

Sketch the graph of the sum of this series on the interval $[-5\pi, 5\pi]$.

10. **(a)** Show that the cosine series for x^2 is

$$x^2 = \frac{\pi^2}{3} + 4 \sum_{j=1}^{\infty} (-1)^j \frac{\cos jx}{j^2} , \qquad -\pi \le x \le \pi .$$

(b) Find the sine series for x^2 and use this expansion together with the formula (5) to obtain the sum

$$1 - \frac{1}{3^3} + \frac{1}{5^3} - \frac{1}{7^3} + - \cdots = \frac{\pi^3}{32} .$$

(c) Denote by s the sum of the reciprocals of the cubes of the odd positive integers:

$$s = \frac{1}{1^3} + \frac{1}{3^3} + \frac{1}{5^3} + \frac{1}{7^3} + \cdots ,$$

and show that then

$$\sum_{j=1}^{\infty} \frac{1}{j^3} = \frac{1}{1^3} + \frac{1}{2^3} + \frac{1}{3^3} + \frac{1}{4^3} + \cdots = \frac{8}{7} \cdot s .$$

The exact numerical value of this last sum has been a matter of great interest since Euler first raised the question in 1736. It is closely related to the Riemann hypothesis. Roger Apéry proved, by an extremely ingenious argument in 1978, that s is irrational.[2]

11. **(a)** Show that the cosine series for x^3 is

$$x^3 = \frac{\pi^3}{4} + 6\pi \sum_{j=1}^{\infty} (-1)^j \frac{\cos jx}{j^2} + \frac{24}{\pi} \sum_{j=1}^{\infty} \frac{\cos(2j-1)x}{(2j-1)^4} , \qquad 0 \le x \le \pi .$$

(b) Use the series in **(a)** to obtain

$$\text{(i)} \ \sum_{j=1}^{\infty} \frac{1}{(2j-1)^4} = \frac{\pi^4}{96} \quad \text{and} \quad \text{(ii)} \ \sum_{j=1}^{\infty} \frac{1}{j^4} = \frac{\pi^4}{90} .$$

[2]The Riemann hypothesis is perhaps the most celebrated open problem in modern mathematics. Originally formulated as a question about the zero set of a complex analytic function, this question has profound implications for number theory and other branches of mathematics. The recent books [DER] and [SAB] discuss the history and substance of the problem.

12. **(a)** Show that the cosine series for x^4 is

$$x^4 = \frac{\pi^4}{5} + 8\sum_{j=1}^{\infty}(-1)^j \frac{\pi^2 j^2 - 6}{j^4} \cos jx\,, \quad -\pi \le x \le \pi\,.$$

(b) Use the series in **(a)** to find a new derivation of the second sum in Exercise 11(b).

13. The functions $\sin^2 x$ and $\cos^2 x$ are both even. Show, without using any calculations, that the identities

$$\sin^2 x = \frac{1}{2}(1 - \cos 2x) = \frac{1}{2} - \frac{1}{2}\cos 2x$$

and

$$\cos^2 x = \frac{1}{2}(1 + \cos 2x) = \frac{1}{2} + \frac{1}{2}\cos 2x$$

are actually the Fourier series expansions of these functions.

14. Find the sine series of the functions in Exercise 13, and verify that these expansions satisfy the identity $\sin^2 x + \cos^2 x = 1$.

15. Prove the trigonometric identities

$$\sin^3 x = \frac{3}{4}\sin x - \frac{1}{4}\sin 3x \quad \text{and} \quad \cos^3 x = \frac{3}{4}x + \frac{1}{4}\cos 3x$$

and show briefly, without calculation, that these are the Fourier series expansions of the functions $\sin^3 x$ and $\cos^3 x$.

7.4 Fourier Series on Arbitrary Intervals

We have developed Fourier analysis on the interval $[-\pi, \pi]$ (resp. the interval $[0, \pi]$) just because it is notationally convenient. In particular,

$$\int_{-\pi}^{\pi} \cos jx \cos kx\, dx = 0 \quad \text{for } j \ne k$$

and

$$\int_{-\pi}^{\pi} \sin jx \sin kx\, dx = 0 \quad \text{for } j \ne k$$

and

$$\int_{-\pi}^{\pi} \sin jx \cos kx\, dx = 0 \quad \text{for all } j, k\,.$$

These facts are special to the interval of length 2π. But many physical problems take place on an interval of some other length. We must therefore be able to adapt our analysis to intervals of any length. This amounts to a straightforward change of scale on the horizontal axis. We treat the matter in the present section.

Now we concentrate our attention on an interval of the form $[-L, L]$. As x runs from $-L$ to L, we shall have a corresponding variable t that runs from $-\pi$ to π. We mediate between these two variables using the formulas

$$t = \frac{\pi x}{L} \qquad \text{and} \qquad x = \frac{Lt}{\pi}.$$

Thus the function $f(x)$ on $[-L, L]$ is transformed to a new function $\widetilde{f}(t) \equiv f(Lt/\pi)$ on $[-\pi, \pi]$.

If f satisfies the conditions for convergence of the Fourier series, then so will $\widetilde{f}$, and vice versa. Thus we may consider the Fourier expansion

$$\widetilde{f}(t) = \frac{1}{2}a_0 + \sum_{n=1}^{\infty}\left(a_n \cos nt + b_n \sin nt\right).$$

Here, of course,

$$a_n = \frac{1}{\pi}\int_{-\pi}^{\pi} \widetilde{f}(t)\cos nt\, dt \quad \text{and} \quad b_n = \frac{1}{\pi}\int_{-\pi}^{\pi} \widetilde{f}(t)\sin nt\, dt.$$

Now let us write out these last two formulas and perform the change of variables $x = Lt/\pi$. We find that

$$\begin{aligned} a_n &= \frac{1}{\pi}\int_{-\pi}^{\pi} f(Lt/\pi)\cos nt\, dt \\ &= \frac{1}{\pi}\int_{-L}^{L} f(x)\cos\frac{n\pi x}{L}\cdot\frac{\pi}{L}\, dx \\ &= \frac{1}{L}\int_{-L}^{L} f(x)\cos\frac{n\pi x}{L}\, dx. \end{aligned}$$

Likewise,

$$b_n = \frac{1}{L}\int_{-L}^{L} f(x)\sin\frac{n\pi x}{L}\, dx.$$

EXAMPLE 7.4.1 Calculate the Fourier series on the interval $[-2, 2]$ of the function

$$f(x) = \begin{cases} 0 & \text{if} \quad -2 \le x < 0 \\ 1 & \text{if} \quad 0 \le x \le 2. \end{cases}$$

Solution: Of course $L = 2$ so we calculate that

$$a_n = \frac{1}{2}\int_0^2 \cos\frac{n\pi x}{2}\, dx = \begin{cases} 1 & \text{if} \quad n = 0 \\ 0 & \text{if} \quad n \ge 1. \end{cases}$$

Also

$$b_n = \frac{1}{2}\int_0^2 \sin\frac{n\pi x}{2}\, dx = \frac{1}{n\pi}[(-1)^n - 1].$$

This may be rewritten as

$$b_{2j} = 0 \qquad \text{and} \qquad b_{2j-1} = \frac{-2}{(2j-1)\pi}\,.$$

In conclusion,

$$\begin{aligned} f(x) = \widetilde{f}(t) &= \frac{1}{2}a_0 + \sum_{n=1}^{\infty}\Big(a_n \cos nt + b_n \sin nt\Big) \\ &= \frac{1}{2} + \sum_{j=1}^{\infty} \frac{-2}{(2j-1)\pi} \sin\left((2j-1)\cdot\frac{\pi x}{2}\right). \qquad \square \end{aligned}$$

EXAMPLE 7.4.2 Calculate the Fourier series of the function $f(x) = \cos x$ on the interval $[-\pi/2, \pi/2]$.

Solution: We calculate that

$$a_0 = \frac{2}{\pi}\int_{-\pi/2}^{\pi/2} \cos x\,dx = \frac{4}{\pi}\,.$$

Also, for $n \geq 1$,

$$\begin{aligned} a_n &= \frac{2}{\pi}\int_{-\pi/2}^{\pi/2} \cos x \cos(2nx)\,dx \\ &= \frac{2}{\pi}\int_{-\pi/2}^{\pi/2} \frac{1}{2}\Big(\cos(2n+1)x + \cos(2n-1)x\Big)\,dx \\ &= \frac{1}{\pi}\left(\frac{\sin(2n+1)x}{2n+1} + \frac{\sin(2n-1)x}{2n-1}\right)\Bigg|_{-\pi/2}^{\pi/2} \\ &= \begin{cases} \dfrac{2}{\pi}\left(\dfrac{-1}{2n+1} + \dfrac{1}{2n-1}\right) = \dfrac{4}{\pi(4n^2-1)} & \text{if } n \text{ is odd}, \\ \dfrac{2}{\pi}\left(\dfrac{1}{2n+1} + \dfrac{-1}{2n-1}\right) = \dfrac{-4}{\pi(4n^2-1)} & \text{if } n \text{ is even}. \end{cases} \end{aligned}$$

A similar calculation shows that

$$\begin{aligned} b_n &= \frac{2}{\pi}\int_{-\pi/2}^{\pi/2} \cos x \sin 2nx\,dx \\ &= \frac{2}{\pi}\int_{-\pi/2}^{\pi/2} \frac{1}{2}\Big(\sin(2n+1)x + \sin(2n-1)x\Big)\,dx \\ &= \frac{1}{\pi}\left(\frac{-\cos(2n+1)x}{2n+1} + \frac{-\cos(2n-1)x}{2n-1}\right)\Bigg|_{-\pi/2}^{\pi/2} \\ &= 0\,. \end{aligned}$$

This last comes as no surprise since the cosine function is even.

As a result, the Fourier series expansion for $\cos x$ on the interval $[-\pi/2, \pi/2]$ is

$$\begin{aligned}\cos x &= f(x)\\ &= \frac{2}{\pi} + \sum_{m=1}^{\infty} \frac{-4}{\pi(4(2m)^2 - 1)} \cos \frac{2m\pi x}{\pi/2}\\ &\quad + \sum_{k=1}^{\infty} \frac{4}{\pi(4(2k-1)^2 - 1)} \cos \frac{(2k-1)\pi x}{\pi/2} .\end{aligned}$$

□

Exercises

1. Calculate the Fourier series for the given function on the given interval.

(a) $f(x) = x, \quad [-1, 1]$

(b) $g(x) = \sin x, \quad [-2, 2]$

(c) $h(x) = e^x, \quad [-3, 3]$

(d) $f(x) = x^2, \quad [-1, 1]$

(e) $g(x) = \cos 2x, \quad [-\pi/3, \pi/3]$

(f) $h(x) = \sin(2x - \pi/3), \quad [-1, 1]$

2. For the functions

$$f(x) \equiv -3, \quad -2 \le x < 0$$

and

$$g(x) \equiv 3, \quad 0 \le x < 2,$$

write down the Fourier expansion directly from Example 6.4.1 in the text—without any calculation.

3. Find the Fourier series for these functions.

(a)

$$f(x) = \begin{cases} 1 + x & \text{if} \quad -1 \le x < 0 \\ 1 - x & \text{if} \quad 0 \le x \le 1 . \end{cases}$$

(b)

$$f(x) = |x|, \qquad -2 \le x \le 2 .$$

4. Show that

$$\frac{L}{2} - x = \frac{L}{\pi} \sum_{j=1}^{\infty} \frac{1}{j} \sin \frac{2j\pi x}{L}, \qquad 0 < x < L .$$

5. Find the cosine series for the function defined on the interval $0 \le x \le 1$ by $f(x) = x^2 - x + 1/6$. This is a special instance of the Bernoulli polynomials.

6. Find the cosine series for the function defined by

$$f(x) = \begin{cases} 2 & \text{if } 0 \le x \le 1 \\ 0 & \text{if } 1 < x \le 2 . \end{cases}$$

7. Expand $f(x) = \cos \pi x$ in a Fourier series on the interval $-1 \le x \le 1$.

8. Find the cosine series for the function defined by

$$f(x) = \begin{cases} \frac{1}{4} - x & \text{if } 0 \le x < \frac{1}{2} \\ x - \frac{3}{4} & \text{if } \frac{1}{2} \le x \le 1 . \end{cases}$$

7.5 Orthogonal Functions

In the classical Euclidean geometry of 3-space, just as we learn in multivariable calculus class, one of the key ideas is that of orthogonality. Let us briefly review it now.

If $\mathbf{v} = \langle v_1, v_2, v_3 \rangle$ and $\mathbf{w} = \langle w_1, w_2, w_3 \rangle$ are vectors in $\mathbb{R}^3$, then we define their *dot product*, or *inner product*, or *scalar product* to be

$$\mathbf{v} \cdot \mathbf{w} = v_1 w_1 + v_2 w_2 + v_3 w_3 .$$

What is the interest of the inner product? There are three answers:

- Two vectors are perpendicular or *orthogonal*, written $\mathbf{v} \perp \mathbf{w}$, if and only if $\mathbf{v} \cdot \mathbf{w} = 0$.

- The *length* of a vector is given by

$$\|\mathbf{v}\| = \sqrt{\mathbf{v} \cdot \mathbf{v}} .$$

- The *angle* θ between two vectors $\mathbf{v}$ and $\mathbf{w}$ is given by

$$\cos \theta = \frac{\mathbf{v} \cdot \mathbf{w}}{\|\mathbf{v}\| \|\mathbf{w}\|} .$$

In fact all of the geometry of 3-space is built on these three facts.

One of the great ideas of twentieth century mathematics is that many other spaces—sometimes abstract spaces, and sometimes infinite dimensional spaces—can be equipped with an inner product that endows that space with a useful geometry. That is the idea that we shall explore in this section.

Let X be a vector space. This means that X is equipped with **(i)** a notion of addition and **(ii)** a notion of scalar multiplication. These two operations are hypothesized to satisfy the expected properties: addition is commutative and associative, scalar multiplication is commutative, associative, and distributive, and so forth. We say that X is equipped with an *inner product* (which we now denote by $\langle \, , \, \rangle$) if there is a binary operation

$$\langle \, , \, \rangle : X \times X \to \mathbb{R}$$

satisfying the following properties for $\mathbf{u}, \mathbf{v}, \mathbf{w} \in X$ and $c \in \mathbb{R}$:

(1) $\langle \mathbf{u}+\mathbf{v}, \mathbf{w}\rangle = \langle \mathbf{u}, \mathbf{w}\rangle + \langle \mathbf{v}, \mathbf{w}\rangle$;

(2) $\langle c\mathbf{u}, \mathbf{v}\rangle = c\langle \mathbf{u}, \mathbf{v}\rangle$;

(3) $\langle \mathbf{u}, \mathbf{u}\rangle \geq 0$ and $\langle \mathbf{u}, \mathbf{u}\rangle = 0$ if and only if $\mathbf{u} = 0$;

(4) $\langle \mathbf{u}, \mathbf{v}\rangle = \langle \mathbf{v}, \mathbf{u}\rangle$.

We shall give some interesting examples of inner products below. Before we do, let us note that an inner product as just defined gives rise to a notion of length, or a *norm*. Namely, we define

$$\|\mathbf{v}\| = \sqrt{\langle \mathbf{v}, \mathbf{v}\rangle}\,.$$

By Property **(3)**, we see that $\|\mathbf{v}\| \geq 0$ and $\|\mathbf{v}\| = 0$ if and only if $\mathbf{v} = 0$.

In fact the two key properties of the inner product and the norm are enunciated in the following proposition:

Proposition 7.5.1 *Let X be a vector space and $\langle\ ,\ \rangle$ an inner product on that space. Let $\|\ \ \|$ be the induced norm. Then*

(1) The Cauchy–Schwarz–Bunyakovsky Inequality: *If* $\mathbf{u}, \mathbf{v} \in X$ *then*

$$|\langle \mathbf{u}, \mathbf{v}\rangle| \leq \|\mathbf{u}\| \cdot \|\mathbf{v}\|\,.$$

(2) The Triangle Inequality: *If* $\mathbf{u}, \mathbf{v} \in X$ *then*

$$\|\mathbf{u}+\mathbf{v}\| \leq \|\mathbf{u}\| + \|\mathbf{v}\|\,.$$

In fact, just as an exercise, we shall derive the Triangle Inequality from the Cauchy–Schwarz–Bunyakovsky Inequality. We have

$$\begin{aligned}
\|\mathbf{u}+\mathbf{v}\|^2 &= \langle(\mathbf{u}+\mathbf{v}), (\mathbf{u}+\mathbf{v})\rangle \\
&= \langle \mathbf{u}, \mathbf{u}\rangle + \langle \mathbf{u}, \mathbf{v}\rangle + \langle \mathbf{v}, \mathbf{u}\rangle + \langle \mathbf{v}, \mathbf{v}\rangle \\
&= \|\mathbf{u}\|^2 + \|\mathbf{v}\|^2 + 2\langle \mathbf{u}, \mathbf{v}\rangle \\
&\leq \|\mathbf{u}\|^2 + \|\mathbf{v}\|^2 + 2\|\mathbf{u}\| \cdot \|\mathbf{v}\| \\
&= (\|\mathbf{u}\| + \|\mathbf{v}\|)^2\,.
\end{aligned}$$

Now taking the square root of both sides completes the argument. We shall explore the proof of the Cauchy–Schwarz–Bunyakovsky Inequality in Exercise 5.

EXAMPLE 7.5.2 Let $X = C[0, 1]$, the continuous functions on the interval $[0, 1]$. This is certainly a vector space with the usual notions of addition of functions and scalar multiplication of functions. We define an inner product by

$$\langle f, g\rangle = \int_0^1 f(x)g(x)\,dx$$

for any $f, g \in X$.

Then it is straightforward to verify that this definition of inner product satisfies all our axioms. Thus we may *define* two functions to be orthogonal if

$$\langle f, g\rangle = 0\,.$$

We say that the angle θ between two functions is given by

$$\cos\theta = \frac{\langle f, g\rangle}{\|f\|\|g\|}.$$

The *length* or *norm* of an element $f \in X$ is given by

$$\|f\| = \sqrt{\langle f, f\rangle} = \left(\int_0^1 f(x)^2\,dx\right)^{1/2}. \qquad \blacksquare$$

EXAMPLE 7.5.3 Let X be the space of all real sequences $\{a_j\}_{j=1}^\infty$ with the property that $\sum_{j=1}^\infty |a_j|^2 < \infty$. This is a vector space with the obvious componentwise notions of addition and scalar multiplication. Define an inner product by

$$\langle \{a_j\}, \{b_j\}\rangle = \sum_{j=1}^\infty a_j b_j\,.$$

Then this inner product satisfies all our axioms. $\blacksquare$

For the purposes of studying Fourier series, the most important inner product space is that which we call $L^2[-\pi, \pi]$. This is the space of real functions f on the interval $[-\pi, \pi]$ with the property that

$$\int_{-\pi}^{\pi} f(x)^2\,dx < \infty\,.$$

The inner product on this space is

$$\langle f, g\rangle = \int_{-\pi}^{\pi} f(x)g(x)\,dx\,.$$

One must note here that, by a variant of the Cauchy–Schwarz–Bunyakovsky inequality, it holds that if $f, g \in L^2$, then the integral $\int f \cdot g\,dx$ exists and is finite. So our inner product makes sense.

Exercises

1. Verify that each pair of functions f, g is orthogonal on the given interval $[a, b]$ using the inner product

$$\langle f, g\rangle = \int_a^b f(x)g(x)\,dx\,.$$

   (a) $f(x) = \sin 2x, \quad g(x) = \cos 3x, \quad [-\pi, \pi]$
   (b) $f(x) = \sin 2x, \quad g(x) = \sin 4x, \quad [0, \pi]$
   (c) $f(x) = x^2, \quad g(x) = x^3, \quad [-1, 1]$
   (d) $f(x) = x, \quad g(x) = \cos 2x, \quad [-2, 2]$

2. Prove the so-called *parallelogram law* in the space L^2:

$$2\|f\|^2 + 2\|g\|^2 = \|f + g\|^2 + \|f - g\|^2\,.$$

   Hint: Expand the right-hand side.

3. Prove the *Pythagorean theorem* and its converse in L^2: The function f is orthogonal to the function g if and only if

$$\|f - g\|^2 = \|f\|^2 + \|g\|^2\,.$$

4. In the space L^2, show that if f is continuous and $\|f\| = 0$ then $f(x) = 0$ for all x.

5. Prove the Cauchy–Schwarz–Bunyakovsky Inequality in L^2: If $f, g \in L^2$ then

$$|\langle f, g\rangle| \le \|f\| \cdot \|g\|\,.$$

   Do this by considering the auxiliary function

$$\varphi(\lambda) = \|f + \lambda g\|^2$$

   and calculating the value of λ for which it is a minimum.

6. *Bessel's inequality* states that, if f is any square-integrable function on $[-\pi, \pi]$ (i.e., $f \in L^2$), then its Fourier coefficients a_j and b_j satisfy

$$\frac{1}{2}a_0^2 + \sum_{j=1}^{\infty}(a_j^2 + b_j^2) \le \frac{1}{\pi}\int_{-\pi}^{\pi} |f(x)|^2\,dx\,.$$

   This inequality is fundamental in the theory of Fourier series.

   (a) For any $n \ge 1$, define

$$s_n(x) = \frac{1}{2}a_0 + \sum_{j=1}^{n}(a_j \cos jx + b_j \sin jx)$$

   and show that

$$\frac{1}{\pi}\int_{-\pi}^{\pi} f(x)s_n(x)\,dx = \frac{1}{2}a_0^2 + \sum_{j=1}^{\infty}(a_j^2 + b_j^2)\,.$$

(b) By considering all possible products in the multiplication of $s_n(x)$ by itself, show that

$$\frac{1}{\pi}\int_{-\pi}^{\pi} |s_n(x)|^2\,dx = \frac{1}{2}a_0^2 + \sum_{k=1}^{n}(a_j^2 + b_j^2)\,.$$

(c) By writing

$$\begin{aligned}&\frac{1}{\pi}\int_{-\pi}^{\pi} |f(x) - s_n(x)|^2\,dx\\ &\quad = \frac{1}{\pi}\int_{-\pi}^{\pi} |f(x)|^2\,dx - \frac{2}{\pi}\int_{-\pi}^{\pi} f(x)s_n(x)\,dx + \frac{1}{\pi}\int_{-\pi}^{\pi} |s_n(x)|^2\,dx\\ &\quad = \frac{1}{\pi}\int_{-\pi}^{\pi} |f(x)|^2\,dx - \frac{1}{2}a_0^2 - \sum_{j=1}^{n}(a_j^2 + b_j^2)\,,\end{aligned}$$

conclude that

$$\frac{1}{2}a_0^2 + \sum_{j=1}^{n}(a_j^2 + b_j^2) \le \frac{1}{\pi}\int_{-\pi}^{\pi} |f(x)|^2\,dx\,.$$

(d) Now complete the proof of Bessel's inequality.

7. Use your symbol manipulation software, such as `Maple` or `Mathematica`, to implement the Gram–Schmidt procedure to orthonormalize a given finite family of functions on the unit interval. Apply your software routine to the family $1, x, x^2, \cdots, x^{10}$.

Historical Note

Riemann

Bernhard Riemann (1826–1866) was the son of a poor country minister in northern Germany. He studied the works of Euler and Legendre while still in high school; indeed, it is said that he mastered Legendre's treatise on number theory in less than a week. Riemann was shy and modest, with little awareness of his own extraordinary powers; thus, at age 19, he went to the University of Göttingen with the aim of pleasing his father by studying theology. Riemann soon tired of this curriculum, and with his father's acquiescence, he turned to mathematics.

Unfortunately, Gauss's austere manner offered little for an apprentice mathematician like Riemann, so he soon moved to Berlin. There he fell in with Dirichlet and Jacobi, and he learned a great deal from both. Two years later he returned to Göttingen and earned his doctorate. During the next eight years, Riemann suffered debilitating poverty and also produced his greatest scientific work. Unfortunately his health was broken. Even after Gauss's death, when Dirichlet took the helm of the Göttingen math institute and did everything in his power to help and advance Riemann, the young man's spirits and health were well in decline. At the age of 39 he died of tuberculosis in Italy, where he had traveled several times to escape the cold and wet of northern Germany.

Riemann made profound contributions to the theory of complex variables. The Cauchy–Riemann equations, the Riemann mapping theorem, Riemann surfaces, the Riemann–Roch theorem, and the Riemann hypothesis all bear his name. Incidentally, these areas are all studied intensely today.

Riemann's theory of the integral, and his accompanying ideas on Fourier series, has made an indelible impression on calculus and real analysis.

At one point in his career, Riemann was required to present a probationary lecture before the great Gauss. In this offering, Riemann developed a theory of geometry that unified and far generalized all existing geometric theories. This is of course the theory of Riemannian manifolds, perhaps the most important idea in modern geometry. Certainly Riemannian curvature plays a major role in mathematical physics, in partial differential equations, and in many other parts of the subject.

Riemann's single paper on number theory (published in 1859), just ten pages, is about the prime number theorem. In it, he develops the so-called Riemann zeta function and formulates a number of statements about that deep and important artifact. All of these statements, save one, have by now been proved. The one exception is the celebrated *Riemann hypothesis*, now thought to be perhaps the most central, the most profound, and the most difficult problem in all of mathematics. The question concerns the location of the zeros of the zeta function, and it harbors profound implications for the distribution of primes and for number theory as a whole.

In a fragmentary note found among his posthumous papers, Riemann wrote that he had a proof of the Riemann hypothesis, and that it followed from a formula for the Riemann zeta function which he had not simplified enough to publish. To this day, nobody has determined what that formula might be, and so Riemann has left a mathematical legacy that has baffled the greatest minds of our time.

Anatomy of an Application

Introduction to the Fourier Transform

Many problems of mathematics and mathematical physics are set on all of Euclidean space—not on an interval. Thus it is appropriate to have analytical tools designed for that setting. The Fourier transform is one of the most important of these devices. In this Anatomy we explore the basic ideas behind the Fourier transform. We shall present the concepts in Euclidean space of any dimension. Throughout, we shall use the standard notation $f \in L^1$ or $f \in L^1(\mathbb{R}^N)$ to mean that f is integrable.

If $t, \xi \in \mathbb{R}^N$ then we let

$$t \cdot \xi \equiv t_1\xi_1 + \cdots + t_N\xi_N.$$

We define the *Fourier transform* of a function $f \in L^1(\mathbb{R}^N)$ by

$$\widehat{f}(\xi) = \int_{\mathbb{R}^N} f(t)e^{it\cdot\xi}\,dt.$$

Here dt denotes N-dimensional volume. We sometimes write the Fourier transform as $\mathcal{F}(f)$ or simply $\mathcal{F}f$.

Many references will insert a factor of 2π in the exponential or in the measure. Others will insert a minus sign in the exponent. There is no agreement on this matter. We have opted for this particular definition because of its simplicity.

Proposition *If $f \in L^1(\mathbb{R}^N)$, then*

$$\sup_{\xi} |\widehat{f}(\xi)| \le \|f\|_{L^1(\mathbb{R}^N)}.$$

Proof Observe that, for any $\xi \in \mathbb{R}^N$,

$$|\widehat{f}(\xi)| \le \int |f(t)|\,dt < \infty\,. \qquad \square$$

In our development of the ideas concerning the Fourier transform, it is frequently useful to restrict attention to certain "testing functions." We define them now. Let us say that $f \in C_c^k$ if f is k-times continuously differentiable and f is identically zero outside of some ball. Figure 7.14 exhibits such a function.

Proposition *If $f \in L^1(\mathbb{R}^N)$, f is differentiable, and $\partial f/\partial x_j \in L^1(\mathbb{R}^N)$, then*

$$\left(\frac{\partial f}{\partial x_j}\right)^{\widehat{}}(\xi) = -i\xi_j\widehat{f}(\xi).$$

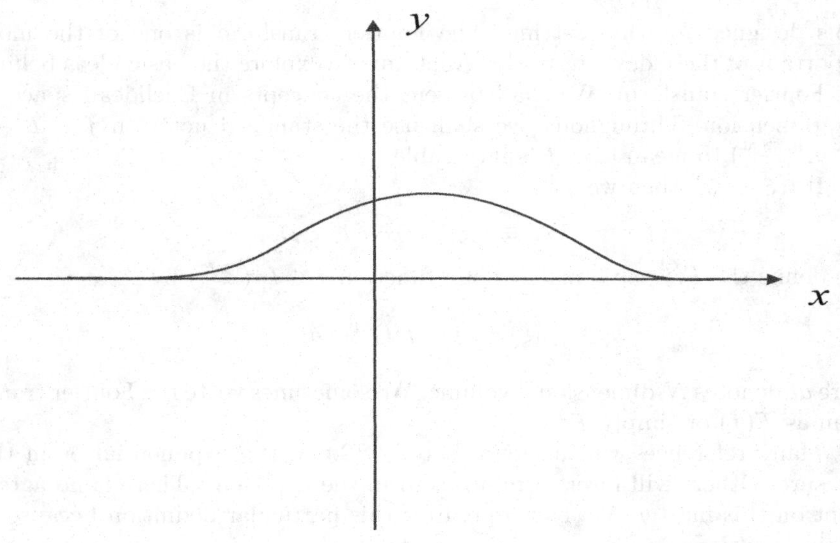

FIGURE 7.14
A C_c^k function.

Proof Integrate by parts: if $f \in C_c^1$, then

$$\begin{aligned}
\widehat{\left(\frac{\partial f}{\partial x_j}\right)}(\xi) &= \int \frac{\partial f}{\partial t_j} e^{it\cdot\xi}\,dt \\
&= \int \cdots \int \left(\int \frac{\partial f}{\partial t_j} e^{it\cdot\xi}\,dt_j\right) dt_1 \ldots dt_{j-1}dt_{j+1} \ldots dt_N \\
&= -\int \cdots \int f(t) \left(\frac{\partial}{\partial t_j} e^{it\cdot\xi}\right) dt_j dt_1 \ldots dt_{j-1}dt_{j+1} \ldots dt_N \\
&= -i\xi_j \int \cdots \int f(t) e^{it\cdot\xi}\,dt \\
&= -i\xi_j \widehat{f}(\xi).
\end{aligned}$$

(Of course the "boundary terms" in the integration by parts vanish since $f \in C_c^\infty$.) The general case follows from a limiting argument. □

Proposition *If $f \in L^1(\mathbb{R}^N)$ and $ix_j f \in L^1(\mathbb{R}^N)$, then*

$$\widehat{(ix_j f)} = \frac{\partial}{\partial \xi_j} \widehat{f}.$$

Proof Differentiate under the integral sign. □

Proposition (The Riemann–Lebesgue Lemma) *If* $f \in L^1(\mathbb{R}^N)$, *then*

$$\lim_{\xi \to \infty} |\widehat{f}(\xi)| = 0.$$

Proof First assume that $g \in C_c^2(\mathbb{R}^N)$. We know that

$$\|\widehat{g}\|_{L^\infty} \leq \|g\|_{L^1} \leq C$$

and, for each j,

$$\sup_\xi |\xi_j^2 \widehat{g}(\xi)| = \sup \left| \left(\left(\frac{\partial^2}{\partial x_j^2} \right) g \right)\widehat{\ } \right| \leq \int \left| \left(\frac{\partial^2}{\partial x_j^2} \right) g(x) \right| \, dx \equiv C_j \, .$$

Then $(1 + |\xi|^2)\widehat{g}$ is bounded. Therefore

$$|\widehat{g}(\xi)| \leq \frac{C''}{1 + |\xi|^2} \stackrel{|\xi| \to \infty}{\longrightarrow} 0.$$

This proves the result for $g \in C_c^2$.

Now let $f \in L^1$ be arbitrary. It is easy to see that there is a C_c^2 function ψ such that $\int |f - \psi\| \, dx < \epsilon/2$.

Choose M so large that when $|\xi| > M$ then $|\widehat{\psi}(\xi)| < \epsilon/2$. Then, for $|\xi| > M$, we have

$$\begin{aligned} |\widehat{f}(\xi)| &= |(f - \psi)\widehat{\ }(\xi) + \widehat{\psi}(\xi)| \\ &\leq |(f - \psi)\widehat{\ }(\xi)| + |\widehat{\psi}(\xi)| \\ &\leq \|f - \psi\|_{L^1} + \frac{\epsilon}{2} \\ &< \frac{\epsilon}{2} + \frac{\epsilon}{2} = \epsilon. \end{aligned}$$

This proves the result. □

Remark: The Riemann–Lebesgue lemma is intuitively clear when viewed in the following way. Fix an L^1 function f. An L^1 function is well-approximated by a continuous function, so we may as well suppose that f is continuous. But a continuous function is well-approximated by a smooth function, so we may as well suppose that f is smooth. On a small interval I—say of length $1/M$—a smooth function is nearly constant. So, if we let $|\xi| \gg 2\pi M$, then the character $e^{i\xi \cdot x}$ will oscillate at least M times on I, and will therefore integrate against a constant to a value that is very nearly zero. As M becomes larger, this statement becomes more and more accurate. That is the Riemann–Lebesgue lemma.

The three Euclidean groups that act naturally on $\mathbb{R}^N$ are

- rotations
- dilations
- translations

Certainly a large part of the utility of the Fourier transform is that it has natural invariance properties under the actions of these three groups. We shall now explicitly describe those properties.

We begin with the orthogonal group $O(N)$; an $N \times N$ matrix is *orthogonal* if it has real entries and its rows form an orthonormal system of vectors. A *rotation* is an orthogonal matrix with determinant 1 (also called a *special orthogonal* matrix).

Proposition *Let ρ be a rotation of $\mathbb{R}^N$. We define $\rho f(x) = f(\rho(x))$. Then we have the formula*

$$\widehat{\rho f} = \rho \widehat{f}.$$

Proof Remembering that ρ is orthogonal and has determinant 1, we calculate that

$$\begin{aligned}
\widehat{\rho f}(\xi) &= \int (\rho f)(t) e^{it\cdot\xi}\, dt = \int f(\rho(t)) e^{it\cdot\xi} dt \\
&\overset{(s=\rho(t))}{=} \int f(s) e^{i\rho^{-1}(s)\cdot\xi}\, ds = \int f(s) e^{is\cdot\rho\xi}\, ds \\
&= \widehat{f}(\rho\xi) = \rho\widehat{f}(\xi).
\end{aligned}$$

Here we have used the fact that $\rho^{-1} = {}^t\rho$ for an orthogonal matrix. The proof is complete. □

Definition For $\delta > 0$ and $f \in L^1(\mathbb{R}^N)$ we set $\alpha_\delta f(x) = f(\delta x)$ and $\alpha^\delta f(x) = \delta^{-N} f(x/\delta)$. These are the dual *dilation operators* of Euclidean analysis.

Proposition *The dilation operators interact with the Fourier transform as follows:*

$$\begin{aligned}
(\alpha_\delta f)\widehat{\ } &= \alpha^\delta \left(\widehat{f}\right) \\
\widehat{\alpha^\delta f} &= \alpha_\delta \left(\widehat{f}\right).
\end{aligned}$$

Proof We calculate that

$$\begin{aligned}
(\alpha_\delta f)\widehat{\ }(\xi) &= \int (\alpha_\delta f)(t) e^{it\cdot\xi}\, dt \\
&= \int f(\delta t) e^{it\cdot\xi}\, dt \\
&\overset{(s=\delta t)}{=} \int f(s) e^{i(s/\delta)\cdot\xi} \delta^{-N}\, ds \\
&= \delta^{-N} \widehat{f}(\xi/\delta) \\
&= \left(\alpha^\delta(\widehat{f})\right)(\xi).
\end{aligned}$$

That proves the first assertion. The proof of the second is similar. □

For any function f on $\mathbb{R}^N$ and $a \in \mathbb{R}^N$ we define $\tau_a f(x) = f(x - a)$. Clearly τ_a is a *translation operator.*

> **Proposition** *If $f \in L^1(\mathbb{R}^N)$ then*
>
> $$\widehat{\tau_a f}(\xi) = e^{ia\cdot\xi}\widehat{f}(\xi)$$
>
> *and*
>
> $$\left(\tau_a\{\widehat{f}\}\right)(\xi) = \left[e^{-ia\cdot t} f(t)\right]\widehat{\ }(\xi).$$

Proof For the first equality, we calculate that

$$\begin{aligned}
\widehat{\tau_a f}(\xi) &= \int_{\mathbb{R}^N} e^{ix\cdot\xi}(\tau_a f)(x)\, dx \\
&= \int_{\mathbb{R}^N} e^{ix\cdot\xi} f(x-a)\, dx \\
&\overset{(x-a)=t}{=} \int_{\mathbb{R}^N} e^{i(t+a)\cdot\xi} f(t)\, dt \\
&= e^{ia\cdot\xi} \int_{\mathbb{R}^N} e^{it\cdot\xi} f(t)\, dt \\
&= e^{ia\cdot\xi}\widehat{f}(\xi).
\end{aligned}$$

The second identity is proved similarly. □

Much of the theory of classical harmonic analysis—especially in this century—concentrates on translation-invariant operators. An operator T on functions is called *translation-invariant* if

$$T(\tau_a f)(x) = (\tau_a T f)(x)$$

for every x.[3] It is a basic fact that any translation-invariant integral operator is given by convolution with a kernel k.

Proposition *For $f \in L^1(\mathbb{R}^N)$ we let $\widetilde{f}(x) = f(-x)$. Then $\widehat{\widetilde{f}} = \widetilde{\widehat{f}}$.*

Proof We calculate that

$$\begin{aligned}\widehat{\widetilde{f}}(\xi) &= \int \widetilde{f}(t)e^{it\cdot\xi}\,dt = \int f(-t)e^{it\cdot\xi}\,dt \\ &= \int f(t)e^{-it\cdot\xi}\,dt = \widehat{f}(-\xi) = \widetilde{\widehat{f}}(\xi).\end{aligned}$$

□

Proposition *We have*

$$\widehat{\overline{f}} = \overline{\widetilde{\widehat{f}}}.$$

Proof We calculate that

$$\widehat{\overline{f}}(\xi) = \int \overline{f}(t)e^{it\cdot\xi}\,dt = \overline{\int f(t)e^{-it\cdot\xi}\,dt} = \overline{\widehat{f}(-\xi)} = \overline{\widetilde{\widehat{f}}}(\xi).$$

□

Proposition *If $f, g \in L^1$, then*

$$\int \widehat{f}(\xi)g(\xi)\,d\xi = \int f(\xi)\widehat{g}(\xi)\,d\xi.$$

[3] It is perhaps more accurate to say that such an operator *commutes with translations.* However, the terminology "translation-invariant" is standard.

Proof This is a straightforward change in the order of integration:

$$\begin{aligned}\int \widehat{f}(\xi)g(\xi)\,d\xi &= \int\int f(t)e^{it\cdot\xi}\,dt\,g(\xi)\,d\xi \\ &= \int\int g(\xi)e^{it\cdot\xi}\,d\xi\,f(t)\,dt \\ &= \int \widehat{g}(t)f(t)\,dt\,.\end{aligned}$$

□

Convolution and Fourier Inversion

If f and g are integrable functions then we define their *convolution* to be

$$f * g(x) = \int f(x-t)g(t)\,dt = \int f(t)g(x-t)\,dt\,.$$

Note that a simple change of variable confirms the second equality.

Proposition *If $f, g \in L^1$, then*

$$\widehat{f * g} = \widehat{f} \cdot \widehat{g}.$$

Proof We calculate that

$$\begin{aligned}\widehat{f * g}(\xi) &= \int (f * g)(t)e^{it\cdot\xi}\,dt = \int\int f(t-s)g(s)\,ds\,e^{it\cdot\xi}\,dt \\ &= \int\int f(t-s)e^{i(t-s)\cdot\xi}\,dt\,g(s)e^{is\cdot\xi}\,ds = \widehat{f}(\xi)\cdot\widehat{g}(\xi)\,.\end{aligned}$$

□

The Inverse Fourier Transform

Our goal is to be able to recover f from $\widehat{f}$. This program entails several technical difficulties. First, we need to know that the Fourier transform is one-to-one in order to have any hope of success. Secondly, we would like to say that

$$f(t) = c \cdot \int \widehat{f}(\xi)e^{-it\cdot\xi}\,d\xi. \tag{7.6}$$

But in general the Fourier transform $\widehat{f}$ of an L^1 function f is not integrable (just calculate the Fourier transform of $\chi_{[0,1]}$)—so the expression on the right of (7) does not necessarily make any sense.

Theorem *If $f, \widehat{f} \in L^1$ (and both are continuous), then*

$$f(0) = (2\pi)^{-N} \int \widehat{f}(\xi)\, d\xi. \tag{7.7}$$

Of course there is nothing special about the point $0 \in \mathbb{R}^N$. We now exploit the compatibility of the Fourier transform with translations to obtain a more general formula. We apply formula (8) in our theorem to $\tau_{-h}f$: The result is

$$(\tau_{-h}f)(0) = (2\pi)^{-N} \int (\tau_{-h}f)\widehat{\ }(\xi)\, d\xi$$

or

Theorem (The Fourier Inversion Formula) *If $f, \widehat{f} \in L^1$ (and if both $f, \widehat{f}$ are continuous), then for any $y \in \mathbb{R}^N$ we have*

$$f(y) = (2\pi)^{-N} \int \widehat{f}(\xi) e^{-iy\cdot\xi}\, d\xi.$$

The proof of the Fourier inversion theorem is rather involved, and requires a number of ancillary ideas. We cannot explore it here, but see [KRA3].

Plancherel's Formula

We now give a treatment of the quadratic Fourier theory.

Proposition (Plancherel) *If $f \in C_c^\infty(\mathbb{R}^N)$, then*

$$(2\pi)^{-N} \int |\widehat{f}(\xi)|^2\, d\xi = \int |f(x)|^2\, dx.$$

Proof Define $g = f * \widetilde{\overline{f}} \in C_c^\infty(\mathbb{R}^N)$. Then

$$\widehat{g} = \widehat{f} \cdot \widehat{\widetilde{\overline{f}}} = \widehat{f} \cdot \widehat{\overline{\widetilde{f}}} = \widehat{f} \cdot \overline{\widetilde{\widehat{f}}} = \widehat{f} \cdot \overline{\widetilde{\widetilde{\widehat{f}}}} = \widehat{f} \cdot \overline{\widehat{f}} = |\widehat{f}|^2. \tag{7.8}$$

Now

$$g(0) = f * \widetilde{\overline{f}}\,(0) = \int f(-t)\overline{f}(-t)\, dt = \int f(t)\overline{f}(t)\, dt = \int |f(t)|^2\, dt.$$

By Fourier inversion and formula (9) we may now conclude that

$$\int |f(t)|^2\, dt = g(0) = (2\pi)^{-N} \int \widehat{g}(\xi)\, d\xi = (2\pi)^{-N} \int |\widehat{f}(\xi)|^2\, d\xi.$$

That is the desired formula. □

Definition For any square integrable function f, the Fourier transform of f can be defined in the following fashion: Let $f_j \in C_c^\infty$ satisfy $f_j \to f$ in the L^2 topology. It follows from the Proposition that $\{\widehat{f_j}\}$ is Cauchy in L^2. Let g be the L^2 limit of this latter sequence. We set $\widehat{f} = g$.

It is easy to check that this definition of $\widehat{f}$ is independent of the choice of sequence $f_j \in C_c^\infty$ and that

$$(2\pi)^{-N} \int |\widehat{f}(\xi)|^2\, d\xi = \int |f(x)|^2\, dx.$$

Problems for Review and Discovery

A. Drill Exercises

1. Find the Fourier series for each of these functions.

(a) $f(x) = x^2\,, -\pi \le x \le \pi$

(b) $g(x) = x - |x|\,, \quad -\pi \le x \le \pi$

(c) $h(x) = x + |x|\,, \quad -\pi \le x \le \pi$

(d) $f(x) = |x|\,, \quad -\pi \le x \le \pi$

(e) $g(x) = \begin{cases} -x^2 & \text{if} \quad -\pi \le x \le 0 \\ x^2 & \text{if} \quad 0 < x \le \pi \end{cases}$

(f) $h(x) = \begin{cases} |x| & \text{if} \quad -\pi \le x \le 1/2 \\ 1/2 & \text{if} \quad 1/2 < x \le \pi \end{cases}$

2. Calculate the Fourier series for each of these functions.

(a) $f(x) = \begin{cases} \sin x & \text{if} \quad -\pi \le x \le \pi/2 \\ 0 & \text{if} \quad \pi/2 < x \le \pi \end{cases}$

(b) $g(x) = \begin{cases} \cos x & \text{if} \quad -\pi \le x \le -\pi/2 \\ 0 & \text{if} \quad -\pi/2 < x \le \pi \end{cases}$

(c) $h(x) = \begin{cases} \cos x & \text{if} \quad -\pi \le x \le 0 \\ 1 & \text{if} \quad 0 < x \le \pi \end{cases}$

(d) $f(x) = \begin{cases} 1 & \text{if} \quad -\pi \le x \le 0 \\ \sin x & \text{if} \quad 0 < x \le \pi \end{cases}$

3. Sketch the graphs of the first three partial sums of the Fourier series for each of the functions in Exercise 2.

4. Calculate the sine series of each of these functions.

(a) $f(x) = \cos 2x\,, \quad 0 \le x \le \pi$

(b) $g(x) = x^2\,, \quad 0 \le x \le \pi$

(c) $h(x) = x - |x - 1/2|/2\,, \quad 0 \le x \le \pi$

(d) $f(x) = x^2 + |x + 1/4|\,, \quad 0 \le x \le \pi$

5. Calculate the cosine series of each of these functions.

(a) $f(x) = \sin 3x\,, \quad 0 \le x \le \pi$

(b) $g(x) = x^2\,, \quad 0 \le x \le \pi$

(c) $h(x) = x - |x - 1/2|/2\,, \quad 0 \le x \le \pi$

(d) $f(x) = x^2 + |x + 1/4|\,, \quad 0 \le x \le \pi$

6. Find the Fourier series expansion for the given function on the given interval.

(a) $f(x) = x^2 - x\,, \quad -1 \le x \le 1$

(b) $g(x) = \sin x\,, \quad -2 \le x \le 2$

(c) $h(x) = \cos x\,, \quad -3 \le x \le 3$

(d) $f(x) = |x|\,, \quad -1 \le x \le 1$

(e) $g(x) = |x - 1/2|\,, \quad -2 \le x \le 2$

(f) $h(x) = |x + 1/2|/2\,, \quad -3 \le x \le 3$

B. Challenge Problems

1. In Section 4.3 we learned about the Legendre polynomials. The first three Legendre polynomials are

$$P_0(x) \equiv 1\ ,\ P_1(x) = x\ ,\ P_2(x) = \frac{3}{2}x^2 - \frac{1}{2}\,.$$

Verify that P_0, P_1, P_2 are mutually orthogonal on the interval $[-1, 1]$. Let

$$f(x) = \begin{cases} -1 & \text{if} \quad -1 \le x \le 0 \\ 2 & \text{if} \quad 0 < x \le 1\,. \end{cases}$$

Find the first three coefficients in the expansion

$$f(x) = a_0P_0(x) + a_1P_1(x) + a_2P_2(x) + \cdots\,.$$

2. Repeat Exercise 1 for the function $f(x) = x + |x|$.

3. The first three Hermite polynomials are

$$H_0(x) \equiv 1\ ,\ H_1(x) = 2x\ ,\ H_2(x) = 4x^2 - 2\,.$$

Verify that these functions are mutually orthogonal on the interval $(-\infty, \infty)$ with respect to the weight e^{-x^2} (this means that

$$\int_{-\infty}^{\infty} H_j(x)H_k(x)e^{-x^2}\,dx = 0$$

if $j \ne k$). Calculate the first three coefficients in the expansion

$$f(x) = b_0H_0(x) + b_1H_1(x) + b_2H_2(x) + \cdots$$

for the function $f(x) = x - |x|$.

4. The first three Chebyshev polynomials are

$$T_0(x) \equiv 1 \ , \ T_1(x) = x \ , \ T_2(x) = 2x^2 - 1 \, .$$

Verify that these functions are mutually orthogonal on the interval $[-1, 1]$ with respect to the weight $(1-x^2)^{-1/2}$ (refer to Exercise 3 for the meaning of this concept). Calculate the first three coefficients in the expansion

$$f(x) = c_0 T_0(x) + c_1 T_1(x) + c_2 T_2(x) + \cdots$$

for the function $f(x) = x$.

C. Problems for Discussion and Exploration

1. Refer to Fejér's Theorem 6.2.5 about convergence of the Cesàro means. Confirm this result by direct calculation for these functions.

(a) $f(x) = \begin{cases} -1 & \text{if} \quad -\pi \le x \le 0 \\ 1 & \text{if} \quad 0 < x \le \pi \end{cases}$

(b) $g(x) = \begin{cases} 0 & \text{if} \quad -\pi \le x \le 0 \\ \cos x & \text{if} \quad 0 < x \le \pi \end{cases}$

(c) $h(x) = \begin{cases} \sin x & \text{if} \quad -\pi \le x \le 0 \\ 2 & \text{if} \quad 0 < x \le \pi \end{cases}$

(d) $f(x) = \begin{cases} |x+1|/2 & \text{if} \quad -\pi \le x \le 0 \\ 0 & \text{if} \quad 0 < x \le \pi \end{cases}$

2. A celebrated result of classical Fourier analysis states that if f is continuously differentiable on $[-\pi, \pi]$, then its Fourier series converges absolutely. Confirm this assertion (at all points except the endpoints of the interval) in the following specific examples.

(a) $f(x) = x$

(b) $g(x) = x^2$

(c) $h(x) = e^x$

3. In other expositions, it is convenient to define Fourier series using the language of complex numbers. Specifically, instead of expanding in terms of $\cos jx$ and $\sin jx$, we expand in terms of e^{ijx}. Specifically, we work with a function f on $[-\pi, \pi]$ and set

$$c_j = \frac{1}{2\pi} \int_{-\pi}^{\pi} f(t) e^{-ijt} \, dt \, .$$

We define the formal Fourier expansion of f to be

$$Sf \sim \sum_{j=-\infty}^{\infty} c_j e^{ijt} \, .$$

Explain why this new formulation of Fourier series is equivalent to that presented in Section 5.1 (i.e., explain how to pass back and forth from one language to the other).

What are the advantages and disadvantages of this new, complex form of the Fourier series?

8

Laplace Transforms

- The idea of the Laplace transform
- The Laplace transform and differential equations
- Derivatives and the Laplace transform
- Integrals and the Laplace transform
- Convolutions
- Step and impulse functions
- Discontinuous input

8.0 Introduction

The idea of the Laplace transform has had a profound influence on the development of mathematical analysis. It also plays a significant role in mathematical applications. More generally, the overall theory of transforms has become an important part of modern mathematics.

The concept of a *transform* is that it turns a given function into another function. We are already acquainted with several transforms:

I. The derivative D takes a differentiable function f (defined on some interval (a, b)) and assigns to it a new function $Df = f'$.

II. The integral I takes a continuous function f (defined on some interval $[a, b]$ and assigns to it a new function

$$If(x) = \int_a^x f(t)\, dt\,.$$

III. The multiplication operator M_φ, which multiplies any given function f on the interval $[a, b]$ by a fixed function φ on $[a, b]$, is a transform:

$$M_\varphi f(x) = \varphi(x) \cdot f(x)\,.$$

DOI: 10.1201/9781003214526-8

We are particularly interested in transforms that are linear. A transform T is *linear* if

$$T[\alpha f + \beta g] = \alpha T(f) + \beta T(g)$$

for any real constants α, β. In particular (taking $\alpha = \beta = 1$),

$$T[f + g] = T(f) + T(g)$$

and (taking $\beta = 0$)

$$T(\alpha f) = \alpha T(f)\,.$$

We are especially interested in linear transformations that are given by integration. Let f be a function with domain $[0, \infty)$. The *Laplace transform* of f is defined by

$$L[f](p) = F(p) = \int_0^\infty e^{-px} f(x)\,dx \qquad \text{for } p > 0\,.$$

Notice that we begin with a function f of x, and the Laplace transform L produces a new function $L[f]$ of p. We sometimes write the Laplace transform as $F(p)$. Notice that the Laplace transform is an improper integral; it exists precisely when

$$\int_0^\infty e^{-px} f(x)\,dx = \lim_{N\to\infty} \int_0^N e^{-px} f(x)\,dx$$

exists.

Let us now calculate some Laplace transforms:

Function f	**Laplace transform F**
$f(x) \equiv 1$	$F(p) = \int_0^\infty e^{-px}\,dx = \frac{1}{p}$
$f(x) = x$	$F(p) = \int_0^\infty e^{-px} x\,dx = \frac{1}{p^2}$
$f(x) = x^n$	$F(p) = \int_0^\infty e^{-px} x^n\,dx = \frac{n!}{p^{n+1}}$
$f(x) = e^{ax}$	$F(p) = \int_0^\infty e^{-px} e^{ax}\,dx = \frac{1}{p-a}$
$f(x) = \sin ax$	$F(p) = \int_0^\infty e^{-px} \sin ax\,dx = \frac{a}{p^2+a^2}$
$f(x) = \cos ax$	$F(p) = \int_0^\infty e^{-px} \cos ax\,dx = \frac{p}{p^2+a^2}$
$f(x) = \sinh ax$	$F(p) = \int_0^\infty e^{-px} \sinh ax\,dx = \frac{a}{p^2-a^2}$
$f(x) = \cosh ax$	$F(p) = \int_0^\infty e^{-px} \cosh ax\,dx = \frac{p}{p^2-a^2}$

We shall not actually perform all these integrations. We content ourselves with the third one, just to illustrate the idea. The student should definitely perform the others, just to get the feel of Laplace transform calculations.

Now

$$\begin{aligned} L[x^n] &= \int_0^\infty e^{-px}x^n\,dx \\ &= -\frac{x^n e^{-px}}{p}\bigg|_0^\infty + \frac{n}{p}\int_0^\infty e^{-px}x^{n-1}\,dx \\ &= \frac{n}{p}L[x^{n-1}] \\ &= \frac{n}{p}\left(\frac{n-1}{p}\right)L[x^{n-2}] \\ &= \cdots \\ &= \frac{n!}{p^n}L[1] \\ &= \frac{n!}{p^{n+1}}\,. \end{aligned}$$

The reader will find, as we just have, that integration by parts is eminently useful in the calculation of Laplace transforms.

It may be noted that the Laplace transform is a linear operator. Thus Laplace transforms of some compound functions may be readily calculated from the table just given:

$$L(5x^3 - 2e^x] = \frac{5\cdot 3!}{p^4} - \frac{2}{p-1}$$

and

$$L(4\sin 2x + 6x] = \frac{4\cdot 2}{p^2+2^2} + \frac{6}{p^2}\,.$$

Exercises

1. Evaluate all the Laplace transform integrals for the table in this section.
2. Without actually integrating, show that
   (a) $L[\sinh ax] = \dfrac{a}{p^2-a^2}$
   (b) $L[\cosh ax] = \dfrac{p}{p^2-a^2}$
3. Find $L[\sin^2 ax]$ and $L[\cos^2 ax]$ without integrating. How are these two transforms related to one another?

4. Use the formulas given in the text to find the transform of each of the following functions.

(a) 10 (d) $4\sin x\cos x + 2e^{-x}$
(b) $x^5 + \cos 2x$ (e) $x^6\sin^2 3x + x^6\cos^2 3x$
(c) $2e^{3x} - \sin 5x$

5. Find a function f whose Laplace transform is:

(a) $\frac{30}{p^4}$ (d) $\frac{1}{p^2+p}$
(b) $\frac{2}{p+3}$ (e) $\frac{1}{p^4+p^2}$
(c) $\frac{4}{p^3} + \frac{6}{p^2+4}$

Hint: The method of partial fractions will prove useful.

6. Give a plausible definition of $\frac{1}{2}!$ (i.e., the factorial of the number 1/2).

7. Use your symbol manipulation software, such as `Maple` or `Mathematica`, to calculate the Laplace transforms of each of these functions.

(a) $\sin(e^t)$
(b) $\ln(1 + \sin^2 t)$
(c) $\sin[\ln t]$
(d) $e^{\cos t}$

8.1 Applications to Differential Equations

The key to our use of Laplace transform theory in the subject of differential equations is the way that L treats derivatives. Let us calculate

$$\begin{aligned} L[y'] &= \int_0^\infty e^{-px} y'(x)\,dx \\ &= y(x)e^{-px}\Big|_0^\infty + p\int_0^\infty e^{-px} y(x)\,dx \\ &= -y(0) + p \cdot L[y]\,. \end{aligned}$$

In summary,

$$L[y'] = p \cdot L[y] - y(0)\,.$$

Likewise,

$$\begin{aligned} L[y''] &= L[(y')'] = p \cdot L[y'] - y'(0) \\ &= p\{p \cdot L[y] - y(0)\} - y'(0) \\ &= p^2 \cdot L[y] - py(0) - y'(0)\,. \end{aligned}$$

Now let us examine the differential equation

$$y'' + ay' + by = f(x), \tag{8.1}$$

with the initial conditions $y(0) = y_0$ and $y'(0) = y_1$. Here a and b are real constants. We apply the Laplace transform L to both sides of (1), of course using the linearity of L. The result is

$$L[y''] + aL[y'] + bL[y] = L[f].$$

Writing out what each term is, we find that

$$\{p^2 \cdot L[y] - py(0) - y'(0)\} + a\{p \cdot L[y] - y(0)\} + bL[y] = L[f].$$

Now we can plug in what $y(0)$ and $y'(0)$ are. We may also gather like terms together. The result is

$$\{p^2 + ap + b\}L[y] = (p + a)y_0 + y_1 + L[f]$$

or

$$L[y] = \frac{(p + a)y_0 + y_1 + L[f]}{p^2 + ap + b}. \tag{8.2}$$

What we see here is a remarkable thing: The Laplace transform changes solving a differential equation from a rather complicated calculus problem to a simple algebra problem. The only thing that remains, in order to find an explicit solution to the original differential equation (1), is to find the inverse Laplace transform of the right-hand side of (2). In practice we shall find that we can often perform this operation in a straightforward fashion. The following examples will illustrate the idea.

EXAMPLE 8.1.1 Use the Laplace transform to solve the differential equation

$$y'' + 4y = 4x \tag{8.3}$$

with initial conditions $y(0) = 1$ and $y'(0) = 5$.

Solution: We proceed mechanically, by applying the Laplace transform to both sides of (3). Thus

$$L[y''] + L[4y] = L[4x].$$

We can use our various Laplace transform formulas to write this out more explicitly:

$$\{p^2 L[y] - py(0) - y'(0)\} + 4L[y] = \frac{4}{p^2}$$

or

$$p^2 L[y] - p \cdot 1 - 5 + 4L[y] = \frac{4}{p^2}$$

or

$$(p^2+4)L[y] = p + 5 + \frac{4}{p^2}\,.$$

It is convenient to write this as

$$L[y] = \frac{p}{p^2+4} + \frac{5}{p^2+4} + \frac{4}{p^2\cdot(p^2+4)} = \frac{p}{p^2+4} + \frac{5}{p^2+4} + \frac{1}{p^2} - \frac{1}{p^2+4}\,,$$

where we have used a partial fractions decomposition in the last step. Simplifying, we have

$$L[y] = \frac{p}{p^2+4} + \frac{4}{p^2+4} + \frac{1}{p^2}\,.$$

Referring to our table of Laplace transforms, we may now deduce what y must be:

$$L[y] = L[\cos 2x] + L[2\sin 2x] + L[x] = L[\cos 2x + 2\sin 2x + x]\,.$$

We deduce then that

$$y = \cos 2x + 2\sin 2x + x\,,$$

and this is the solution of our initial value problem. □

Remark 8.1.2 It is useful to note that our formulas for the Laplace transform of the first and second derivative incorporated automatically the values $y(0)$ and $y'(0)$. Thus our initial conditions got built in during the course of our solution process.

A useful general property of the Laplace transform concerns its interaction with translations. Indeed, we have

$$L[e^{ax}f(x)] = F(p-a)\,. \qquad (8.4)$$

To see this, we calculate

$$\begin{aligned} L[e^{ax}f(x)] &= \int_0^\infty e^{-px}e^{ax}f(x)\,dx \\ &= \int_0^\infty e^{-(p-a)x}f(x)\,dx \\ &= F(p-a)\,. \end{aligned}$$

We frequently find it useful to use the notation L^{-1} to denote the inverse operation to the Laplace transform.[1] For example, since

$$L[x^2] = \frac{2!}{p^3}\,,$$

[1]We tacitly use here the fact that the Laplace transform L is one-to-one: if $L[f] = L[g]$ then $f = g$. Thus L is invertible on its image. We are able to verify this assertion empirically through our calculations; the general result is proved rigorously in a more advanced treatment.

we may write

$$L^{-1}\left(\frac{2!}{p^3}\right) = x^2 .$$

Since

$$L[\sin x - e^{2x}] = \frac{1}{p^2+1} - \frac{1}{p-2},$$

we may write

$$L^{-1}\left(\frac{1}{p^2+1} - \frac{1}{p-2}\right) = \sin x - e^{2x} .$$

EXAMPLE 8.1.3 Since

$$L[\sin bx] = \frac{b}{p^2+b^2},$$

we conclude that

$$L[e^{ax}\sin bx] = \frac{b}{(p-a)^2+b^2} .$$

Since

$$L^{-1}\left(\frac{1}{p^2}\right) = x ,$$

we conclude that

$$L^{-1}\left(\frac{1}{(p-a)^2}\right) = e^{ax}x .$$

■

EXAMPLE 8.1.4 Use the Laplace transform to solve the differential equation

$$y'' + 2y' + 5y = 3e^{-x}\sin x \tag{8.5}$$

with initial conditions $y(0) = 0$ and $y'(0) = 3$.

Solution: We calculate the Laplace transform of both sides, using our new formula (4) on the right-hand side, to obtain

$$\{p^2L[y] - py(0) - y'(0)\} + 2\{pL[y] - y(0)\} + 5L[y] = 3 \cdot \frac{1}{(p+1)^2+1} .$$

Plugging in the initial conditions, and organizing like terms, we find that

$$(p^2+2p+5)L[y] = 3 + \frac{3}{(p+1)^2+1}$$

or

$$\begin{aligned} L[y] &= \frac{3}{p^2+2p+5} + \frac{3}{(p^2+2p+2)(p^2+2p+5)} \\ &= \frac{3}{p^2+2p+5} + \frac{1}{p^2+2p+2} - \frac{1}{p^2+2p+5} \\ &= \frac{2}{(p+1)^2+4} + \frac{1}{(p+1)^2+1} . \end{aligned}$$

We see therefore that

$$y = e^{-x} \sin 2x + e^{-x} \sin x\,.$$

This is the solution of our initial value problem. □

Exercises

1. Find the Laplace transforms of

(a) $x^5 e^{-2x}$ (e) $e^{3x} \cos 2x$

(b) $(1 - x^2)e^{-x}$ (f) xe^x

(c) $e^{-x} \sin x$ (g) $x^2 \cos x$

(d) $x \sin 3x$ (h) $\sin x \cos x$

2. Find the inverse Laplace transform of

(a) $\dfrac{6}{(p+2)^2+9}$ (d) $\dfrac{12}{(p+3)^4}$

(b) $\dfrac{p+3}{p^2+2p+5}$ (e) $\dfrac{1}{p^4+p^2+1}$

(c) $\dfrac{p}{4p^2+1}$ (f) $\dfrac{6}{(p-1)^3}$

3. Solve each of the following differential equations with initial values using the Laplace transform.

(a) $y' + y = e^{2x}, \quad y(0) = 0$

(b) $y'' - 4y' + 4y = 0, \quad y(0) = 0$ and $y'(0) = 3$

(c) $y'' + 2y' + 2y = 2, \quad y(0) = 0$ and $y'(0) = 1$

(d) $y'' + y' = 3x^2, \quad y(0) = 0$ and $y'(0) = 1$

(e) $y'' + 2y' + 5y = 3e^{-x} \sin x, \quad y(0) = 0$ and $y'(0) = 3$

4. Find the solution of $y'' - 2ay' + a^2 y = 0$ in which the initial conditions $y(0) = y_0$ and $y'(0) = y_0'$ are left unrestricted. (This provides an additional derivation of our earlier solution for the case in which the auxiliary equation has a double root.)

5. Apply the formula $L[y'] = pL[y] - y(0)$ to establish the formula for the Laplace transform of an integral:

$$L\left(\int_0^x f(t)\, dt\right) = \frac{F(p)}{p}\,.$$

Do so by finding

$$L^{-1}\left(\frac{1}{p(p+1)}\right)$$

in two different ways.

6. Solve the equation

$$y' + 4y + 5\int_0^x y dx = e^{-x} , \qquad y(0) = 0 .$$

8.2 Derivatives and Integrals of Laplace Transforms

In some contexts it is useful to calculate the derivative of the Laplace transform of a function (when the corresponding integral makes sense). For instance, consider

$$F(p) = \int_0^\infty e^{-px} f(x)\, dx .$$

Then

$$\begin{aligned} \frac{d}{dp} F(p) &= \frac{d}{dp}\int_0^\infty e^{-px} f(x)\, dx \\ &= \int_0^\infty \frac{d}{dp}\left[e^{-px} f(x)\right]\, dx \\ &= \int_0^\infty e^{-px}\{-xf(x)\}\, dx = L[-xf(x)] . \end{aligned}$$

We see that the derivative[2] of $F(p)$ is the Laplace transform of $-xf(x)$. More generally, the same calculation shows us that

$$\frac{d^2}{dp^2} F(p) = L[x^2 f(x)](p)$$

and

$$\frac{d^j}{dp^j} F(p) = L[(-1)^j x^j f(x)](p) .$$

EXAMPLE 8.2.1 Calculate

$$L[x \sin ax] .$$

Solution: We have

$$L[x\sin ax] = -L[-x\sin ax] = -\frac{d}{dp}L[\sin ax] = -\frac{d}{dp}\frac{a}{p^2+a^2} = \frac{2ap}{(p^2+a^2)^2} .$$

□

EXAMPLE 8.2.2 Calculate the Laplace transform of $\sqrt{x}$.

[2]The passage of the derivative under the integral sign in this calculation requires advanced ideas from real analysis which we cannot treat here—see [KRA2].

Solution: This calculation actually involves some tricky integration. We first note that

$$L[\sqrt{x}] = L[x^{1/2}] = -L[-x \cdot x^{-1/2}] = -\frac{d}{dp}L[x^{-1/2}]. \tag{8.6}$$

Thus we must find the Laplace transform of $x^{-1/2}$.

Now

$$L[x^{-1/2}] = \int_0^\infty e^{-px}x^{-1/2}\,dx.$$

The change of variables $px = t$ yields

$$= p^{-1/2}\int_0^\infty e^{-t}t^{-1/2}\,dt.$$

The further change of variables $t = s^2$ gives the integral

$$L[x^{-1/2}] = 2p^{-1/2}\int_0^\infty e^{-s^2}\,ds. \tag{8.7}$$

Now we must evaluate the integral $I = \int_0^\infty e^{-s^2}\,ds$. Observe that

$$I \cdot I = \int_0^\infty e^{-s^2}\,ds \cdot \int_0^\infty e^{-u^2}\,du = \int_0^\infty \int_0^{\pi/2} e^{-r^2} \cdot r\,d\theta dr.$$

Here we have introduced polar coordinates in the standard way.

Now the last integral is easily evaluated and we find that

$$I^2 = \frac{\pi}{4}$$

hence $I = \sqrt{\pi}/2$. Thus $L[x^{-1/2}](p) = 2p^{-1/2}\{\sqrt{\pi}/2\} = \sqrt{\pi/p}$. Finally,

$$L[\sqrt{x}] = -\frac{d}{dp}\sqrt{\frac{\pi}{p}} = \frac{1}{2p}\sqrt{\frac{\pi}{p}}. \qquad \square$$

We now derive some additional formulas that will be useful in solving differential equations. We let $y = f(x)$ be our function and $Y = L[f]$ be its Laplace transform. Then

$$L[xy] = -\frac{d}{dp}L[y] = -\frac{dY}{dp}. \tag{8.8}$$

Also

$$L[xy'] = -\frac{d}{dp}L[y'] = -\frac{d}{dp}[pY - y(0)] = -\frac{d}{dp}[pY] \qquad (8.9)$$

and

$$L[xy''] = -\frac{d}{dp}L[y''] = -\frac{d}{dp}[p^2Y - py(0) - y'(0)] = -\frac{d}{dp}[p^2Y] + y(0) \,. \qquad (8.10)$$

EXAMPLE 8.2.3 Use the Laplace transform to analyze Bessel's equation

$$xy'' + y' + xy = 0$$

with the single initial condition $y(0) = 1$.

Solution: Apply the Laplace transform to both sides of the equation. Thus

$$L[xy''] + L[y'] + L[xy] = L[0] = 0 \,.$$

We can apply our new formulas (10) and (8) to the first and third terms on the left. And of course we apply the usual formula for the Laplace transform of the derivative to the second term on the left. The result is

$$-\frac{d}{dp}[p^2Y] + 1 + \{pY - 1\} - \frac{dY}{dp} = 0 \,.$$

We may simplify this equation to

$$(p^2 + 1)\frac{dY}{dp} = -pY \,.$$

This is a *new* differential equation, and we may solve it by separation of variables. Now

$$\frac{dY}{Y} = -\frac{p\,dp}{p^2 + 1}$$

so

$$\ln Y = -\frac{1}{2}\ln(p^2 + 1) + C \,.$$

Exponentiating both sides gives

$$Y = D \cdot \frac{1}{\sqrt{p^2 + 1}} \,.$$

It is useful (with a view to calculating the inverse Laplace transform) to write this solution as

$$Y = \frac{D}{p} \cdot \left(1 + \frac{1}{p^2}\right)^{-1/2} . \tag{8.11}$$

Recall the binomial expansion (Section 6.2):

$$\begin{aligned}(1+z)^a &= 1 + az + \frac{a(a-1)}{2!}z^2 + \frac{a(a-1)(a-2)}{3!}z^3 \\ &\quad + \cdots + \frac{a(a-1)\cdots(a-n+1)}{n!}z^n + \cdots .\end{aligned}$$

We apply this formula to the second term on the right of (11). Thus

$$\begin{aligned} Y &= \frac{D}{p} \cdot \left(1 - \frac{1}{2} \cdot \frac{1}{p^2} + \frac{1}{2!} \cdot \frac{1}{2} \cdot \frac{3}{2} \cdot \frac{1}{p^4} - \frac{1}{3!} \cdot \frac{1}{2} \cdot \frac{3}{2} \cdot \frac{5}{2} \cdot \frac{1}{p^6} \right. \\ &\quad \left. + \cdots + \frac{1 \cdot 3 \cdot 5 \cdots (2n-1)}{2^n n!} \frac{(-1)^n}{p^{2n}} + \cdots \right) \\ &= D \cdot \sum_{j=0}^{\infty} \frac{(2j)!}{2^{2j}(j!)^2} \cdot \frac{(-1)^j}{p^{2j+1}} . \end{aligned}$$

The good news is that we can now calculate L^{-1} of Y (thus obtaining y) by just calculating the inverse Laplace transform of each term of this series. The result is

$$\begin{aligned} y(x) &= D \cdot \sum_{j=0}^{\infty} \frac{(-1)^j}{2^{2j}(j!)^2} \cdot x^{2j} \\ &= D \cdot \left(1 - \frac{x^2}{2^2} + \frac{x^4}{2^2 \cdot 4^2} - \frac{x^6}{2^2 \cdot 4^2 \cdot 6^2} + \cdots \right) . \end{aligned}$$

Since $y(0) = 1$ (the initial condition), we see that $D = 1$ and

$$y(x) = 1 - \frac{x^2}{2^2} + \frac{x^4}{2^2 \cdot 4^2} - \frac{x^6}{2^2 \cdot 4^2 \cdot 6^2} + \cdots .$$

It should be noted that the origin $x_0 = 0$ is a regular singular point for this differential equation, and the values of m are $m = 0, 0$. That explains why we do *not* have two undetermined constants in our solution.

The series we have just derived defines the celebrated and important *Bessel function* J_0. We have learned that the Laplace transform of J_0 is $1/\sqrt{p^2+1}$. □

It is also a matter of some interest to integrate the Laplace transform. We can anticipate how this will go by running the differentiation formulas in reverse. Our main result is

$$L\left[\frac{f(x)}{x}\right] = \int_p^{\infty} F(s)\, ds . \tag{8.12}$$

In fact

$$\begin{aligned}
\int_p^\infty F(s)\,ds &= \int_p^\infty \left(\int_0^\infty e^{-sx} f(x)\,dx\right) ds \\
&= \int_0^\infty f(x) \int_p^\infty e^{-sx}\,dsdx \\
&= \int_0^\infty f(x) \left(\frac{e^{-sx}}{-x}\right)_p^\infty dx \\
&= \int_0^\infty f(x) \cdot \frac{e^{-px}}{x}\,dx \\
&= \int_0^\infty \left(\frac{f(x)}{x}\right) \cdot e^{-px}\,dx \\
&= L\left[\frac{f(x)}{x}\right] .
\end{aligned}$$

Remark 8.2.4 From formula (12) we have

$$L\left[\frac{f(x)}{x}\right] = \int_0^\infty e^{-px} f(x)\,dx = \int_p^\infty F(s)\,ds\,.$$

Thus taking $p = 0$ yields

$$\int_0^\infty \frac{f(x)}{x}\,dx = \int_0^\infty F(s)\,ds\,.$$

EXAMPLE 8.2.5 Use the fact that $L[\sin x] = 1/(p^2+1)$ to calculate $\int_0^\infty (\sin x)/x\,dx$.

Solution: By formula (12) (with $f(x) = \sin x$),

$$\int_0^\infty \frac{\sin x}{x}\,dx = \int_0^\infty \frac{dp}{p^2+1} = \arctan p\Big|_0^\infty = \frac{\pi}{2}\,. \qquad \square$$

We conclude this section by summarizing the chief properties of the Laplace transform in a table. The last property listed concerns convolution, and we shall treat that topic in the next section.

Properties of the Laplace Transform

$$L[\alpha f(x) + \beta g(x)] = \alpha F(p) + \beta G(p)$$

$$L[e^{ax} f(x)] = F(p-a)$$

$$L[f'(x)] = pF(p) - f(0)$$

$$L[f''(x)] = p^2 F(p) - pf(0) - f'(0)$$

$$L\left[\int_0^x f(t)\,dt\right] = \frac{F(p)}{p}$$

$$L[-xf(x)] = F'(p)$$

$$L[(-1)^n x^n f(x)] = F^{(n)}(p)$$

$$L\left[\frac{f(x)}{x}\right] = \int_p^\infty F(p)\,dp$$

$$L\left[\int_0^x f(x-t)g(t)\,dt\right] = F(p)G(p)$$

Exercises

1. Verify that

$$L[x \cos ax] = \frac{p^2 - a^2}{(p^2+a^2)^2} .$$

Use this result to find

$$L^{-1}\left(\frac{1}{(p^2+a^2)^2}\right) .$$

2. Calculate each of the following Laplace transforms.
   (a) $L[x^2 \sin ax]$
   (b) $L[x^{3/2}]$
   (c) $L[x \cos ax]$
   (d) $L[xe^x]$
3. Solve each of the following differential equations.
   (a) $xy'' + (3x-1)y' - (4x+9)y = 0$
   (b) $xy'' + (2x+3)y' + (x+3)y = 3e^{-x}$
4. If a and b are positive constants, then evaluate the following integrals.

(a) $\displaystyle\int_0^\infty \frac{e^{-ax} - e^{-bx}}{x}\,dx$

(b) $\displaystyle\int_0^\infty \frac{e^{-ax}\sin bx}{x}\,dx$

5. Without worrying about convergence issues, verify that

(a) $\displaystyle\int_0^\infty J_0(x)\,dx = 1$

(b) $\displaystyle J_0(x) = \frac{1}{\pi}\int_0^\pi \cos(x\cos t)\,dt$

6. Without worrying about convergence issues, and assuming $x > 0$, show that

(a) $\displaystyle f(x) = \int_0^\infty \frac{\sin xt}{t}\,dt \equiv \frac{\pi}{2}$

(b) $\displaystyle f(x) = \int_0^\infty \frac{\cos xt}{1+t^2}\,dt = \frac{\pi}{2}e^{-x}$

7. (a) If f is periodic with period a, so that $f(x+a) = f(x)$, then show that

$$F(p) = \frac{1}{1-e^{-ap}}\int_0^a e^{-px} f(x)\,dx\,.$$

(b) Find $F(p)$ if $f(x) = 1$ in the intervals $[0,1]$, $[2,3]$, $[4,5]$, etc., and $f \equiv 0$ in the remaining intervals.

8. If y satisfies the differential equation

$$y'' + x^2 y = 0\,,$$

where $y(0) = y_0$ and $y'(0) = y_0'$, then show that its Laplace transform $Y(p)$ satisfies the equation

$$Y'' + p^2 Y = py_0 + y_0'\,.$$

Observe that the new equation is of the same type as the original equation, so that no real progress has been made. The method of Example 8.2.3 is effective only when the coefficients are first-degree polynomials.

8.3 Convolutions

An interesting question, that occurs frequently with the use of the Laplace transform, is this: Let f and g be functions and F and G their Laplace transforms; what is $L^{-1}[F\cdot G]$? To discover the answer, we write

$$\begin{aligned} F(p)\cdot G(p) &= \left(\int_0^\infty e^{-ps} f(s)\,ds\right)\cdot\left(\int_0^\infty e^{-pt} g(t)\,dt\right) \\ &= \int_0^\infty\int_0^\infty e^{-p(s+t)} f(s)g(t)\,dsdt \\ &= \int_0^\infty\left(\int_0^\infty e^{-p(s+t)} f(s)\,ds\right) g(t)\,dt\,. \end{aligned}$$

Now we perform the change of variable $s = x - t$ in the inner integral. The result is

$$\begin{aligned} F(p) \cdot G(p) &= \int_0^\infty \left(\int_t^\infty e^{-px} f(x-t)\, dx \right) g(t)\, dt \\ &= \int_0^\infty \int_t^\infty e^{-px} f(x-t) g(t)\, dx dt . \end{aligned}$$

Reversing the order of integration, we may finally write

$$\begin{aligned} F(p) \cdot G(p) &= \int_0^\infty \left(\int_0^x e^{-px} f(x-t) g(t)\, dt \right) dx \\ &= \int_0^\infty e^{-px} \left(\int_0^x f(x-t) g(t)\, dt \right) dx \\ &= L\left[\int_0^x f(x-t) g(t)\, dt \right] . \end{aligned}$$

We call the expression $\int_0^x f(x-t)g(t)\,dt$ the *convolution* of f and g. Many texts write

$$f * g(x) = \int_0^x f(x-t) g(t)\, dt . \tag{8.13}$$

Our calculation shows that

$$L[f * g](p) = F \cdot G = L[f] \cdot L[g] .$$

The convolution formula is particularly useful in calculating inverse Laplace transforms.

EXAMPLE 8.3.1 Calculate

$$L^{-1}\left(\frac{1}{p^2(p^2+1)} \right) .$$

Solution: We write

$$\begin{aligned} L^{-1}\left(\frac{1}{p^2(p^2+1)} \right) &= L^{-1}\left(\frac{1}{p^2} \cdot \frac{1}{p^2+1} \right) \\ &= \int_0^x (x-t) \cdot \sin t\, dt . \end{aligned}$$

Notice that we have recognized that $1/p^2$ is the Laplace transform of x and $1/(p^2+1)$ is the Laplace transform of $\sin x$, and then applied the convolution result.

Now the last integral is easily evaluated (just integrate by parts) and seen to equal

$$x - \sin x\,.$$

We have thus discovered, rather painlessly, that

$$L^{-1}\left(\frac{1}{p^2(p^2+1)}\right) = x - \sin x\,. \qquad \square$$

An entire area of mathematics is devoted to the study of integral equations of the form

$$f(x) = y(x) + \int_0^x k(x-t)y(t)\,dt\,. \tag{8.14}$$

Here f is a given forcing function, and k is a given function known as the *kernel.* Usually k is a mathematical model for the physical process being studied. The object is to solve for y. As you can see, the integral equation involves a convolution. And, not surprisingly, the Laplace transform comes to our aid in unraveling the equation.

In fact we apply the Laplace transform to both sides of (14). The result is

$$L[f] = L[y] + L[k]\cdot L[y]$$

hence

$$L[y] = \frac{L[f]}{1+L[k]}\,.$$

Let us look at an example in which this paradigm occurs.

EXAMPLE 8.3.2 Use the Laplace transform to solve the integral equation

$$y(x) = x^3 + \int_0^x \sin(x-t)y(t)\,dt\,.$$

Solution: We apply the Laplace transform to both sides (using the convolution formula):

$$L[y] = L[x^3] + L[\sin x]\cdot L[y]\,.$$

Solving for $L[y]$, we see that

$$L[y] = \frac{L[x^3]}{1-L[\sin x]} = \frac{3!/p^4}{1-1/(p^2+1)}\,.$$

We may simplify the right-hand side to obtain

$$L[y] = \frac{3!}{p^4} + \frac{3!}{p^6}\,.$$

Of course it is easy to determine the inverse Laplace transform of the right-hand side. The result is

$$y(x) = x^3 + \frac{x^5}{20}\,.$$

$\square$

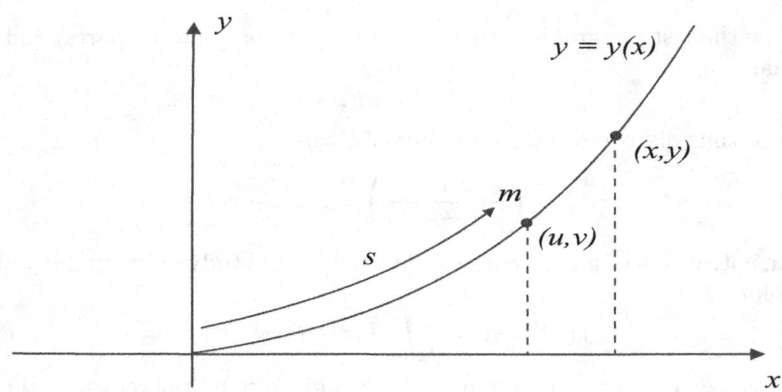

FIGURE 8.1
A smooth curve which a bead will slide down.

8.3.1 Abel's Mechanics Problem

We now study an old problem from mechanics that goes back to Niels Henrik Abel (1802–1829). Imagine a wire bent into a smooth curve (Figure 8.1). The curve terminates at the origin. Imagine a bead sliding from the top of the wire, without friction, down to the origin. The only force acting on the bead is gravity, depending only on the weight of the bead. Say that the wire is the graph of a function $y = y(x)$. Then the total time for the descent of the bead is some number $T(y)$ that depends on the shape of the wire and on the initial height y. Abel's problem is to run the process in reverse: Suppose that we are given a function T. Then find the shape y of a wire that will result in this time-of-descent function T.

What is interesting about this problem, from the point of view of the present section, is that its mathematical formulation leads to an integral equation of the sort that we have just been discussing. And we shall be able to solve it using the Laplace transform.

We begin our analysis with the principle of conservation of energy. Namely,

$$\frac{1}{2}m\left(\frac{ds}{dt}\right)^2 = m \cdot g \cdot (y - v)\,.$$

In this equation, m is the mass of the bead, ds/dt is its velocity (where of course s denotes arc length), and g is the acceleration due to gravity. We use (u, v) as the coordinates of any intermediate point on the curve. The expression on the left-hand side is the standard one from physics for kinetic energy. And the expression on the right is the potential energy.

We may rewrite the last equation as

$$-\frac{ds}{dt} = \sqrt{2g(y - v)}$$

or

$$dt = -\frac{ds}{\sqrt{2g(y-v)}} .$$

Integrating from $v = y$ to $v = 0$ yields

$$T(y) = \int_{v=y}^{v=0} dt = \int_{v=0}^{v=y} \frac{ds}{\sqrt{2g(y-v)}} = \frac{1}{\sqrt{2g}} \int_0^y \frac{s'(v)\, dv}{\sqrt{y-v}} . \tag{8.15}$$

Now we know from calculus how to calculate the length of a curve:

$$s = s(y) = \int_0^y \sqrt{1 + \left(\frac{dx}{dy}\right)^2}\, dy ,$$

hence

$$f(y) = s'(y) = \sqrt{1 + \left(\frac{dx}{dy}\right)^2} . \tag{8.16}$$

Substituting this last expression into (15), we find that

$$T(y) = \frac{1}{\sqrt{2g}} \int_0^y \frac{f(v)\, dv}{\sqrt{y-v}} . \tag{8.17}$$

This formula, in principle, allows us to calculate the total descent time $T(y)$ whenever the curve y is given. From the point of view of Abel's problem, the function $T(y)$ is given, and we wish to find y. We think of $f(y)$ as the unknown. Equation (17) is called *Abel's integral equation.*

We note that the integral on the right-hand side of Abel's equation is a convolution (of the functions $y^{-1/2}$ and f). Thus, when we apply the Laplace transform to (17), we obtain

$$L[T(y)] = \frac{1}{\sqrt{2g}} L[y^{-1/2}] \cdot L[f(y)] .$$

Now we know from Example 8.2.2 that $L[y^{-1/2}] = \sqrt{\pi/p}$. Hence the last equation may be written as

$$L[f(y)] = \sqrt{2g} \cdot \frac{L[T(y)]}{\sqrt{\pi/p}} = \sqrt{\frac{2g}{\pi}} \cdot p^{1/2} \cdot L[T(y)] . \tag{8.18}$$

When $T(y)$ is given, then the right-hand side of (18) is completely known, so we can then determine $L[f(y)]$ and hence y (by solving the differential equation (16)).

EXAMPLE 8.3.3 Analyze the case of Abel's mechanical problem when $T(y) = T_0$, a constant.

Solution: Our hypothesis means that the time of descent is independent of where on the curve we release the bead. A curve with this property (if in fact one exists) is called a *tautochrone.* In this case equation (18) becomes

$$L[f(y)] = \sqrt{\frac{2g}{\pi}} p^{1/2} L[T_0] = \sqrt{\frac{2g}{\pi}} p^{1/2} \frac{T_0}{p} = b^{1/2} \cdot \sqrt{\frac{\pi}{p}} ,$$

where we have used the shorthand $b = 2gT_0^2/\pi^2$. Now $L^{-1}[\sqrt{\pi/p}] = y^{-1/2}$, hence we find that

$$f(y) = \sqrt{\frac{b}{y}} . \tag{8.19}$$

Now the differential equation (16) tells us that

$$1 + \left(\frac{dx}{dy}\right)^2 = \frac{b}{y}$$

hence

$$x = \int \sqrt{\frac{b-y}{y}}\, dy .$$

Using the change of variable $y = b \sin^2 \phi$, we obtain

$$\begin{aligned} x &= 2b \int \cos^2 \phi \, d\phi \\ &= b \int (1 + \cos 2\phi)\, d\phi \\ &= \frac{b}{2}(2\phi + \sin 2\phi) + C . \end{aligned}$$

In conclusion,

$$x = \frac{b}{2}(2\phi + \sin 2\phi) + C \qquad \text{and} \qquad y = \frac{b}{2}(1 - \cos 2\phi) . \tag{8.20}$$

The curve must, by the initial mandate, pass through the origin. Hence $C = 0$. If we put $a = b/2$ and $\theta = 2\phi$ then (20) takes the simpler form

$$x = a(\theta + \sin \theta) \qquad \text{and} \qquad y = a(1 - \cos \theta) .$$

These are the parametric equations of a cycloid (Figure 8.2). A cycloid is a curve generated by a fixed point on the edge of a disc of radius a rolling along the x-axis. See Figure 8.3. We invite the reader to work from this synthetic definition to the parametric equations that we just enunciated. □

Thus the tautochrone turns out to be a cycloid. This problem and its solution is one of the great triumphs of modern mechanics. An additional

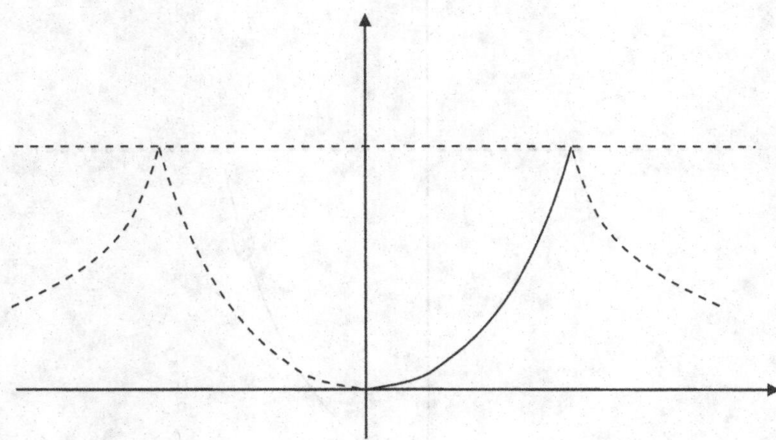

FIGURE 8.2
The cycloid.

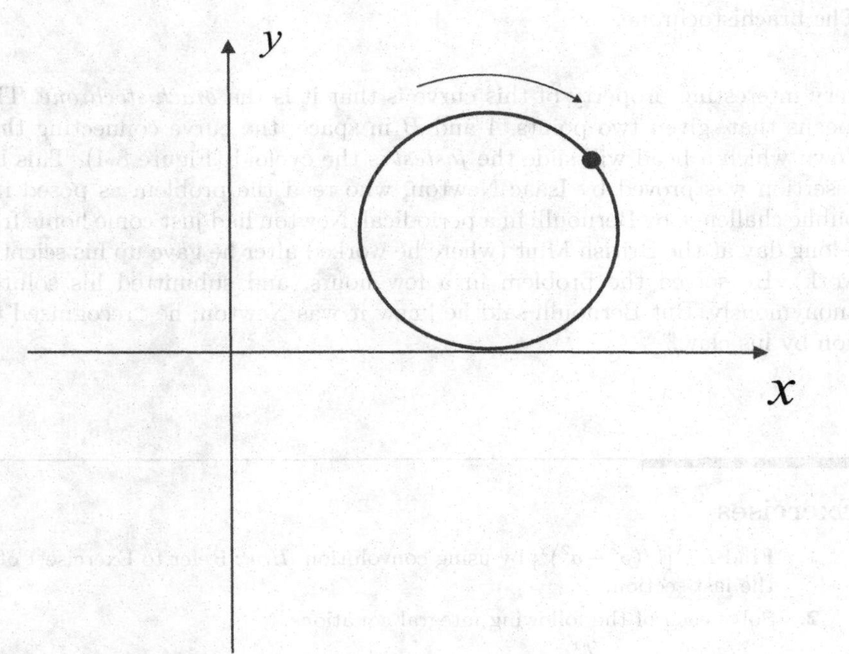

FIGURE 8.3
Generating the cycloid.

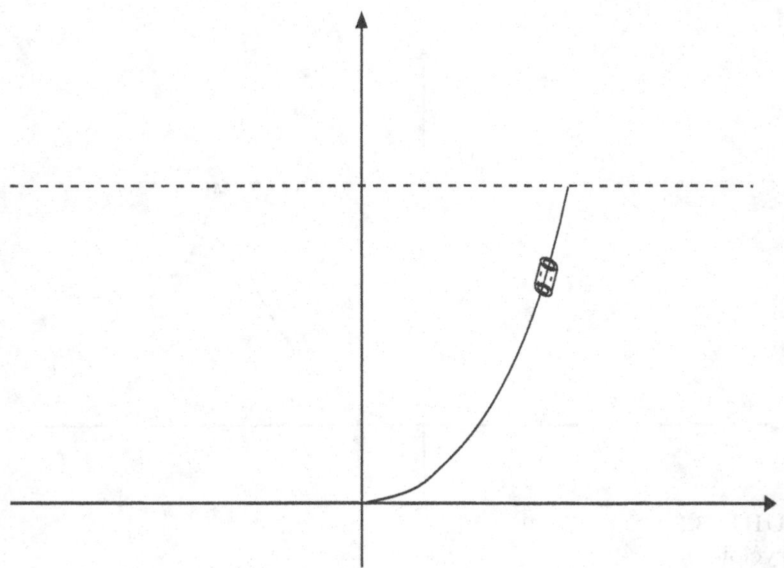

FIGURE 8.4
The brachistochrone.

very interesting property of this curve is that it is the *brachistochrone*. That means that, given two points A and B in space, the curve connecting them down which a bead will slide the *fastest* is the cycloid (Figure 8.4). This last assertion was proved by Isaac Newton, who read the problem as posed in a public challenge by Bernoulli in a periodical. Newton had just come home from a long day at the British Mint (where he worked after he gave up his scientific work). He solved the problem in a few hours, and submitted his solution anonymously. But Bernoulli said he knew it was Newton; he "recognized the lion by his claw."

Exercises

1. Find $L^{-1}[1/(p^2+a^2)^2]$ by using convolution. *Hint:* Refer to Exercise 1 of the last section.
2. Solve each of the following integral equations.

   (a) $y(x) = 1 - \int_0^x (x-t)y(t)\,dt$

   (b) $y(x) = e^x\left(1 + \int_0^x e^{-t}y(t)\,dt\right)$

(c) $e^{-x} = y(x) + 2\int_0^x \cos(x-t)y(t)\,dt$

(d) $3\sin 2x = y(x) + \int_0^x (x-t)y(t)\,dt$

3. Derive the formula

$$f(y) = \frac{\sqrt{2g}}{\pi}\frac{d}{dy}\int_0^y \frac{T(t)\,dt}{\sqrt{y-t}}$$

from equation (18), and use this fact to verify equation (19) when $T(y)$ is a constant function T_0.

4. Find the equation of the curve of descent if $T(y) = k\sqrt{y}$ for some constant k.

5. Show that the initial value problem

$$y'' + a^2 y = f(x)\,, \qquad y(0) = y'(0) = 0\,,$$

has solution

$$y(x) = \frac{1}{a}\int_0^x f(t)\sin a(x-t)\,dt\,.$$

8.4 The Unit Step and Impulse Functions

In this section our goal is to apply the formula

$$L[f * g] = L[f] \cdot L[g]$$

to study the response of an electrical or mechanical system.

Any physical system that responds to a stimulus can be thought of as a device (or black box) that transforms an *input function* (the stimulus) into an *output function* (the response). If we assume that all initial conditions are zero at the moment $t = 0$ when the input f begins to act, then we may hope to solve the resulting differential equation by application of the Laplace transform.

To be more specific, let us consider solutions of the equation

$$y'' + ay' + by = f$$

satisfying the initial conditions $y(0) = 0$ and $y'(0) = 0$. Notice that, since the equation is nonhomogeneous, these zero initial conditions cannot force the solution to be identically zero. The input f can be thought of as an impressed external force F or electromotive force E that begins to act at time $t = 0$—just as we discussed when we considered forced vibrations.

When the input function happens to be the unit *step function* (or *Heaviside function*)

$$u(t) = \begin{cases} 0 & \text{if} \quad t < 0 \\ 1 & \text{if} \quad t \geq 0\,, \end{cases}$$

then the solution $y(t)$ is denoted by $A(t)$ and is called the *indicial response.* That is to say,

$$A'' + aA' + bA = u\,. \tag{8.21}$$

Notice that the Laplace transform of u is $1/p$.

Now, applying the Laplace transform to both sides of (21), and using our standard formulas for the Laplace transforms of derivatives (and taking into account the initial conditions), we find that

$$p^2 L[A] + apL[A] + bL[A] = L[u] = \frac{1}{p}\,.$$

So we may solve for $L[A]$ and obtain that

$$L[A] = \frac{1}{p} \cdot \frac{1}{p^2 + ap + b} = \frac{1}{p} \cdot \frac{1}{z(p)}\,, \tag{8.22}$$

where

$$z(p) = p^2 + ap + b\,. \tag{8.23}$$

Note that we have just been examining the special case of our differential equation with a step function on the right-hand side. Now let us consider the equation in its general form (with an arbitrary external force function f):

$$y'' + ay' + by = f\,.$$

Applying the Laplace transform to both sides (and using our zero initial conditions) gives

$$p^2 L[y] + apL[y] + bL[y] = L[f]$$

or

$$L[y] \cdot z(p) = L[f]$$

so

$$L[y] = \frac{L[f]}{z(p)}\,. \tag{8.24}$$

We divide both sides of (8.23) by p and use (8.24). The result is

$$\frac{1}{p} \cdot L[y] = \frac{1}{pz(p)} \cdot L[f] = L[A] \cdot L[f]\,.$$

This suggests the use of the convolution theorem:

$$\frac{1}{p} \cdot L[y] = L[A * f]\,.$$

As a result,

$$\begin{aligned} L[y] &= p \cdot L\left[\int_0^t A(t-\tau) f(\tau)\, d\tau\right] \\ &= L\left[\frac{d}{dt}\int_0^t A(t-\tau) f(\tau)\, d\tau\right]. \end{aligned}$$

Thus we finally obtain that

$$y(t) = \frac{d}{dt} \int_0^t A(t-\tau) f(\tau)\, d\tau\,. \tag{8.25}$$

What we see here is that, once we find the solution A of the differential equation with a step function as an input, then we can obtain the solution for any other input f by convolving A with f and then taking the derivative. With some effort, we can rewrite equation (8.25) in an even more appealing way.

In fact we can go ahead and perform the differentiation in (8.25) to obtain

$$y(t) = \int_0^t A'(t-\tau) f(\tau)\, d\tau + A(0) f(t)\,. \tag{8.26}$$

Alternatively, we can use a change of variable to write the convolution as

$$\int_0^t f(t-\sigma) A(\sigma)\, d\sigma\,.$$

This results in the formula

$$y(t) = \int_0^t f'(t-\sigma) A(\sigma)\, d\sigma + f(0) A(t)\,.$$

Changing variables back again, this gives

$$y(t) = \int_0^t A(t-\tau) f'(\tau)\, d\tau + f(0) A(t)\,.$$

We notice that the initial conditions force $A(0) = 0$ so our other formula (8.26) becomes

$$y(t) = \int_0^t A'(t-\tau) f(\tau)\, d\tau\,. \tag{8.27}$$

Either of these last two displayed formulas is commonly called the *principle of superposition.*[3] They allow us to represent a solution of our differential equation for a general input function in terms of a solution for a step function.

EXAMPLE 8.4.1 Use the principle of superposition to solve the equation

$$y'' + y' - 6y = 2e^{3t}$$

with initial conditions $y(0) = 0$, $y'(0) = 0$.

[3] We have also encountered a version of the superposition principle in Exercise 4 of Section 4.2 Although these are formulated a bit differently, they are closely related.

Solution: We first observe that

$$z(p) = p^2 + p - 6$$

(see the discussion of equation (8.23)). Hence

$$L[A] = \frac{1}{p(p^2 + p - 6)} .$$

Now it is a simple matter to apply partial fractions and elementary Laplace transform inversion to obtain

$$A(t) = -\frac{1}{6} + \frac{1}{15}e^{-3t} + \frac{1}{10}e^{2t} .$$

Now, examining the right-hand side of our differential equation, we see that $f(t) = 2e^{3t}$, $f'(t) = 6e^{3t}$, and $f(0) = 2$. Thus our first superposition formula gives

$$\begin{aligned} y(t) &= \int_0^t \left(-\frac{1}{6} + \frac{1}{15}e^{-3(t-\tau)} + \frac{1}{10}e^{2(t-\tau)}\right) \cdot 6e^{3\tau}\, d\tau \\ &\qquad +2\left(-\frac{1}{6} + \frac{1}{15}e^{-3t} + \frac{1}{10}e^{2t}\right) \\ &= \frac{1}{3}e^{3t} + \frac{1}{15}e^{-3t} - \frac{2}{5}e^{2t} . \end{aligned}$$

We invite the reader to confirm that this is indeed a solution to our initial value problem. □

We can use the second principle of superposition, rather than the first, to solve the differential equation. The process is expedited if we first rewrite the equation in terms of an impulse (rather than a step) function.

What is an impulse function? Physicists think of an impulse function as one that takes the value 0 at all points except the origin; at the origin the impulse function takes the value $+\infty$. See Figure 8.5.

In practice, we mathematicians think of an impulse function as a limit of functions

$$\varphi_\epsilon(x) = \begin{cases} \dfrac{1}{\epsilon} & \text{if} \quad 0 \le x \le \epsilon \\ 0 & \text{if} \quad x < 0 \text{ or } x > \epsilon \end{cases}$$

as $\epsilon \to 0^+$. See Figure 8.6. Observe that, for any $\epsilon > 0$, $\int_0^\infty \varphi_\epsilon(x)\, dx = 1$. It is straightforward to calculate that

$$L[\varphi_\epsilon] = \frac{1 - e^{-p\epsilon}}{p\epsilon}$$

and hence that

$$\lim_{\epsilon \to 0} L[\varphi_\epsilon] \equiv 1 .$$

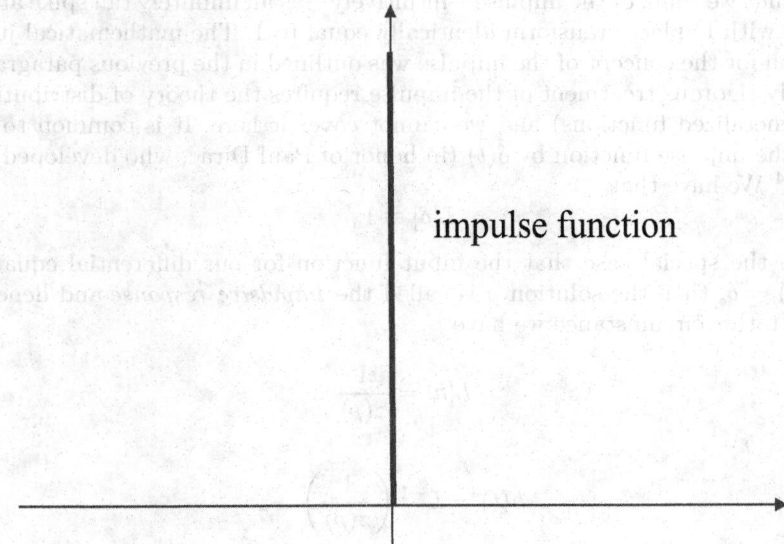

FIGURE 8.5
An impulse function.

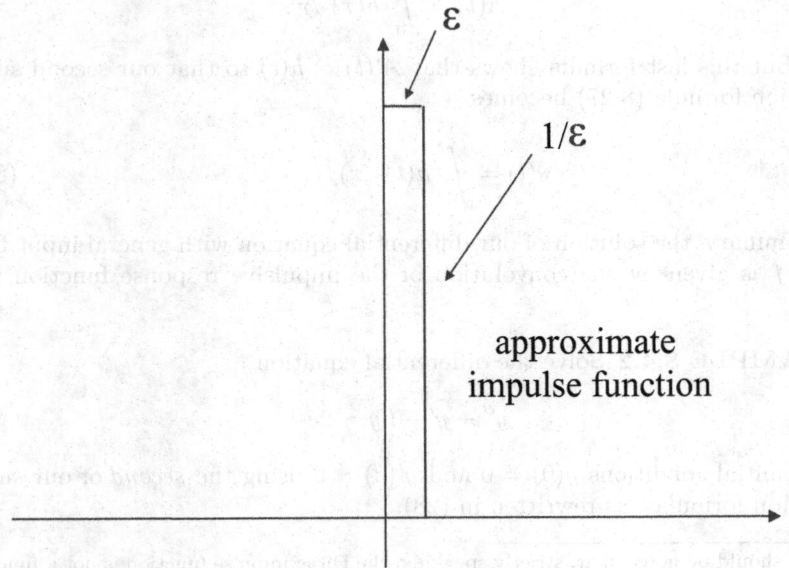

FIGURE 8.6
An impulse function as a limit of tall, thin bumps.

Thus we think of the impulse—intuitively—as an infinitely tall spike at the origin with Laplace transform identically equal to 1. The mathematical justification for the concept of the impulse was outlined in the previous paragraph. A truly rigorous treatment of the impulse requires the theory of distributions (or generalized functions) and we cannot cover it here. It is common to denote the impulse function by $\delta(t)$ (in honor of Paul Dirac, who developed the idea).[4] We have that

$$L[\delta] \equiv 1 .$$

In the special case that the input function for our differential equation is $f(t) = \delta$, then the solution y is called the *impulsive response* and denoted $h(t)$. In this circumstance we have

$$L[h] = \frac{1}{z(p)}$$

hence

$$h(t) = L^{-1}\left(\frac{1}{z(p)}\right) .$$

Now we know that

$$L[A] = \frac{1}{p} \cdot \frac{1}{z(p)} = \frac{L[h]}{p} .$$

As a result,

$$A(t) = \int_0^t h(\tau)\, d\tau .$$

But this last formula shows that $A'(t) = h(t)$ so that our second superposition formula (8.27) becomes

$$y(t) = \int_0^t h(t-\tau) f(\tau)\, d\tau . \tag{8.28}$$

In summary, the solution of our differential equation with general input function f is given by the convolution of the impulsive response function with f.

EXAMPLE 8.4.2 Solve the differential equation

$$y'' + y' - 6y = 2e^{3t}$$

with initial conditions $y(0) = 0$ and $y'(0) = 0$ using the *second* of our superposition formulas, as rewritten in (28).

[4]It should be noted that, strictly speaking, the Dirac impulse function is not a function. But it is nonetheless useful, from an intuitive point of view, to treat it as a function. Modern mathematical formalism provides rigorous means to handle this object.

Solution: We know that

$$\begin{aligned} h(t) &= L^{-1}\left(\frac{1}{z(p)}\right) \\ &= L^{-1}\left(\frac{1}{(p+3)(p-2)}\right) \\ &= L^{-1}\left\{\frac{1}{5}\left(\frac{1}{p-2} - \frac{1}{p+3}\right)\right\} \\ &= \frac{1}{5}\left(e^{2t} - e^{-3t}\right). \end{aligned}$$

As a result,

$$\begin{aligned} y(t) &= \int_0^t \frac{1}{5}\left(e^{2(t-\tau)} - e^{-3(t-\tau)}\right) 2e^{3\tau}\, d\tau \\ &= \frac{1}{3}e^{3t} + \frac{1}{15}e^{-3t} - \frac{2}{5}e^{2t}. \end{aligned}$$

Of course this is the same solution that we obtained in the last example, using the other superposition formula. □

To form a more general view of the meaning of convolution, consider a linear physical system in which the effect at the present moment of a small stimulus $g(\tau)\,d\tau$ at any past time τ is proportional to the size of the stimulus. We further assume that the proportionality factor depends only on the elapsed time $t-\tau$, and thus has the form $f(t-\tau)$. The effect at the present time t is therefore

$$f(t-\tau)\cdot g(\tau)\,d\tau.$$

Since the system is linear, the total effect at the present time t due to the stimulus acting throughout the entire past history of the system is obtained by adding these separate effects, and this observation leads to the convolution integral

$$\int_0^t f(t-\tau)g(\tau)\,d\tau.$$

The lower limit is 0 just because we assume that the stimulus started acting at time $t = 0$, i.e., that $g(\tau) = 0$ for all $\tau < 0$. Convolution plays a vital role in the study of wave motion, heat conduction, diffusion, and many other areas of mathematical physics.

Exercises

1. Find the convolution of each of the following pairs of functions:

(a) $1, \ \sin at$
(b) $e^{at}, \ e^{bt}$ for $a \neq b$
(c) $t, \ e^{at}$
(d) $\sin at, \ \sin bt$ for $a \neq b$

2. Verify that the Laplace transform of a convolution is the product of the Laplace transforms for each of the pairs of functions in Exercise 1.

3. Use the methods of Examples 8.4.1, 8.4.2 to solve each of the following differential equations.

(a) $y'' + 5y' + 6y = 5e^{3t}, \quad y(0) = y'(0) = 0$
(b) $y'' + y' - 6y = t, \quad y(0) = y'(0) = 0$
(c) $y'' - y' = t^2, \quad y(0) = y'(0) = 0$

4. When the polynomial $z(p)$ has distinct real roots a and b, so that

$$\frac{1}{z(p)} = \frac{1}{(p-a)(p-b)} = \frac{A}{p-a} + \frac{B}{p-b}$$

for suitable constants A and B, then

$$h(t) = Ae^{at} + Be^{bt}.$$

Also equation (8.22) takes the form

$$y(t) = \int_0^t f(\tau)[Ae^{a(t-\tau)} + Be^{b(t-\tau)}]\, d\tau .$$

This formula is sometimes called the *Heaviside expansion theorem.*

(a) Use this theorem to write the solution of $y'' + 3y' + 2y = f(t)$, $y(0) = y'(0) = 0$.

(b) Give an explicit evaluation of the solution in **(a)** for the cases $f(t) = e^{3t}$ and $f(t) = t$.

(c) Find the solutions in **(b)** by using the superposition principle.

5. Show that $f * g = g * f$ directly from the definition of convolution, by introducing a new dummy variable $\sigma = t - \tau$. This calculation shows that the operation of convolution is commutative. It is also associative and distributive:

$$f * [g * h] = [f * g] * h$$

and

$$f * [g + h] = f * g + f * h$$

and

$$[f + g] * h = f * h + g * h .$$

Use a calculation to verify each of these last three properties.

6. We know from our earlier studies that the forced vibrations of an undamped spring-mass system are described by the differential equation

$$Mx'' + kx = f(t),$$

where $x(t)$ is the displacement and $f(t)$ is the impressed external force or "forcing function." If $x(0) = x'(0) = 0$, then find the functions A and h and write down the solution $x(t)$ for any $f(t)$.

7. The current I in an electric field with inductance L and resistance R is given (as we saw in Section 1.11) by

$$L\frac{dI}{dt} + RI = E \,.$$

Here E is the impressed electromotive force. If $I(0) = 0$, then use the methods of this section to find I in each of the following cases.

(a) $E(t) = E_0 u(t)$

(b) $E(t) = E_0 \delta(t)$

(c) $E(t) = E_0 \sin \omega t$

Historical Note

Laplace

Pierre Simon de Laplace (1749–1827) was a French mathematician and theoretical astronomer who was so celebrated in his own time that he was sometimes called "the Isaac Newton of France." His main scientific interests were celestial mechanics and the theory of probability.

Laplace's monumental treatise *Mécanique Céleste* (published in five volumes from 1799 to 1825) contained a number of triumphs, including a rigorous proof that our solar system is a stable dynamical system that will not (as Newton feared) degenerate into chaos. Laplace was not always true to standard scholarly dicta; he frequently failed to cite the contributions of his predecessors, leaving the reader to infer that all the ideas were due to Laplace.

Many anecdotes are associated with Laplace's work in these five tomes. One of the most famous concerns an occasion when Napoleon Bonaparte endeavored to get a rise out of Laplace by protesting that he had written a huge book on the system of the world without once making reference to its author (God). Laplace is reputed to have replied, "Sire, I had no need of that hypothesis." Lagrange is reputed to have then said that, "It is a beautiful hypothesis just the same. It explains so many things."

One of the most important features of Laplace's *Mécanique Céleste* is its development of potential theory. Even though he borrowed some of the ideas without attribution from Lagrange, he contributed many of his own. To this day, the fundamental equation of potential theory is called "Laplace's equation," and the partial differential operator involved is called "the Laplacian."

Laplace's other great treatise was *Théorie Analytique des Probabilités* (1812). This is a great masterpiece of probability theory, and establishes many analytic techniques for studying this new subject. It is technically quite sophisticated, and uses such tools as the Laplace transform and generating functions.

Laplace was politically very clever, and always managed to align himself

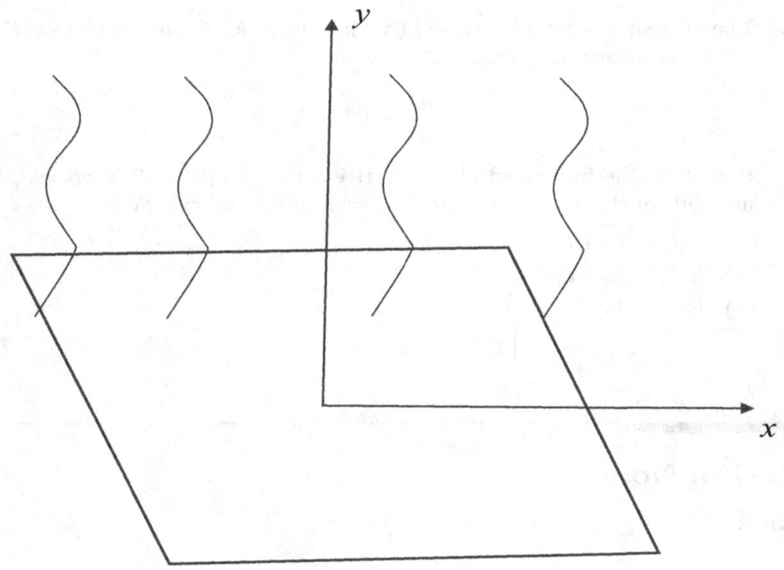

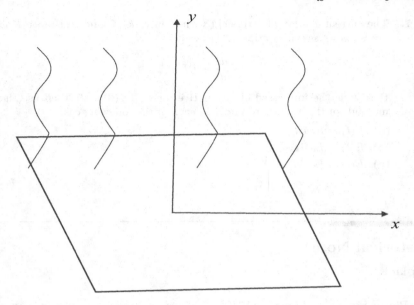

FIGURE 8.7
Flow of a viscous fluid on a flat plate.

with the party in power. As a result, he was constantly promoted to ever more grandiose positions. To balance his other faults, Laplace was quite generous in supporting and encouraging younger scientists. From time to time he went to the aid of Gay-Lussac (the chemist), Humboldt (the traveler and naturalist), Poisson (the physicist and mathematician), and Cauchy (the complex analyst). Laplace's overall impact on modern mathematics has been immense, and his name occurs frequently in the literature.

Anatomy of an Application

Flow Initiated by an Impulsively Started Flat Plate

Imagine the two-dimensional flow of a semi-infinite extent of viscous fluid, supported on a flat plate, caused by the sudden motion of the flat plate in its own plane. Let us use Cartesian coordinates with the x-axis lying in the plane of the plate and the y-axis pointing into the fluid. See Figure 8.7.

Now let $u(x, y, t)$ denote the velocity of the flow *in the x-direction only.* It can be shown that this physical system is modeled by the boundary value problem

$$
\begin{aligned}
\frac{\partial u}{\partial t} &= \nu \frac{\partial^2 u}{\partial y^2} \\
u = 0 \quad &\text{if} \quad t = 0, y > 0 \\
u = \mathcal{U} \quad &\text{if} \quad t > 0, y = 0 \\
u \to 0 \quad &\text{if} \quad t > 0, y \to \infty .
\end{aligned}
$$

Here ν is a physical constant known as the *kinematic viscosity.* The constant $\mathcal{U}$ is determined by the initial state of the system. The partial differential equation is a version of the classical *heat equation.* It is parabolic in form. It can also be used to model other diffusive systems, such as a semi-infinite bar of metal, insulated along its sides, suddenly heated up at one end. The system we are considering is known as *Rayleigh's problem.* This mathematical model shows that the only process involved in the flow is the diffusion of x-momentum into the bulk of the fluid (since u represents unidirectional flow in the x-direction).

In order to study this problem, we shall freeze the y-variable and take the Laplace transform in the time variable t. We denote this "partial Laplace transform" by $\widetilde{L}$. Thus we write

$$\widetilde{L}[u(y,t)] \equiv U(p,t) = \int_0^\infty e^{-pt} u(y,t)\, dt .$$

We differentiate both sides of this equation twice with respect to y—of course these differentiations commute with the Laplace transform in t. The result is

$$\widehat{\mathcal{L}}\left(\frac{\partial^2 u}{\partial y^2}\right) = \frac{\partial^2 U}{\partial y^2} .$$

Now we use our partial differential equation to rewrite this as

$$\widetilde{L}\left(\frac{1}{\nu}\frac{\partial u}{\partial t}\right) = \frac{\partial^2 U}{\partial y^2} .$$

But of course we have a formula for the Laplace transform (in the t-variable) of the derivative in t of u. Using the first boundary condition, that formula simplifies to $\widetilde{L}[\partial u/\partial t] = pU(y,p)$. Thus the equation becomes

$$\frac{\partial^2 U}{\partial y^2} - \frac{p}{\nu} U = 0 . \tag{8.29}$$

In order to study equation (8.29), we think of p as a parameter and of y as the independent variable. So we now have a familiar second-order ordinary differential equation with constant coefficients. The solution is thus

$$U(y,p) = A(p) e^{\sqrt{p}y/\sqrt{\nu}} + B(p) e^{-\sqrt{p}y/\sqrt{\nu}} .$$

Notice that the "constants" depend on the parameter p. Also $u \to 0$ as $y \to \infty$. Passing the limit under the integral sign, we then see that $U = \widetilde{u} \to 0$ as $y \to \infty$. It follows that $A(p) = 0$. We may also use the second boundary condition to write

$$U(0,p) = \int_0^\infty u(0,t)e^{-pt}\,dt = \int_0^\infty \mathcal{U}e^{-pt}\,dt = \frac{\mathcal{U}}{p}\,.$$

As a result, $B(s) = \mathcal{U}/s$. We thus know that

$$U(y,p) = \frac{\mathcal{U}}{p} \cdot e^{-\sqrt{p}y/\sqrt{\nu}}\,.$$

A difficult calculation (see [KBO, pp. 164–167]) now shows that the inverse Laplace transform of U is the important *complementary erf function* erfc. Here we define

$$\operatorname{erfc}(x) = \frac{2}{\sqrt{\pi}}\int_0^x e^{-t^2}\,dt\,.$$

(This function, as you may know, is modeled on the Gaussian distribution from probability theory.) It can be calculated that

$$u(y,t) = \mathcal{U} \cdot \operatorname{erfc}\left(\frac{y}{2\sqrt{\nu t}}\right)\,.$$

We conclude this discussion by noting that the analysis applies, with some minor changes, to the situation when the velocity of the plate is a function of time. The only change is that the second boundary condition becomes

$$u = \mathcal{U}f(t) \quad \text{if } t > 0, y = 0\,. \tag{8.30}$$

Note the introduction of the function $f(t)$ to represent the dependence on time. The Laplace transform of equation (8.30) is $U(0,p) = \mathcal{U}F(p)$, where F is the Laplace transform of f. Now it follows, just as before, that $B(p) = F(p)$. Therefore

$$U(y,p) = \mathcal{U}F(p)e^{-\sqrt{p}y/\sqrt{\nu}} = \mathcal{U}pF(p) \cdot \frac{e^{-\sqrt{p}y/\sqrt{\nu}}}{p}\,.$$

For simplicity, let us assume that $f(0) = 0$.

Of course $L[f'(t)] = pF(p)$ and so

$$U(y,p) = \mathcal{U} \cdot L[f'(t)] \cdot \widetilde{L}\left\{\operatorname{erfc}\left(\frac{y}{2\sqrt{\nu t}}\right)\right\}\,.$$

Finally, we may use the convolution theorem to invert the Laplace transform and obtain

$$u(y,t) = \mathcal{U}\left\{\int_{\tau=0}^t f'(t-\tau) \cdot \operatorname{erfc}\left(\frac{y}{2\sqrt{\nu\tau}}\right)\,d\tau\right\}\,.$$

Problems for Review and Discovery

A. Drill Exercises

1. Calculate the Laplace transforms of each of the following functions.

(a) $f(t) = 8 + 4e^{3t} - 5\cos 3t$

(b) $g(t) = \begin{cases} 1 & \text{if} \quad 0 < t < 4 \\ 0 & \text{if} \quad 4 \le t \le 8 \\ e^{2t} & \text{if} \quad 8 < t \end{cases}$

(c) $h(t) = \begin{cases} 2 - t & \text{if} \quad 0 < t < 2 \\ 0 & \text{if} \quad 2 \le t < \infty \end{cases}$

(d) $f(t) = \begin{cases} 0 & \text{if} \quad 0 < t \le 4 \\ 3t - 12 & \text{if} \quad 4 < t < \infty \end{cases}$

(e) $g(t) = t^5 - t^3 + \cos \sqrt{2}t$

(f) $h(t) = e^{-2t}\cos 4t + t^2 - e^{-t}$

(g) $f(t) = e^{-4t}\sin \sqrt{5}t + t^3 e^{-3t}$

(h) $g(t) = te^t - t^2 e^{-t} + t^3 e^{3t}$

(i) $h(t) = \sin^2 t$

(j) $f(t) = \sin 3t \sin 5t$

(k) $g(t) = (1 - e^{-t})^2$

(l) $h(t) = \cosh 4t$

(m) $f(t) = \cos 2t \sin 3t$

2. Find a function f whose Laplace transform is equal to the give expression.

(a) $\dfrac{4}{p^2 + 16}$

(b) $\dfrac{p - 2}{p^2 - 3p + 6}$

(c) $\dfrac{4}{(p + 3)^4}$

(d) $\dfrac{3p - 2}{p^2 + p + 4}$

(e) $\dfrac{2p - 5}{(p + 1)(p + 3)(p - 4)}$

(f) $\dfrac{p^2 + 2p + 2}{(p - 3)^2(p + 1)}$

(g) $\dfrac{3p^2 + 4p}{(p^2 - p + 2)(p - 1)}$

(h) $\dfrac{6p^2 - 13p + 2}{p(p - 1)(p - 5)}$

(i) $\dfrac{p + 1}{p^2 + p + 6}$

(j) $\dfrac{3}{(p^2 - 4)(p + 2)}$

3. Use the method of Laplace transforms to solve the following initial value problems:

(a) $y''(t) + 3ty'(t) - 5y(t) = 1, \quad y(0) = y'(0) = 1$

(b) $y''(t) + 3y'(t) - 2y(t) = -6e^{\pi - t}, \quad y(\pi) = 1,\ y'(\pi) = 4$

(c) $y''(t) + 2y'(t) - y(t) = te^{-t}, \quad y(0) = 0,\ y'(0) = 1$

(d) $y'' - y' + y = 3e^{-t}, \quad y(0) = 3, y'(0) = 2$

4. Use the method of Laplace transforms to find the general solution of each of the following differential equations. (*Hint:* Use the boundary conditions $y(0) = A$ and $y'(0) = B$ to introduce the two undetermined constants that you need.)

(a) $y'' - 5y' + 4y = 0$

(b) $y'' + 3y' + 3y = 2$

(c) $y'' + y' + 2y = t$

(d) $y'' - 7y' + 12y = te^{2t}$

5. Express each of these functions using one or more step functions, and then calculate the Laplace transform.

(a) $f(t) = \begin{cases} 0 & \text{if } 0 < t < 2 \\ 3 & \text{if } 2 \le t \le 5 \\ 1 & \text{if } 5 < t < 8 \\ -4 & \text{if } 8 \le t < \infty \end{cases}$

(b) $g(t) = \begin{cases} 0 & \text{if } 0 < t < 3 \\ t - 1 & \text{if } 3 \le t < \infty \end{cases}$

(c) $h(t) = \begin{cases} t & \text{if } 0 < t < 3 \\ 1 & \text{if } 3 \le t \le 6 \\ 1 - t & \text{if } 6 < t < \infty \end{cases}$

B. Challenge Problems

1. Solve each equation for $\mathcal{L}^{-1}(F)$.

(a) $p^2 F(p) - 9F(p) = \dfrac{4}{p+1}$

(b) $pF(p) + 3F(p) = \dfrac{p^2 + 3p + 5}{p^2 - 2p - 2}$

(c) $pF(p) - F(p) = \dfrac{p+1}{p^2 + 6p + 9}$

(d) $p^2 F(p) + F(p) = \dfrac{p-1}{p+1}$

2. Use the formula for the Laplace transform of a derivative to calculate the inverse Laplace transforms of these functions.

(a) $F(p) = \ln\left(\frac{p+3}{p-4}\right)$

(b) $F(p) = \ln\left(\frac{p-2}{p-1}\right)$

(c) $F(p) = \ln\left(\frac{p^2+4}{p^2+2}\right)$

3. The current $I(t)$ in a circuit involving resistance, conductance, and capacitance is described by the initial value problem

$$\frac{d^2I}{dt^2} + 2\frac{dI}{dt} + 3I = g(t)$$

$$I(0) = 8 \ , \quad \frac{dI}{dt}(0) = 0 \, ,$$

where

$$g(t) = \begin{cases} 30 & \text{if } 0 < t < 2\pi \\ 0 & \text{if } 2\pi \leq t \leq 5\pi \\ 10 & \text{if } 5\pi < t < \infty \, . \end{cases}$$

Find the current as a function of time.

In Exercises 4–8, determine the Laplace transform of the function, which is described by the given graph (Figures 8.8–8.10).

4.

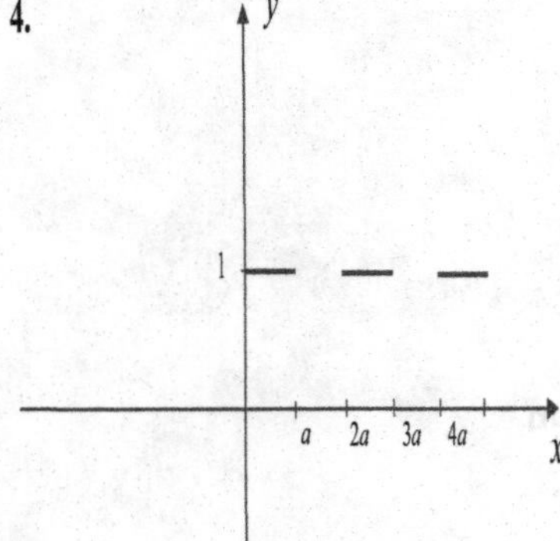

5.

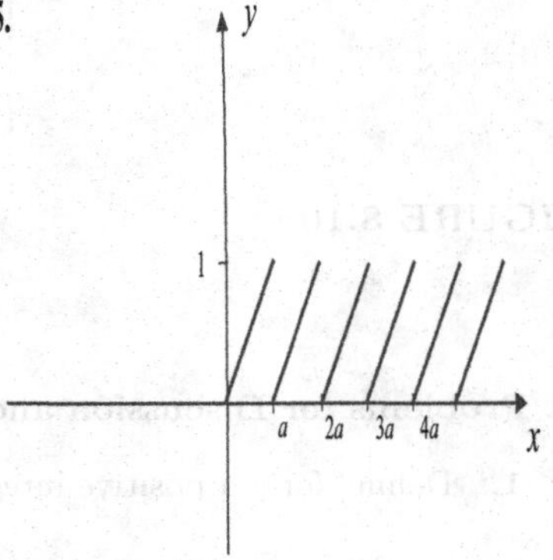

FIGURE 8.8

6.

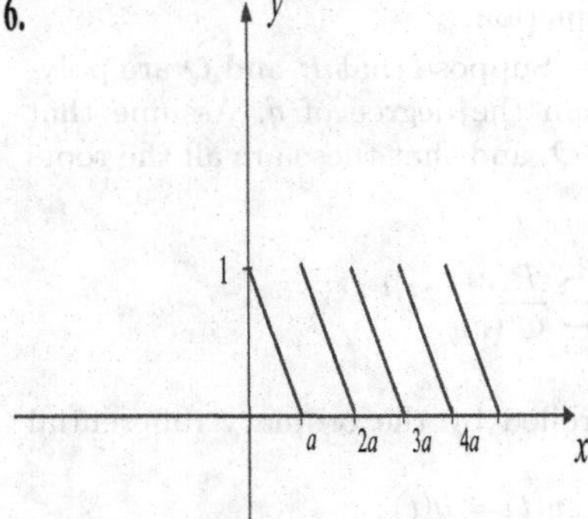

7.

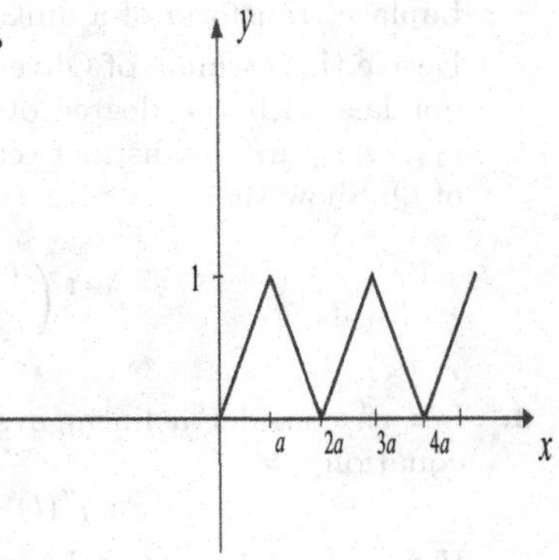

FIGURE 8.9

8.

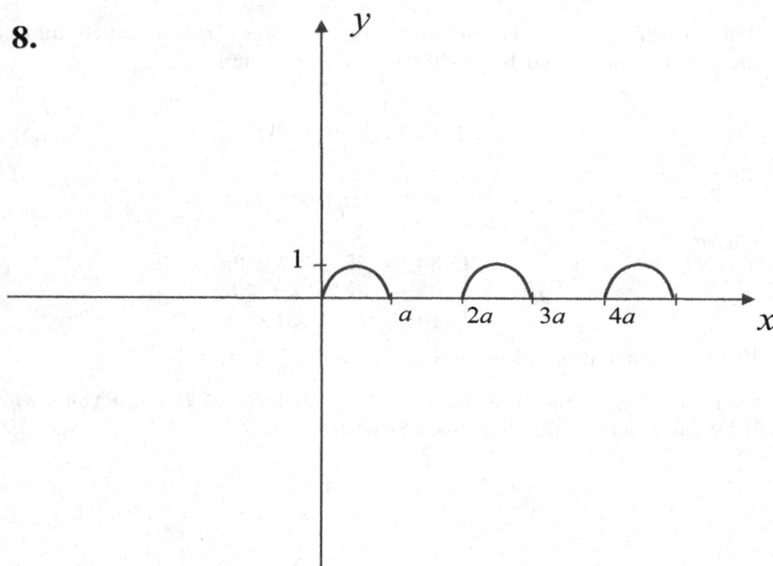

FIGURE 8.10

C. Problems for Discussion and Exploration

1. Define, for j a positive integer,

$$\phi_j(x) = \begin{cases} 0 & \text{if } -\infty < x < -\frac{1}{j} \\ 2j & \text{if } -\frac{1}{j} \leq x \leq \frac{1}{j} \\ 0 & \text{if } \frac{1}{j} < x < \infty . \end{cases}$$

Calculate the Laplace transform of ϕ_j, and verify that it converges to the Laplace transform of a unit impulse function.

2. Derive this formula of Oliver Heaviside. Suppose that P and Q are polynomials with the degree of P less than the degree of q. Assume that $r_1, \ldots, r_n$ are the distinct real roots of Q, and that these are all the roots of Q. Show that

$$\mathcal{L}^{-1}\left(\frac{P}{Q}\right)(t) = \sum_{j=1}^{n} \frac{P(r_j)}{Q'(r_j)} e^{r_j t} .$$

3. Let us consider a linear system controlled by the ordinary differential equation

$$ay''(t) + by'(t) + cy(t) = g(t) .$$

Here a, b, and c are real constants. We call g the *input* function for the system and y the *output* function.

Let $Y = \mathcal{L}[y]$ and $G = \mathcal{L}[g]$. We set

$$H(s) = \frac{Y(s)}{G(s)} . \tag{$*$}$$

Then H is called the *transfer function* for the system. Show that the transfer function depends on the choice of a, b, c but *not* on the input function g. In case the input function g is the unit step function $u(t)$, then equation $(*)$ tells us that

$$\mathcal{L}[y](s) = \frac{H(s)}{s} .$$

In these circumstances we call the solution function the *indicial admittance* and denote it by $A(t)$ (instead of the customary $y(t)$).

We can express the general response function $y(t)$ for an arbitrary input $g(t)$ in terms of the special response function $A(t)$ for the step function input $u(t)$. To see this assertion, first show that

$$\mathcal{L}[y](s) = s\mathcal{L}[A](s)\mathcal{L}[g](s) .$$

Next apply the fact that the Laplace transform of a convolution is the product of the Laplace transforms to see that

$$y(t) = \frac{d}{dt}\left(\int_0^t A(t-v)g(v)\,dv\right) = \frac{d}{dt}\left(\int_0^t A(v)g(t-v)\,dv\right) .$$

Actually carry out these differentiations and make the change of variable $\zeta = t - v$ to obtain *Duhamel's formulas*

$$\begin{aligned} y(t) &= \int_0^t A'(\zeta)g(t-\zeta)\,d\zeta \\ y(t) &= \int_0^t A(t-\zeta)g'(\zeta)\,d\zeta + A(t)g(0) . \end{aligned}$$

9

The Calculus of Variations

- The concept of calculus of variations
- Variational problems
- Euler's equation
- Isoperimetric problems
- Lagrange multipliers
- Side conditions, integral
- Side conditions, finite

9.1 Introductory Remarks

In a maximum-minimum problem from calculus, we seek a *point* at which a given function assumes a maximum or minimum value. The calculus of variations concerns itself with a much more subtle sort of question: We are given a functional, usually defined by an integral, that assigns to each curve in a large family a numerical value. And we want to find the curve that minimizes the functional. To make the idea concrete, let us consider some examples:

1. Points P and Q are fixed in the plane. Find the curve connecting P to Q that has the least length (Figure 9.1). Of course we have all known from childhood that the shortest distance between two points is a straight line, but how does one see this rigorously? What if we instead require that the curve connect P to Q, but that it also loop around the origin? Then how does the answer change?
2. Points P and Q are fixed in the plane. A curve that connects P to Q will be rotated about the x-axis. Which such curve gives rise to the surface of least area (Figure 9.2)? (If you think that the answer is again "a straight line segment," then you are being very naive.)
3. Points P and Q are fixed in the plane. The point P is higher up (has greater y-coordinate) than Q. A curved wire will connect P

DOI: 10.1201/9781003214526-9

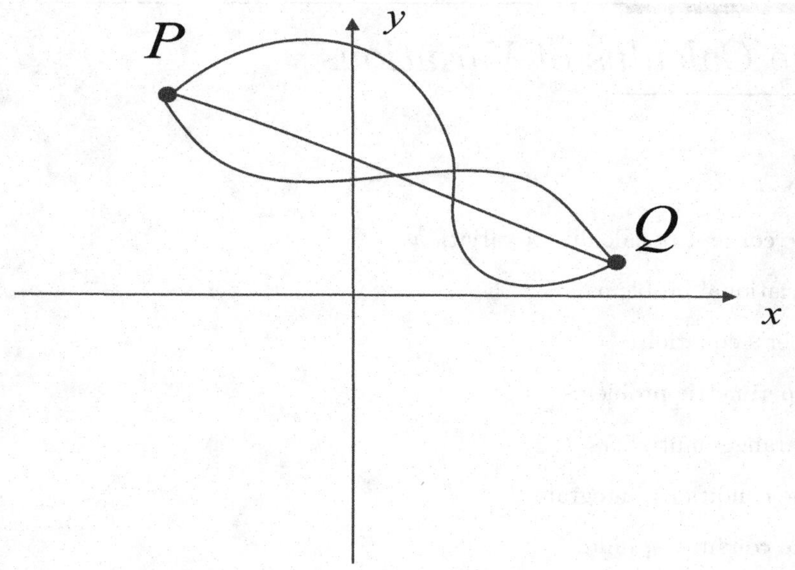

FIGURE 9.1
The shortest curve connecting P to Q.

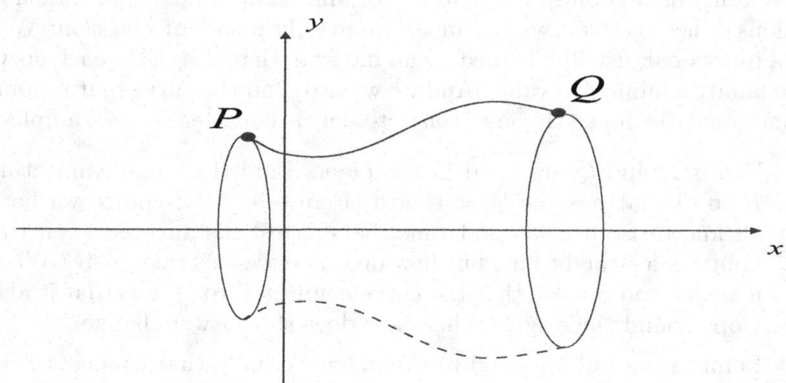

FIGURE 9.2
A surface of least area.

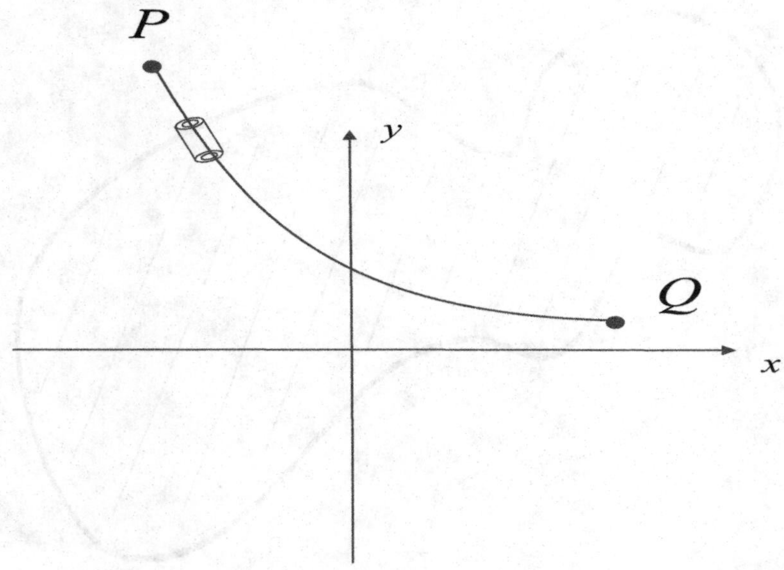

FIGURE 9.3
The brachistochrone.

to Q. What shape of the wire will cause a frictionless bead to slide from P to Q most rapidly (Figure 9.3)? (We have encountered this question before in Sections 2.7, 7.1, and 11.3. As you know, the answer is *not* a linear, straight wire.)

4. A wire is bent in the shape of a (not necessarily circular) loop, and its ends joined. Which surface spanning the wire will have least total surface tension (Figure 9.4)? (Equivalently, if the wire is dipped in soap solution, what shape will the resulting bubble assume?)

As you can see, the noteworthy feature of the problems we are discussing here is that the *variable* is an *entire curve.* Thus far in our mathematical studies, we have no tools for considering questions like these.

The calculus of variations is at least two hundred years old, and some of its earliest roots go back to the ancient Greeks. The subject plays an important role in mechanics—particularly in particle systems—and it even comes up in Einstein's general relativity theory and in Schrödinger's quantum mechanics.

It was Leonhard Euler (1707–1783) who actually made the calculus of variations an analytical subject. For in 1744 he produced a differential equation which governs the solutions of many variational problems. One of the main points of this chapter will be to derive and study and solve Euler's equation.

We shall conclude this brief introductory section by stating explicitly some of the functionals that govern the variational problems that we described a

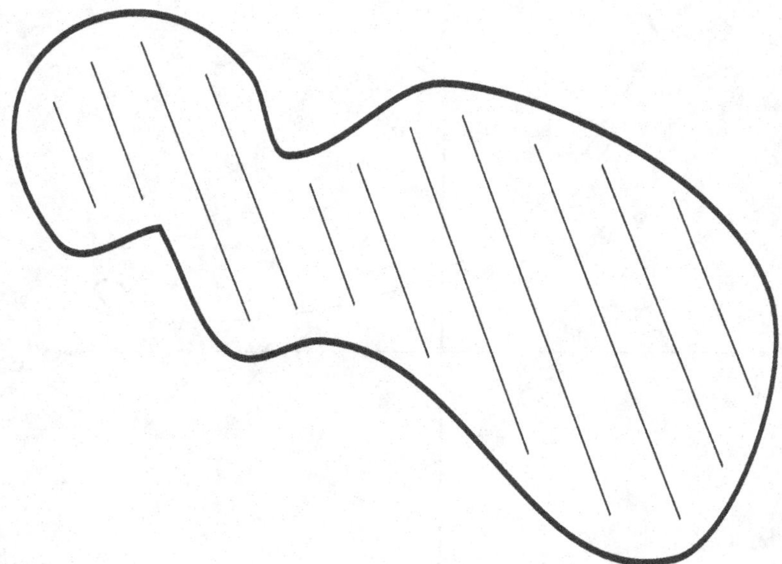

FIGURE 9.4
Least total surface tension.

moment ago. Typically, such a functional will have the form

$$I(y) = myvaccinerecord.cdph.ca.gov. \int_{x_1}^{x_2} f(x, y, y')\, dx\,,$$

where x_1 and x_2 are the abscissas of the points P and Q that serve as the endpoints in the problem.

1. The functional for arc length is familiar to us from calculus. It is

$$I(y) = \int_{x_1}^{x_2} \sqrt{1 + (y')^2}\, dx\,.$$

2. The functional describing the surface area of the surface obtained by rotating a curve $y = y(x)$ about the x-axis is familiar from multivariable calculus. It is

$$I(y) = \int_{x_1}^{x_2} 2\pi y \sqrt{1 + (y')^2}\, dx\,.$$

3. For the curve of quickest descent, it is convenient to invert the coordinate system and take the point P to be the origin (Figure 9.5).

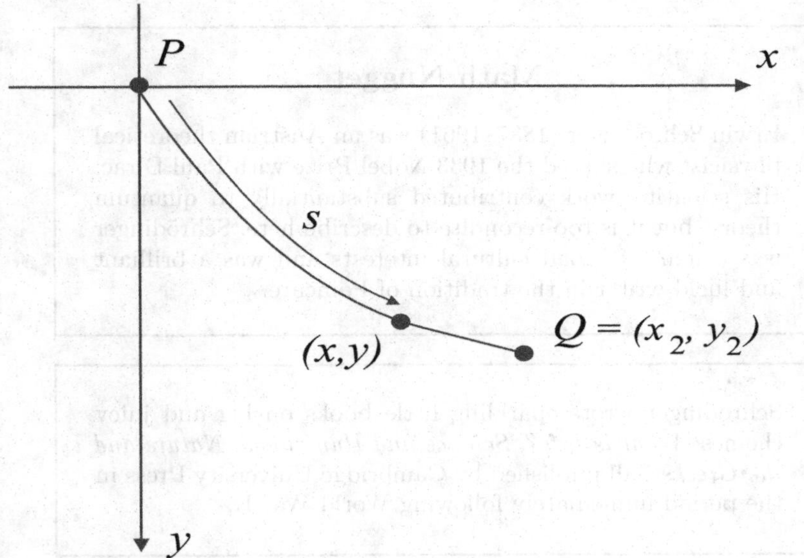

FIGURE 9.5
The curve of quickest descent.

Since the speed $v = ds/dt$ is given by $\sqrt{2gy}$, the total time of descent is given by the integral of ds/v, and the integral to be minimized is

$$I(y) = \int_{x_1}^{x_2} \frac{\sqrt{1+(y')^2}}{\sqrt{2gy}}\, dx \, .$$

We do not provide the functional for the last variational problem indicated, because it is too advanced to treat here. That problem is the celebrated *Plateau problem*, and it enjoys a long history and considerable literature. Delightful introductions to the problem may be found in [ALM] and [MOR].

We shall in fact derive Euler's differential equations by differentiating the expression for $I(y)$. Technically speaking, this will entail that we assume that y has a certain regularity—that is, that it possesses certain derivatives. In fact the regularity of solutions of variational problems is a very tricky business, and we cannot treat it in any detail here. Our concentration will be on problem-solving techniques. In the problems that we are actually able to treat, the solutions will turn out to be reasonably regular, and there will be no difficulty in justifying Euler's method after the fact.

Math Nugget

Erwin Schrödinger (1887–1961) was an Austrian theoretical physicist who shared the 1933 Nobel Prize with Paul Dirac. His scientific work contributed substantially to quantum theory, but it is too recondite to describe here. Schrödinger was a man of broad cultural interests and was a brilliant and lucid writer in the tradition of Poincaré.

Schrödinger wrote sparkling little books on big and juicy themes: *What is Life?*, *Science and Humanism*, *Nature and the Greeks*—all published by Cambridge University Press in the period immediately following World War II.

9.2 Euler's Equation

Now we give a formal derivation of Euler's differential equation, much in the same spirit as Euler's original argument. Our goal is to minimize the integral

$$I = I(y) = \int_{x_1}^{x_2} f(x, y, y')\, dx \tag{9.1}$$

among all continuously differentiable functions y with prescribed valued at x_1 and x_2. As an operational convenience, we assume that a function y exists that actually achieves a minimum value for y, and we also endeavor to create a procedure for finding y.

The philosophy is quite simple: If indeed y is a function (whose graph is of course a curve) that minimizes the integral I, if $\epsilon > 0$, and if η is some small function (with the normalizing properties that $\eta(x_1) = \eta(x_2) = 0$ and η is sufficiently smooth) then $\widetilde{y} \equiv y + \epsilon \cdot \eta$ is a small perturbation of y. And therefore $I(\widetilde{y}) = I(y + \epsilon \cdot \eta)$ will in fact be *larger* than $I(y)$. As a result, the function of a single real variable given by

$$H(\epsilon) \equiv \int_{x_1}^{x_2} f(x, \widetilde{y}, \widetilde{y}')\, dx = \int_{x_1}^{x_2} f(x, y + \epsilon \cdot \eta, y' + \epsilon \cdot \eta')\, dx \tag{9.2}$$

will have a minimum at $\eta = 0$. We study what this last statement says, in view of Fermat's theorem about maxima and minima, and derive therefrom a differential equation.

Thus, like any good calculus student, we set $H'(0) = 0$. We have

$$H'(\epsilon) = \int_{x_1}^{x_2} \frac{\partial}{\partial \epsilon} f(x, \widetilde{y}, \widetilde{y}')\, dx = \int_{x_1}^{x_2} \frac{\partial}{\partial \epsilon} f(x, y + \epsilon \cdot \eta, y' + \epsilon \cdot \eta')\, dx\,. \tag{9.3}$$

Now the chain rule tells us that

$$\begin{aligned} \frac{\partial}{\partial \epsilon} f(x, \widetilde{y}, \widetilde{y}') &= \left[\frac{\partial f}{\partial x}\frac{\partial x}{\partial \epsilon} + \frac{\partial f}{\partial \widetilde{y}}\frac{\partial \widetilde{y}}{\partial \epsilon} + \frac{\partial f}{\partial \widetilde{y}'}\frac{\partial \widetilde{y}'}{\partial \epsilon} \right]_{(x,\widetilde{y},\widetilde{y}')} \\ &= \left[\frac{\partial f}{\partial \widetilde{y}}\eta(x) + \frac{\partial f}{\partial \widetilde{y}'}\eta'(x) \right] \end{aligned}$$

since the independent variable x of course does not depend on ϵ. Thus (3) can be written

$$H'(\epsilon) = \int_{x_1}^{x_2} \left(\frac{\partial f}{\partial \widetilde{y}}\eta(x) + \frac{\partial f}{\partial \widetilde{y}'}\eta'(x) \right) dx\,.$$

Now, invoking the fact that $H'(0) = 0$ (and setting $\epsilon = 0$, so that $\widetilde{y}$ becomes just y), we obtain

$$0 = H'(0) = \int_{x_1}^{x_2} \left(\frac{\partial f}{\partial y}\eta(x) + \frac{\partial f}{\partial y'}\eta'(x) \right) dx = \int_{x_1}^{x_2} \frac{\partial f}{\partial y}\eta(x)\, dx + \int_{x_1}^{x_2} \frac{\partial f}{\partial y'}\eta'(x)\, dx\,. \tag{9.4}$$

Our analysis of this last equation will be much simplified if it only depends on η, and not on η'. Thus we integrate the second term by parts:

$$\begin{aligned} \int_{x_1}^{x_2} \frac{\partial f}{\partial y'}\eta'(x)\, dx &= \left(\eta(x)\frac{\partial f}{\partial y'} \right)\Bigg|_{x_1}^{x_2} - \int_{x_1}^{x_2} \eta(x)\frac{d}{dx}\left(\frac{\partial f}{\partial y'} \right) dx \\ &= -\int_{x_1}^{x_2} \eta(x)\frac{d}{dx}\left(\frac{\partial f}{\partial y'} \right) dx \end{aligned}$$

just because $\eta(x_1) = \eta(x_2) = 0$. Thus (4) can now be written as

$$\int_{x_1}^{x_2} \eta(x) \left(\frac{\partial f}{\partial y} - \frac{d}{dx}\left(\frac{\partial f}{\partial y'} \right) \right) dx = 0\,. \tag{9.5}$$

Now it is a basic fact of life that if g is some continuous function and if

$$\int_{x_1}^{x_2} \eta(x) \cdot g(x)\, dx = 0$$

for *every* choice of η, then it must be that $g \equiv 0$. We apply this observation to equation (5) with $g = \partial f/\partial y - [d/dx](\partial f/\partial y')$ to conclude that

$$\frac{\partial f}{\partial y} - \frac{d}{dx}\left(\frac{\partial f}{\partial y'} \right) = 0\,. \tag{9.6}$$

This is *Euler's equation* for the variational problem (1).

Euler's equation is a beautiful and elegant piece of mathematics, but it is not the answer to all our prayers. There are both philosophical and technical difficulties. Recall that we learned in calculus that, when we endeavor to solve a maximum-minimum problem by taking the derivative and setting it equal to zero, the resulting critical points may or may not actually be the extreme points we are looking for. They could be maxima, or they could be minima, or they could be inflection points. Similar complications occur in the application of Euler's equation.

In ordinary calculus there are second derivative tests and other tests that enable us to positively recognize the extrema that we seek. There are similar tests in the calculus of variations, but they are in fact very complicated and we cannot consider them here. It will turn out for us that, in practice, the geometry or the physics of the problems we are considering will enable us to identify which solutions of Euler's equation are the extrema that we seek. We use the phrase *stationary function* to denote any solution of Euler's equation for the problem we are studying. As noted, additional arguments will be required to determine which stationary functions are extrema.

There are also technical difficulties with studying Euler's equation. It is nonlinear, and in general not solvable in closed form. Fortunately for us, the particular problems that we want to study lead to simplifications in Euler's equation that actually make it tractable. Thus we shall be able to enjoy several illustrative examples of the Euler technique. We shall now spend some time discussing ways to analyze and simplify Euler's equation.

Acknowledging explicitly that x, y, and y' are being treated as independent variables, but also that y and y' depend on x in obvious ways, and using the chain rule, we may rewrite the second term in (6) as

$$\frac{\partial}{\partial x}\left(\frac{\partial f}{\partial y'}\right) + \frac{\partial}{\partial y}\left(\frac{\partial f}{\partial y'}\right)\frac{dy}{dx} + \frac{\partial}{\partial y'}\left(\frac{\partial f}{\partial y'}\right)\frac{dy'}{dx} .$$

We note that

$$\frac{\partial}{\partial y'}\left(\frac{\partial f}{\partial y'}\right)\frac{dy'}{dx} = \frac{\partial^2 f}{\partial y'^2}\frac{d^2 y}{dx^2} .$$

Similar simplifications may be made to the other terms. As a result, Euler's equation may be written as

$$f_{y'y'}\frac{d^2y}{dx^2} + f_{y'y}\frac{dy}{dx} + (f_{y'x} - f_y) = 0 .$$

Here subscripts denote derivatives.

We see immediately that Euler's equation, written in this form, is a second-order, nonlinear equation. The solution set for such an equation is, in general, a two-parameter family of curves. The extremals for the corresponding variational problem will correspond to those choices of the parameters that cause the solution to satisfy the standing boundary conditions. As we have indicated, a second-order, nonlinear equation of this kind is generally impossible to solve. But we shall frequently have simplifying information that does make an explicit solution possible. Some notable examples will now be sketched.

CASE I. If x and y are both missing from the function f (i.e., f does not depend on these parameters), then Euler's equation reduces to

$$f_{y'y'}\frac{d^2y}{dx^2} = 0\,.$$

In case $f_{y'y'} \neq 0$, we may conclude that $d^2y/dx^2 \equiv 0$, and hence that $y = Ax + B$. Thus, for this particularly simple setup, the stationary functions are all straight lines.

CASE II. If y alone is missing from the function f, then Euler's equation becomes

$$\frac{d}{dx}\left(\frac{\partial f}{\partial y'}\right) = 0\,.$$

This equation can be integrated once to yield

$$\frac{\partial f}{\partial y'} = C$$

for the extremals.

CASE III. If x alone is missing from the function f, then we may proceed as follows: Consider the identity

$$\begin{aligned}
\frac{d}{dx}\left(\frac{\partial f}{\partial y'}y' - f\right) &= \frac{d}{dx}\left(\frac{\partial f}{\partial y'}\right)y' + \frac{\partial f}{\partial y'}\frac{dy'}{dx} - \frac{d}{dx}[f] \\
&= \frac{d}{dx}\left(\frac{\partial f}{\partial y'}\right)y' + \frac{\partial f}{\partial y'}\frac{dy'}{dx} - \left(\frac{\partial f}{\partial y}\frac{dy}{dx} + \frac{\partial f}{\partial y'}\frac{dy'}{dx} + \frac{\partial f}{\partial x}\right) \\
&= y'\left\{\frac{d}{dx}\left(\frac{\partial f}{\partial y'}\right) - \frac{\partial f}{\partial y}\right\} - \frac{\partial f}{\partial x}\,.
\end{aligned}$$

Now we use the fact that $\partial f/\partial x = 0$ and also that the expression in parentheses vanishes by Euler's equation. Thus

$$\frac{d}{dx}\left(\frac{\partial f}{\partial y'}y' - f\right) = 0$$

so

$$\frac{\partial f}{\partial y'}y' - f = C\,.$$

Now we may actually attack some of the variational problems formulated at the beginning of this chapter.

EXAMPLE 9.2.1 Find the shortest curve joining two points (x_1, y_1) and (x_2, y_2) in the plane.

Solution: We must minimize the arc-length integral

$$I(y) = \int_{x_1}^{x_2} \sqrt{1 + (y')^2}\, dx\,.$$

The variables x and y are missing from the integrand $f(y') = \sqrt{1 + (y')^2}$, so this problem falls under the rubric of **CASE I** above. Since

$$f_{y'y'} = \frac{\partial^2 f}{\partial y' \partial y'} = \frac{1}{[1 + (y')^2]^{3/2}} \neq 0\,,$$

we may safely conclude that $d^2y/dx^2 \equiv 0$, and hence that the stationary functions are those of the form $y = Ax + B$. The boundary conditions (that the curve pass through the two given points) then yield that

$$y = \left(\frac{y_2 - y_1}{x_2 - x_1} \right) x + \left(y_1 - x_1 \cdot \frac{y_2 - y_1}{x_2 - x_1} \right).$$

We only know for certain at this point that, if I has a stationary value, then that extremal function must be a straight line. However, it is clear that I has no maximizing curve (i.e., there are curves of arbitrary length that connect the two given points). Geometric reasoning can now convince us that the line we have found must be the shortest curve. □

Remark 9.2.2 Although Example 9.2.1 is both satisfying and rewarding, it is not very profound. A much more interesting question, and one that has serious consequences for modern geometry, is to find curves of shortest length that connect two points on a surface (and which lie entirely in that surface). Such a curve is called a *geodesic*. Geodesics are a matter of intense study in modern differential geometry.

EXAMPLE 9.2.3 Find the curve joining two points (x_1, y_1) and (x_2, y_2) in the plane that yields a surface of minimal area when the curve is rotated about the x-axis.

Solution: The integral that represents the surface area of the described surface of revolution is

$$I(y) = \int_{x_1}^{x_2} 2\pi y \sqrt{1 + (y')^2}\, dx\,.$$

We see that the variable x is missing from the integrand f. Thus **CASE III** tells us that Euler's equation is

$$\frac{y(y')^2}{\sqrt{1 + (y')^2}} - y\sqrt{1 + (y')^2} = C\,.$$

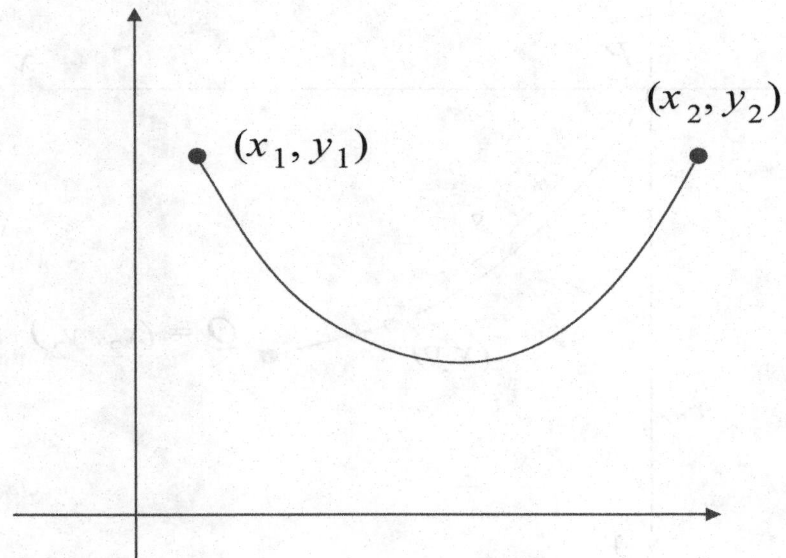

FIGURE 9.6
The catenary.

This simplifies, with some algebra, to

$$Cy' = \sqrt{y^2 - C^2}\,.$$

Separating variables and integrating, we find that

$$x = C\int \frac{dy}{\sqrt{y^2 - C^2}} = C\ln\left(\frac{y + \sqrt{y^2 - C^2}}{C}\right) + D\,.$$

Solving for y gives

$$y = C\cosh\left(\frac{x - D}{C}\right). \tag{9.7}$$

In conclusion, the extremals are catenaries (Figure 9.6). The required minimal surface is obtained by revolving a catenary about the x-axis. It remains to be seen whether the constants C and D may be chosen so that the curve described by (7) actually joins the two given points. This turns out to be algebraically quite complex, and is not our primary interest. Therefore we shall omit the details. □

EXAMPLE 9.2.4 Find the curve of quickest descent (Figure 9.7) that connects given points P and Q in the plane.

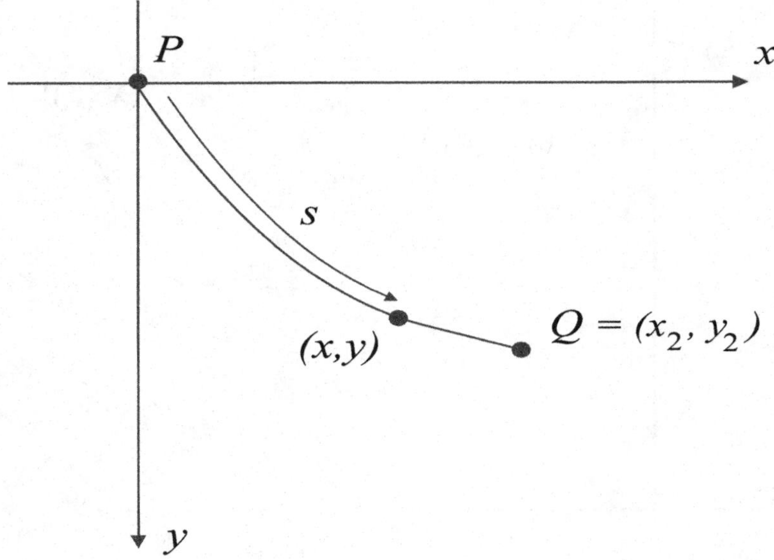

FIGURE 9.7
The curve of quickest descent.

Solution: We must minimize

$$I(y) = \int_{x_1}^{x_2} \frac{\sqrt{1+(y')^2}}{\sqrt{2gy}}\, dx\,.$$

Obviously the variable x is missing from the integrand f. Thus **CASE III** applies and Euler's equation becomes

$$\frac{(y')^2}{\sqrt{y}\sqrt{1+(y')^2}} - \frac{\sqrt{1+(y')^2}}{\sqrt{y}} = C\,.$$

After some algebra, this last equation reduces to

$$y[1+(y')^2] = C\,.$$

This is exactly the same equation we derived in our earlier discussion of the brachistochrone (Section 11.3) from a different point of view. We solved the equation at that time, and found that the resulting stationary curve is the cycloid

$$x = a(\theta - \sin\theta) \quad \text{and} \quad y = a(1-\cos\theta)\,;$$

this curve is generated by a mark on the edge of a disc of radius a rolling along the x-axis (see Figure 8.3), where a is selected so that the first inverted arch passes through the point (x_2, y_2).

It is reasonably clear from physical considerations that there is no maximal curve for this problem. Thus the stationary curve that we have found must be the sought-after minimizing curve. □

We conclude this section by noting that the methodology that led to the Euler equation can be applied to more general circumstances. We illustrate this idea in just one particular case.

Suppose we want to find stationary functions y and z for the integral

$$I(y,z) = I = \int_{x_1}^{x_2} f(x,y,z,y',z')\,dx\,.$$

Here we think of y and z as functions of the independent variable x, and the integrand f depends on x, y, z and on the first derivatives of y and z with respect to x. As in the simpler case that we have already treated, boundary values for y and z are specified in advance.

Just as before, we introduce perturbation functions η_1 and η_2 that are sufficiently smooth and which vanish at the endpoints. Then we form the perturbed functions $\widetilde{y}(x) = y(x)+\epsilon\eta_1(x)$ and $\widetilde{z}(x) = z(x)+\epsilon\eta_2(x)$. We consider the function of ϵ that is defined by

$$H(\epsilon) = \int_{x_1}^{x_2} f(x, y+\epsilon\eta_1, z+\epsilon\eta_2, y'+\epsilon\eta_1', z'+\epsilon\eta_2')\,dx\,.$$

If y and z are stationary functions for the problem, then of course it must be that $H'(0) = 0$. Computing the derivative of H and setting $\epsilon = 0$ then yields

$$\int_{x_1}^{x_2} \left(\frac{\partial f}{\partial y}\eta_1 + \frac{\partial f}{\partial z}\eta_2 + \frac{\partial f}{\partial y'}\eta_1' + \frac{\partial f}{\partial z'}\eta_2'\right) dx = 0\,.$$

If the terms involving η_1' and η_2' are integrated by parts, just as we did before, then the result is

$$\int_{x_1}^{x_2} \left\{\eta_1(x)\left(\frac{\partial f}{\partial y} - \frac{d}{dx}\left(\frac{\partial f}{\partial y'}\right)\right) + \eta_2(x)\left(\frac{\partial f}{\partial z} - \frac{d}{dx}\left(\frac{\partial f}{\partial z'}\right)\right)\right\} dx = 0\,.$$

Finally, since this last equation must hold for all choices of the perturbation functions η_1, η_2, we are led at once to the dual Euler equations

$$\frac{d}{dx}\left(\frac{\partial f}{\partial y'}\right) - \frac{\partial f}{\partial y} = 0 \quad \text{and} \quad \frac{d}{dx}\left(\frac{\partial f}{\partial z'}\right) - \frac{\partial f}{\partial z} = 0\,. \tag{9.8}$$

We see that finding the stationary functions for our problem amounts to solving the system (8). Once the system is solved—and this can be quite difficult—then one might hope to choose the arbitrary constants (four of them!) so that the solution meets the boundary conditions.

Exercises

1. Find the extremals for the integral

$$I = \int_{x_1}^{x_2} f(x, y, y')\, dx$$

if the integrand is given by

(a) $\dfrac{\sqrt{1+(y')^2}}{y}$

(b) $y^2 - (y')^2$

(c) $1 + (y')^2$

(d) $2y^2 - (y')^2$

2. Find the stationary function of

$$I = \int_0^4 [xy' - (y')^2]\, dx$$

which is determined by the boundary conditions $y(0) = 0$ and $y(4) = 3$.

3. When the integrand in Exercise 1 is of the form

$$a(x)(y')^2 + 2b(x)yy' + c(x)y^2 ,$$

show that Euler's equation is a second-order linear differential equation.

4. If P and Q are two points in the plane then, in terms of polar coordinates, the length of a curve from P to Q is

$$\int_P^Q dx = \int_P^Q \sqrt{dr^2 + r^2 d\theta^2} .$$

Find the polar equation of the shortest such path (i.e., a straight line) by minimizing this integral

(a) with θ as the independent variable;

(b) with r as the independent variable.

5. Consider two points P and Q on the surface of the sphere $x^2+y^2+z^2 = a^2$. Coordinatize this surface by means of the standard spherical coordinates θ and ϕ (see [STE, p. 694] for details of this coordinate system), where $x = a \sin\phi \cos\theta$, $y = a \sin\phi \sin\theta$, and $z = a\cos\phi$. Let $\theta = F(\phi)$ be a curve lying on the surface and joining P to Q. Show that the shortest such curve (a *geodesic*) is an arc or a great circle, that is, it lies on the intersection of a plane through the center of the sphere with the sphere itself. *Hint:* Express the length of the curve in the form

$$\begin{aligned} \int_P^Q ds &= \int_P^Q \sqrt{dx^2 + dy^2 + dz^2} \\ &= a \int_P^Q \sqrt{1 + \left(\frac{d\theta}{d\phi}\right)^2 \sin^2\phi}\, d\phi , \end{aligned}$$

solve the corresponding Euler equation for θ, and convert the result back into rectangular coordinates.

6. Demonstrate that any geodesic on the right circular cone $z^2 = a^2(x^2+y^2)$, $z \geq 0$, has the following property: If the cone is cut along a generator and flattened into a plane, then the geodesic becomes a straight line. *Hint:* Represent the cone parametrically by means of the equation

$$x = \frac{r\cos(\theta\sqrt{1+a^2})}{\sqrt{1+a^2}}\,, \quad y = \frac{r\sin(\theta\sqrt{1+a^2})}{\sqrt{1+a^2}}\,, \quad z = \frac{ar}{\sqrt{1+a^2}}\,.$$

Show that the parameters r and θ represent ordinary polar coordinates on the flattened cone. Then show that a geodesic $r = r(\theta)$ is a straight line in these polar coordinates.

7. If the curve $y = g(z)$ is revolved about the z-axis, then the resulting surface of revolution has $x^2 + y^2 = g(z)^2$ as its equation. A convenient parametric representation of this surface is given by

$$x = g(z)\cos\theta\,, \quad y = g(z)\sin\theta\,, z = z\,,$$

where θ is the polar angle in the x-y plane. Show that a geodesic $\theta = \theta(z)$ on this surface has equation

$$\theta = c_1 \int \frac{\sqrt{1+[g'(z)]^2}}{g(z)\sqrt{g(z)^2 - c_1^2}}\,dz + c_2\,.$$

8. If the surface of revolution in Exercise 7 is a right circular cylinder, then show that every geodesic of the form $\theta = \theta(z)$ is a helix or a generator.

9.3 Isoperimetric Problems and the Like

It is a problem that goes back to the ancient Greeks to find a curve of fixed length that encloses the greatest area. The problem was given the name *isoperimetric problem*, and it has a long and important history. The Greeks produced a fairly rigorous proof that the intuitively correct answer—a circle—is the stationary curve for the problem. By today there are dozens of different (rigorous) proofs.

If we parameterize the curve in the isoperimetric problem as $x = x(t)$, $y = y(t)$, and consider the curve to be traversed once counterclockwise as t increases from t_1 to t_2, then the enclosed area is given by

$$A = \frac{1}{2}\int_{t_1}^{t_2}\left(x\frac{dy}{dt} - y\frac{dx}{dt}\right)dt\,; \tag{9.9}$$

this formula is an immediate consequence of Green's theorem, and is part of any multivariable calculus course. Now the length of the curve is

$$L = \int_{t_1}^{t_2}\sqrt{\left(\frac{dx}{dt}\right)^2 + \left(\frac{dy}{dt}\right)^2}\,dt\,. \tag{9.10}$$

Thus the isoperimetric problem is to maximize (9), subject to the constraint that (10) have a fixed, constant value.

Today the terminology has been broadened. In general, the family of problems in mathematics that are termed "isoperimetric problems" involves extremizing one function subject to a constraint defined by some other function. In the problem above, the constraint is expressed by an integral. But it need not be. An example is to consider two points P and Q in a surface S in space defined by

$$G(x, y, z) = 0 .$$

Our job is to find the shortest curve from P to Q that lies entirely in S. Thus we have a curve $\gamma(t) = (x(t), y(t), z(t))$ and we want to minimize the integral

$$\text{length} = \int_{t_1}^{t_2} \sqrt{\left(\frac{dx}{dt}\right)^2 + \left(\frac{dy}{dt}\right)^2 + \left(\frac{dz}{dt}\right)^2}\, dt$$

subject to the constraint

$$G(x(t), y(t), z(t)) = 0 .$$

9.3.1 Lagrange Multipliers

We consider a class of problems from elementary calculus that are similar in spirit to isoperimetric problems. Consider, for example, the problem of finding points $(x, y) \in \mathbb{R}^2$ that yield stationary values for a function $z = f(x, y)$ subject to the constraint

$$g(x, y) = 0 . \tag{9.11}$$

One classical approach to such a problem is to declare one of the variables—say x—to be independent or free, and to consider that y depends on x. With this assumption in place, we calculate from (11) that

$$\frac{\partial g}{\partial x} + \frac{\partial g}{\partial y} \cdot \frac{dy}{dx} = 0 .$$

Since z is now considered to be a function of x alone (as y depends on x), we see that $dz/dx = 0$ is a necessary condition for z to have a stationary value. Thus

$$\frac{dz}{dx} = \frac{\partial f}{\partial x} + \frac{\partial f}{\partial y}\frac{dy}{dx} = 0$$

or

$$\frac{\partial f}{\partial x} - \frac{\partial f}{\partial y}\frac{\partial g/\partial x}{\partial g/\partial y} = 0 . \tag{9.12}$$

Solving (11) and (12) simultaneously, we find the stationary points (x, y).

One drawback—at least a philosophical one—that occurs in this last approach is that x and y have symmetric roles in the original problem, but we end up treating them asymmetrically. We now introduce a different, and more elegant, method that has many practical advantages. It goes back to Joseph Louis Lagrange (1736–1813) and is called *the method of Lagrange multipliers.*

Lagrange's idea was to introduce the function

$$F(x, y, \lambda) = f(x, y) + \lambda g(x, y)$$

and investigate its *unconstrained stationary values* by means of the obvious necessary conditions

$$\begin{aligned}\frac{\partial F}{\partial x} &= \frac{\partial f}{\partial x} + \lambda \frac{\partial g}{\partial x} = 0 \\ \frac{\partial F}{\partial y} &= \frac{\partial f}{\partial y} + \lambda \frac{\partial g}{\partial y} = 0 \\ \frac{\partial F}{\partial \lambda} &= g(x, y) = 0 \,.\end{aligned}$$

If we use elementary algebra to eliminate λ from the first two of these equations, then the system reduces to

$$\frac{\partial f}{\partial x} - \frac{\partial f}{\partial y} \frac{\partial g / \partial x}{\partial g / \partial y} = 0 \qquad \text{and} \qquad g(x, y) = 0 \,.$$

This is the very same system that we obtained earlier by different means. Notice that, in the end, the parameter λ (known as the *Lagrange multiplier*) disappears. Its role is as an aid, but in the end it has no part in the solution.[1]

The advantage of the Lagrange multiplier method over the earlier method is that **(i)** it does not disturb the symmetry of the problem by making an arbitrary choice of variable, and **(ii)** it absorbs the side condition (at the small expense of introducing the spare variable λ). The Lagrange multiplier method works just as well for functions of any number of variables.

9.3.2 Integral Side Conditions

Now we shall consider the problem of finding the differential equation that must be satisfied by a function $y = y(x)$ that is a stationary function for the integral

$$I = I(y) = \int_{x_1}^{x_2} f(x, y, y')\, dx$$

and also satisfies the constraint

$$J = J(y) = \int_{x_1}^{x_2} g(x, y, y')\, dx = C \,, \tag{9.13}$$

[1] In some applications, such as econometrics, there are important interpretations for the value of λ. We shall not explore that topic here.

together with the boundary conditions $y(x_1) = y_1$ and $y(x_2) = y_2$. We shall follow the pattern, used successfully before, of assuming that the stationary function y exists and then perturbing it.

Thus define a function $\widetilde{y}(x) = y(x) + \epsilon_1\eta_1(x) + \epsilon_2\eta_2(x)$. Notice that this is a bit more complex than our earlier approach, because now we must maintain the side condition (13). We require, as usual, that η_1, η_2 be sufficiently smooth and that both functions vanish at the endpoints x_1, x_2. The parameters ϵ_1, ϵ_2 are not entirely independent (as already indicated), because we must require that

$$J(\epsilon_1, \epsilon_2) = \int_{x_1}^{x_2} g(x, \widetilde{y}, \widetilde{y}')\, dx = C\,. \tag{9.14}$$

In summary, our problem is now reduced to finding necessary conditions for the function

$$I(\epsilon_1, \epsilon_2) \equiv \int_{x_1}^{x_2} f(x, \widetilde{y}, \widetilde{y}')\, dx$$

to have a stationary value at $\epsilon_1 = \epsilon_2 = 0$ provided that ϵ_1, ϵ_2 satisfy (14). This situation is natural for the method of Lagrange multipliers. Thus we introduce the auxiliary function

$$\begin{aligned} K(\epsilon_1, \epsilon_2, \lambda) &\equiv I(\epsilon_1, \epsilon_2) + \lambda J(\epsilon_1, \epsilon_2) \\ &= \textstyle\int_{x_1}^{x_2} F(x, \widetilde{y}, \widetilde{y}')\, dx\,, \end{aligned} \tag{9.15}$$

where

$$F \equiv f + \lambda g\,.$$

We investigate the *unconstrained* stationary functions for $\epsilon_1 = \epsilon_2 = 0$ by means of the necessary conditions (from the calculus)

$$\frac{\partial K}{\partial \epsilon_1} = \frac{\partial K}{\partial \epsilon_2} = 0 \quad \text{when } \epsilon_1 = \epsilon_2 = 0\,. \tag{9.16}$$

If we differentiate (15) under the integral sign and use the definition of $\widetilde{y}$, we obtain

$$\frac{\partial K}{\partial \epsilon_j} = \int_{x_1}^{x_2} \left(\frac{\partial F}{\partial \widetilde{y}} \eta_j(x) + \frac{\partial F}{\partial \widetilde{y}'} \eta_j'(x) \right) dx \qquad \text{for } j = 1, 2\,.$$

Setting $\epsilon_1 = \epsilon_2 = 0$, we obtain

$$\int_{x_1}^{x_2} \left(\frac{\partial F}{\partial y} \eta_j(x) + \frac{\partial F}{\partial y'} \eta_j'(x) \right) dx = 0$$

by virtue of (16). After the second term is integrated by parts (just as we usually do in these problems), we obtain the equation

$$\int_{x_1}^{x_2} \eta_j(x) \left[\frac{\partial F}{\partial y} - \frac{d}{dx}\left(\frac{\partial F}{\partial y'} \right) \right] dx = 0 \qquad \text{for } j = 1, 2\,.$$

Since η_1 and η_2 are both arbitrary, the two conditions just enunciated are really just one condition. We conclude then as usual that the expression in brackets must be identically zero. So

$$\frac{d}{dx}\left(\frac{\partial F}{\partial y'}\right) - \frac{\partial F}{\partial y} = 0 .$$

This is the Euler equation, not for f or g but for the combined function F. The solutions of this equation (the extremals of our problem) involve three undetermined parameters: two constants of integration, and the Lagrange multiplier λ. The stationary function is then selected from this three-parameter family of extremals by imposing the two boundary conditions and by prescribing that the integral J takes the value C.

Remark 9.3.1 In the case of integrals that depend on two or more functions, our last result can be extended in the same way as in the previous section. For instance, if

$$I = I(y, z) = \int_{x_1}^{x_2} f(x, y, z, y', z')\, dx$$

has a pair of stationary functions subject to the constraint

$$J = J(y, z) = \int_{x_1}^{x_2} g(x, y, z, y', z')\, dx = C ,$$

then the stationary functions $y(x)$ and $z(x)$ must satisfy the system of equations (with $F = f + \lambda g$) given by

$$\frac{d}{dx}\left(\frac{\partial F}{\partial y'}\right) - \frac{\partial F}{\partial y} = 0 \quad \text{and} \quad \frac{d}{dx}\left(\frac{\partial F}{\partial z'}\right) - \frac{\partial F}{\partial z} = 0 . \tag{9.17}$$

The derivation of this result uses reasoning that we have already presented, and we shall not repeat it.

EXAMPLE 9.3.2 Solve the following restricted version of the isoperimetric problem. Find the curve of fixed length L, joining the points $(0, 0)$ and $(1, 0)$ in the plane, that

(i) lies above the x-axis;

(ii) encloses the maximum area between itself and the x-axis.

Solution: Our problem is to maximize the integral

$$I(y) = I = \int_0^1 y\, dx$$

subject to the constraint

$$J(y) = J = \int_0^1 \sqrt{1 + (y')^2}\, dx = L$$

and of course the boundary conditions $y(0) = 0$ and $y(1) = 0$. Thus

$$F = y + \lambda\sqrt{1 + (y')^2},$$

so Euler's equation is

$$\frac{d}{dx}\left(\frac{\lambda y'}{\sqrt{1 + (y')^2}}\right) - 1 = 0. \tag{9.18}$$

Carrying out the differentiation, we find that our differential equation is

$$\frac{y''}{[1 + (y')^2]^{3/2}} = \frac{1}{\lambda}.$$

Using a formula that we learned in multivariable calculus, we see immediately that the curvature of the stationary curve y is constantly equal to $1/\lambda$. It follows that the extremal curve that we seek is an arc of a circle with radius λ.

An alternative approach would be to integrate (18) to obtain

$$\frac{y'}{\sqrt{1 + (y')^2}} = \frac{x - c}{\lambda}.$$

Solving this equation for y' and then integrating one more time yields

$$(x - c)^2 + (y - d)^2 = \lambda^2.$$

This is of course the equation of a circle with radius λ. □

EXAMPLE 9.3.3 Solve the original isometric problem of finding the curve with given length L that encloses the greatest area.

Solution: In order to avoid the somewhat artificial constraints imposed earlier, we consider curves in the parameteric form

$$x = x(t), \quad y = y(t).$$

Our job is to maximize

$$I = \frac{1}{2}\int_{t_1}^{t_2} (x\dot{y} - y\dot{x})\, dt$$

(where, of course, $\dot{x} = dx/dt$ and $\dot{y} = dy/dt$) with the constraint

$$\int_{t_1}^{t_2} \sqrt{\dot{x}^2 + \dot{y}^2}\, dt = L.$$

Following the Lagrange multiplier paradigm, we set

$$F = \frac{1}{2}(x\dot{y} - y\dot{x}) + \lambda\sqrt{\dot{x}^2 + \dot{y}^2}\,.$$

Thus the Euler equations (17) become

$$\frac{d}{dt}\left(-\frac{1}{2}y + \frac{\lambda\dot{x}}{\sqrt{\dot{x}^2 + \dot{y}^2}}\right) - \frac{1}{2}\dot{y} = 0$$

and

$$\frac{d}{dt}\left(\frac{1}{2}x + \frac{\lambda\dot{y}}{\sqrt{\dot{x}^2 + \dot{y}^2}}\right) + \frac{1}{2}\dot{x} = 0\,.$$

These equations may be integrated directly, and the result is

$$-y + \frac{\lambda\dot{x}}{\sqrt{\dot{x}^2 + \dot{y}^2}} = -C$$

and

$$x + \frac{\lambda\dot{y}}{\sqrt{\dot{x}^2 + \dot{y}^2}} = D\,.$$

If we solve for $x - D$ and $y - C$, square, and add, then the result is

$$(x - D)^2 + (y - C)^2 = \lambda^2\,,$$

so that the maximizing curve is a circle.

There is in fact a *quantitative* way to express the isoperimetric result. If L is the length of a closed, plane curve that encloses an area A, then

$$A \leq \frac{L^2}{4\pi}\,.$$

Moreover, equality occurs if and only if the curve is a circle. A relation of this kind is called an *isoperimetric inequality.* See [OSS], [GAK] for further reading. □

9.3.3 Finite Side Conditions

We have, a few times, mentioned the problem of finding geodesics on a given surface

$$G(x, y, z) = 0\,. \tag{9.19}$$

We now consider instead the slightly more general problem of finding a space curve $x = x(t)$, $y = y(t)$, $z = z(t)$ that gives a stationary value to an integral of the form

$$I = \int_{t_1}^{t_2} f(\dot{x}, \dot{y}, \dot{z})\,dt\,, \tag{9.20}$$

where the curve is required (as a side, or constraint, condition) to *lie in the surface.*

Our strategy is to eliminate the explicit use of the constraint condition (19). We may assume[2] that the curve we seek lies in a part of the surface where $G_z \neq 0$. With this assumption, we may solve (19) for z (by the implicit function theorem), which gives a formula of the type

$$z = g(x, y)$$

and

$$\dot{z} = \frac{\partial g}{\partial x}\dot{x} + \frac{\partial g}{\partial y}\dot{y}\,. \tag{9.21}$$

When this last equation is inserted in (20), our problem is reduced to that of finding unconstrained stationary functions for the integral

$$\int_{t_1}^{t_2} f\left(\dot{x}, \dot{y}, \frac{\partial g}{\partial x}\dot{x} + \frac{\partial g}{\partial y}\dot{y}\right)\, dt\,.$$

We know from the previous section that the Euler equations for such an integral are

$$\frac{d}{dt}\left(\frac{\partial f}{\partial \dot{x}} + \frac{\partial f}{\partial \dot{z}}\frac{\partial g}{\partial x}\right) - \frac{\partial f}{\partial \dot{z}}\frac{\partial \dot{z}}{\partial x} = 0$$

and

$$\frac{d}{dt}\left(\frac{\partial f}{\partial \dot{y}} + \frac{\partial f}{\partial \dot{z}}\frac{\partial g}{\partial y}\right) - \frac{\partial f}{\partial \dot{z}}\frac{\partial \dot{z}}{\partial y} = 0\,.$$

It follows from (21) that

$$\frac{\partial \dot{z}}{\partial x} = \frac{d}{dt}\left(\frac{\partial g}{\partial x}\right) \quad \text{and} \quad \frac{\partial \dot{z}}{\partial y} = \frac{d}{dt}\left(\frac{\partial g}{\partial y}\right)\,.$$

As a result, the Euler equations can be written in the form

$$\frac{d}{dt}\left(\frac{\partial f}{\partial \dot{x}}\right) + \frac{\partial g}{\partial x}\frac{d}{dt}\left(\frac{\partial f}{\partial \dot{z}}\right) = 0 \quad \text{and} \quad \frac{d}{dt}\left(\frac{\partial f}{\partial \dot{y}}\right) + \frac{\partial g}{\partial y}\frac{d}{dt}\left(\frac{\partial f}{\partial \dot{z}}\right) = 0\,.$$

Now we define a new function $\lambda(t)$ by

$$\frac{d}{dt}\left(\frac{\partial f}{\partial \dot{z}}\right) = \lambda(t) G_z\,. \tag{9.22}$$

We use the relations $\partial g/\partial x = -G_x/G_z$, $\partial g/\partial y = -G_y/G_z$ (once again here the normalizing assumption $G_z \neq 0$ is useful). Then Euler's equation becomes

$$\frac{d}{dt}\left(\frac{\partial f}{\partial \dot{x}}\right) = \lambda(t) G_x \tag{9.23}$$

[2]This assumption requires a moment's thought. It means that we are restricting attention to a part of the surface on which the normal to the surface has a vertical component. This can be arranged by rotating the surface in space.

and

$$\frac{d}{dt}\left(\frac{\partial f}{\partial \dot{y}}\right) = \lambda(t)G_y . \tag{9.24}$$

Thus a necessary condition for the existence of a stationary function is the existence of a function $\lambda(t)$ satisfying equations (22), (23), and (24). Eliminating $\lambda(t)$ from these equations, we obtain the symmetric system

$$\frac{(d/dt)(\partial f/\partial \dot{x})}{G_x} = \frac{(d/dt)(\partial f/\partial \dot{y})}{G_y} = \frac{(d/dt)(\partial f/\partial \dot{z})}{G_z} . \tag{9.25}$$

Together with equation (19), this new system determines the extremals of the problem. We note that equations (22), (23), and (24) can be regarded as the Euler equations for the problem of finding unconstrained stationary functions for the integral

$$I = \int_{t_1}^{t_2} [f(\dot{x}, \dot{y}, \dot{z}) + \lambda(t)G(x, y, z)]\, dt .$$

This is very similar to our conclusion earlier for integral side conditions, except that in the present circumstance the multiplier is an undetermined function of t instead of an undetermined constant.

Let us now summarize all of these ideas for the particular problem of finding geodesics. In this case, we have

$$f = \sqrt{\dot{x}^2 + \dot{y}^2 + \dot{z}^2} .$$

Equation (25) becomes

$$\frac{(d/dt)(\dot{x}/f)}{G_x} = \frac{(d/dt)(\dot{y}/f)}{G_y} = \frac{(d/dt)(\dot{z}/f)}{G_z} , \tag{9.26}$$

and the problem is to extract information from this system of equations.

EXAMPLE 9.3.4 Find the geodesics on the sphere

$$x^2 + y^2 + z^2 = 1 .$$

Solution: The surface for this problem is

$$G(x, y, z) = x^2 + y^2 + z^2 = 1$$

and equation (26) becomes

$$\frac{f\ddot{x} - \dot{x}\dot{f}}{2xf^2} = \frac{f\ddot{y} - \dot{y}\dot{f}}{2yf^2} = \frac{f\ddot{z} - \dot{z}\dot{f}}{2zf^2} .$$

This can be rewritten in the form

$$\frac{x\ddot{y} - y\ddot{x}}{x\dot{y} - y\dot{x}} = \frac{\dot{f}}{f} = \frac{y\ddot{z} - z\ddot{y}}{y\dot{z} - z\dot{y}} .$$

This is just

$$\frac{(d/dt)(x\dot{y} - y\dot{x})}{x\dot{y} - y\dot{x}} = \frac{(d/dt)(y\dot{z} - z\dot{y})}{y\dot{z} - z\dot{y}}.$$

A simple integration now gives

$$x\dot{y} - y\dot{x} = C(y\dot{z} - z\dot{y})$$

or

$$\frac{\dot{x} + C\dot{z}}{x + Cz} = \frac{\dot{y}}{y}.$$

Another integration yields

$$x + Cz = Dy.$$

This is just the equation of a plane through the origin. The intersection of that plane with our surface $G = 0$ is nothing other than a great circle. We have discovered that the geodesics on a sphere are great circles. □

Remark 9.3.5 In the last example we were lucky: it was not difficult to solve equation (26). In general, the problem of solving these equations can be intractable. The main significance of these equations lies in their connection with the following very important result in mathematical physics: If a particle glides without friction along a surface, free from the action of any external force, then it will follow the path of a geodesic. We shall study the details of this assertion later. In studying this question, it is most convenient to parameterize the curve by arc length, so that $t = s$, $f \equiv 1$, and equation (26) become

$$\frac{d^2x/ds^2}{G_x} = \frac{d^2y/ds^2}{G_y} = \frac{d^2z/ds^2}{G_z}. \tag{9.27}$$

Exercises

1. Solve the following problems by the method of Lagrange multipliers.

   (a) Find the point on the plane $ax + by + cz = d$ that is nearest to the origin. *Hint:* Minimize $w = x^2 + y^2 + z^2$ with the side condition $ax + by + cz - d = 0$.

   (b) Show that the triangle with greatest area A for a given perimeter is equilateral. *Hint:* If $x, y,$ and z are the sides, then $A = \sqrt{s(s - x)(s - y)(s - z)}$, where $s = (x + y + z)/2$.

(c) If the sum of n positive numbers $x_1, x_2, \ldots, x_n$ has a fixed value s, then prove that their product $x_1 \cdot x_2 \cdots x_n$ has s^n/n^n as its maximum value. Conclude therefore that the geometric mean of n positive numbers can never exceed their arithmetic mean:

$$\sqrt[n]{x_1 \cdot x_2 \cdots x_n} \leq \frac{x_1 + x_2 + \cdots + x_n}{n} .$$

2. A curve in the first quadrant in the plane joins $(0,0)$ and $(1,0)$ and has a given area A beneath it. Show that the shortest such curve is an arc of a circle.

3. A uniform, flexible chain of given length hangs between two points. Find its shape if it hangs in such a way as to minimize its potential energy.

4. Solve the original isoperimetric problem, as described in the text, by using polar coordinates. *Hint:* Choose the origin to be any point on the curve and the polar axis to be the tangent line at that point. Then maximize

$$I = \frac{1}{2} \int_0^\pi r^2 \, d\theta$$

with the side condition that

$$\int_0^\pi \sqrt{\left(\frac{dr}{d\theta}\right)^2 + r^2 \, d\theta}$$

is constant.

5. Show that the geodesics on any cylinder of the form $g(x, z) = 0$ make a constant angle with the y-axis.

6. Give a convincing argument for the validity of the formula

$$A = \frac{1}{2} \int_{t_1}^{t_2} \left(x \frac{dy}{dt} - y \frac{dx}{dt} \right) dt$$

for a closed, convex curve like that shown in Figure 9.8. *Hint:* What is the geometric meaning of

$$\int_P^Q y \, dx + \int_Q^P y \, dx \, ,$$

where the first integral is taken from right to left along the upper part of the curve and the second from left to right along the lower part of the curve?

7. Verify the formula for A in the last exercise for the circle whose parametric equations are $x = a \cos t$ and $y = a \sin t$, $0 \leq t \leq 2\pi$.

8. Use your symbol manipulation software, such as `Maple` or `Mathematica`, to confirm the solution of the isoperimetric problem. Set up a routine for calculating the area inside various curves of length 2π, including ellipses, squares, and some irregular curves as well. Conclude by calculating the area inside the unit circle and observing that it contains the greatest area.

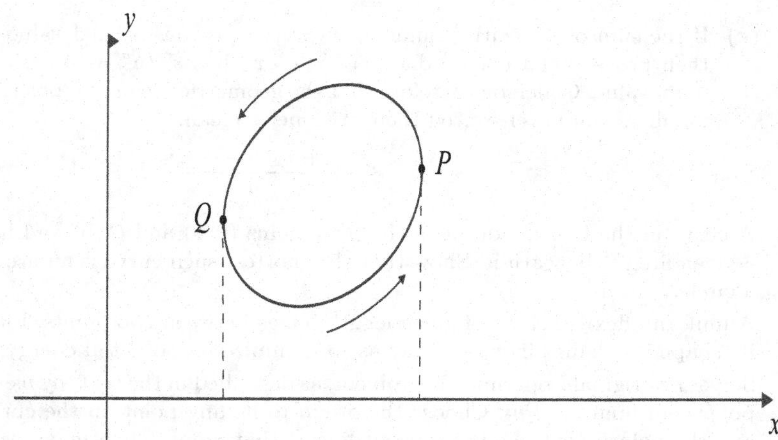

FIGURE 9.8
The area inside a curve.

Historical Note

Newton

Isaac Newton (1642–1727) was born to a farm family in the village of Woolsthorpe in northern England. Almost nothing is known of his early life, and his undergraduate years at Cambridge University were relatively undistinguished. In 1665, an outbreak of the plague caused the universities to close; to avoid contagion, Newton returned to his home in the country, where he remained until 1667. During those two years—from ages 22 to 24—Newton experienced a flood of creativity that has been unmatched in the history of the human race. He invented the binomial series for negative and fractional exponents, the differential and integral calculus, the law of universal gravitation, and the resolution of sunlight into the visual spectrum (by means of a prism). It has been said that Newton virtually created physical science, and he did this in just two years while still a callow youth.

Newton was an inward and secretive man, and by all accounts also an irascible one. He refrained from publishing many of his monumental discoveries, and this practice later contributed to several priority disputes. In fact the rivalry between Newton and Leibniz, and more generally between England and Germany, over the discovery of calculus was long and intense and did a great deal of damage to international relations and to the progress of science.

One of the legends of Isaac Newton is that his talents were so evident, and so remarkable, that his teacher Isaac Barrow in 1669 resigned his professorship at Cambridge so that Newton could have the post. Newton settled into Cambridge life and spent the next 27 years studying science.

Many of Newton's mathematical discoveries were never really published in connected form; they only became known in a limited way, and virtually by accident—through conversations or replies to questions put to him in correspondence. He seems to have regarded his mathematics mainly as a fruitful tool for the study of scientific problems, and of comparatively little intrinsic interest.

Little is known of the early years of Newton's professorship at Cambridge. It is clear that optics and the construction of telescopes were among his main interests. He experimented with different techniques for grinding lenses, using tools which he made himself. In 1670 he built the first reflecting telescope, a model for some of the great modern instruments (such as the one in the Mount Palomar Observatory). He developed his prismatic analysis of sunlight and continued to explore aspects of the theory of optics. He endeavored to publish some of his ideas, but the leading scientists of the day were unprepared to appreciate Newton's work. Disappointed, Newton retreated back into his shell and determined to do future scientific work only for his own satisfaction.

Newton had periodic lapses into disinterest or distaste for science. He was roused from one of the deepest of these by his friend Edmund Halley (of Halley's Comet), who had been discussing with Christopher Wren (the noted British architect) and Robert Hooke (of Hooke's Law—see elsewhere in this book) the motions of the planets, the nature of gravitation, and the shapes of the planetary orbits. One day Halley visited Newton in Cambridge and asked, "What would be the curve described by the planets on the supposition that gravity diminishes as the square of the distance?" Newton immediately answered that it would be an ellipse. Struck with joy and amazement, Halley asked how Newton could know this. "Because I have calculated it." This was just what Halley, Hooke, and Wren had sought for so long. Halley demanded to see the calculations, but Newton could not find them.

To make a long story short, it was this interchange that led Newton to put in the immense effort to write his treatise *Principia Mathematica*. This volume, still recognized as one of the premiere scientific works of all time, contains the basic principles of theoretical mechanics and fluid dynamics, the first mathematical treatment of wave motion, the deduction of Kepler's laws of planetary motion from the inverse square law of gravitation, an explanation of the orbits of comets, calculations of the masses of the earth and the planets, an explanation for the flattened shape of the earth, a theory of the precession of the equinoxes, and the basics of the theory of tides. This is only a sample of the volcano of ideas that lies in the *Principia*; this book truly lays the foundations of modern physical science.

In 1696, Newton left Cambridge for London to become the Warden (and soon the Master) of the British Mint. He withdrew from scientific pursuits, and even began to enjoy a bit of society and some of his celebrity as a premiere scientist. Late one afternoon, at the end of a hard day at the Mint, Newton learned of a now-famous problem that the Swiss scientist Johann Bernoulli had posed as a challenge "to the most acute mathematicians of the entire world." Bernoulli published the problem in a well-known periodical, and offered a reward for its solution. The problem is this: Suppose two nails are driven at random into the wall, and let the upper nail be connected to the lower one by a wire in the shape of a smooth curve. What is the shape of the wire which will allow a bead to slide down it (under the influence of gravity alone, without any friction) in the least possible time? This problem has become known as the *brachistochrone* (or "shortest time") problem. We have discussed this famous problem in Sections 7.1 and 11.3. In spite of being out of the habit of scientific thought, Newton solved the problem that evening before going to bed. He published his solution anonymously; when Bernoulli saw it, he said, "I recognize the lion by his claw."

Newton published his *Opticks* in 1704. On the one hand, this work collected all of Newton's important ideas about the nature of light and color. On the other hand, the work followed some speculative paths into alchemy and the nature of matter. An appendix to the book contains a list of queries that are far-ranging and often prescient. For example, he speculates that the gravitational field of a body can bend light rays and that mass and light can be converted one to the other (thus anticipating some of Einstein's ideas).

Newton is remembered as the greatest scientist who ever lived. He was a very strange individual—tormented by paranoia, psychosis, and a strange relationship with women (especially his mother). He had dreadful dealings with many of his professional contemporaries. But his legacy of scientific achievement is unmatched in history. His ideas and his deep insights have shaped the science that we have today. His achievements are unmatched in all of history.

Anatomy of an Application

Hamilton's Principle and its Implications

Mathematicians of the nineteenth century sought to discover a general principle from which Newtonian mechanics could be deduced. In searching for clues, they noticed a simple unifying principle that explained a number of different physical phenomena:

- A ray of light follows the quickest path through an optical medium.
- The equilibrium shape of a hanging chain minimizes its potential energy.
- Soap bubbles assume a shape having the least surface area for a given boundary.

- The configuration of blood vessels and arteries is determined by the most efficient path for pumping the blood.

There are many other examples (this general principle of economy in nature is explored in a delightful manner in [HIT]). This circle of ideas suggested to Leonhard Euler that hidden simplicities, and the pursuit of economy, govern many of the apparently chaotic phenomena in nature. This rather vague idea led him to create the calculus of variations for studying such questions. Euler's dream, of finding a *mathematical* principle to unify these many phenomena, was realized a century later by William Rowan Hamilton (1805–1865). In this Anatomy we shall learn about Hamilton's Principle.

We consider a particle of mass m moving through space under the influence of a force field

$$\mathbf{F} = F_1\mathbf{i} + F_2\mathbf{j} + F_3\mathbf{k}\,.$$

We make the physically natural assumption that the force is *conservative*, in the sense that the work it does in moving a particle from one point to another is independent of the path chosen. It is easy to show (and you learned the details in multivariable calculus) that, for such a force, there is a potential function $U(x, y, z)$ such that

$$\frac{\partial U}{\partial x} = F_1\,, \frac{\partial U}{\partial y} = F_2\,, \frac{\partial U}{\partial z} = F_3\,.$$

The function $V = -U$ is called the *potential energy* of the particle, since the change in its value from point A to point B is precisely equal to the work done by $\mathbf{F}$ in moving the particle from A to B. Moreover, if $\mathbf{r}(t) = x(t)\mathbf{i}+y(t)\mathbf{j}+z(t)\mathbf{k}$ is the position vector of the particle so that

$$\mathbf{v} = \frac{dx}{dt}\mathbf{i} + \frac{dy}{dt}\mathbf{j} + \frac{dz}{dt}\mathbf{k} \quad \text{and} \quad v = \sqrt{\left(\frac{dx}{dt}\right)^2 + \left(\frac{dy}{dt}\right)^2 + \left(\frac{dz}{dt}\right)^2}$$

are its velocity and speed, respectively, then $T = mv^2/2$ is its *kinetic energy*.

If the particle is at point P_1 at time t_1 and at point P_2 at time t_2, then we are interested in the path it traverses in moving from P_1 to P_2. The *action* (or *Hamilton's integral*) is defined to be

$$A = \int_{t_1}^{t_2} (T - V)\,dt\,.$$

In general, the value of A will depend on the particular path along which the particle moves from P_1 to P_2. We shall show that the actual path that the particle will choose to take is one that yields a stationary value for the functional A. (This corresponds to a critical point in the study of maximum/minimum problems.)

The function $L \equiv T - V$ is called the *Lagrangian.* In the case we have been describing the Lagrangian is given by

$$L = \frac{1}{2}m\left(\left(\frac{dx}{dt}\right)^2 + \left(\frac{dy}{dt}\right)^2 + \left(\frac{dz}{dt}\right)^2\right) - V(x, y, z).$$

The integrand of the action is therefore a function that has the form

$$f(x, y, z, dx/dt, dy/dt, dz/dt);$$

if the action has a stationary value, then Euler's equations must be satisfied.

In the present circumstance, Euler's equations become

$$m\frac{d^2x}{dt^2} + \frac{\partial V}{\partial x} = 0, \quad m\frac{d^2y}{dt^2} + \frac{\partial V}{\partial y} = 0, \quad m\frac{d^2z}{dt^2} + \frac{\partial V}{\partial z} = 0.$$

These equations can be written in the more compact form

$$m\frac{d^2\mathbf{r}}{dt^2} = -\frac{\partial V}{\partial x}\mathbf{i} - \frac{\partial V}{\partial y}\mathbf{j} - \frac{\partial V}{\partial z}\mathbf{k} = \mathbf{F}.$$

This equation is precisely *Newton's second law of motion.* Thus Newton's law is a necessary condition for the action of the particle to have a stationary value. Since Newton's law also governs the motion of the particle, we may draw the following conclusion:

Hamilton's Principle. If a particle moves from a point P_1 to a point P_2 in a time interval $t_1 \leq t \leq t_2$, then the actual path it follows is one for which the action assumes a stationary value.

There are examples of the motion of a particle in which, over a long period of time, the path actually maximizes the action (this is, of course, a stationary value—as predicted). However, over a sufficiently short time interval, it can be shown that the action is actually a minimum. In this form, Hamilton's principle is sometimes called the *principle of least action.* Loosely interpreted, it says that nature tends to equalize the kinetic and potential energies throughout the motion.

The preceding discussion shows that Newton's law implies Hamilton's principle. It can also be shown that Hamilton's principle implies Newton's law. Thus these two approaches to the dynamics of a particle—the vectorial and the variational—are equivalent. This result is an instance of a very important principle in physics: that pertinent physical laws can be expressed in terms of energy alone, without reference to any coordinate system.

The reasoning that we have presented thus far can be generalized to a system of n particles with masses $m_1, \ldots, m_n$. Suppose that these particles have position vectors $\mathbf{r}_j(t) = x_j(t)\mathbf{i} + y_j(t)\mathbf{k} + z_j(t)\mathbf{k}$. Each of these particles moves under the influence of conservative forces $\mathbf{F}_j = F_{j1}\mathbf{i} + F_{j2}\mathbf{j} + F_{j3}\mathbf{k}$. Then the potential energy of the system is a function $V(x_1, y_1, z_1, \ldots, x_n, y_n, z_n)$ such that

$$\frac{\partial V}{\partial x_j} = -F_{j1}, \quad \frac{\partial V}{\partial y_j} = -F_{j2}, \quad \frac{\partial V}{\partial z_j} = -F_{j3}\,.$$

The kinetic energy is then

$$T = \frac{1}{2}\sum_{j=1}^{n} m_j \left\{ \left(\frac{dx_j}{dt}\right)^2 + \left(\frac{dy_j}{dt}\right)^2 + \left(\frac{dz_j}{dt}\right)^2 \right\}.$$

The action over a time interval $t_1 \leq t \leq t_2$ is

$$A = \int_{t_1}^{t_2} (T - V)\, dt\,.$$

Just as before, we conclude (you should provide the details) that Newton's equations of motion for the system are

$$m_j \frac{d^2\mathbf{r}_j}{dt^2} = \mathbf{F}_j\,, \qquad j = 1, \ldots, n\,.$$

These equations are a necessary condition for the action to have a stationary value. Hamilton's principle thus holds for any finite system of particles in which the forces are conservative. It applies just as well to more general dynamical systems involving constraints and rigid bodies, as well as to continuous media.

Hamilton's principle can also be used to derive the basic laws of electricity and magnetism, quantum theory, and relativity. Many scientists regard the idea to be the most important unifying idea in mathematical physics. We close this Anatomy with an example:

EXAMPLE 9.3.6 Suppose that a particle of mass m is constrained to move on a given surface $G(x, y, z) = 0$, and that no force acts on it. Show then that the particle glides along a geodesic.

Solution: First notice that, since no force is present in the system, we have $V = 0$ so the Lagrangian $L = T - V$ reduces to T, where

$$T = \frac{1}{2}m \left\{ \left(\frac{dx}{dt}\right)^2 + \left(\frac{dy}{dt}\right)^2 + \left(\frac{dz}{dt}\right)^2 \right\}.$$

Now we apply Hamilton's principle, requiring the action

$$\int_{t_1}^{t_2} L\, dt = \int_{t_1}^{t_2} T\, dt$$

(subject, of course, to the side condition $G(x, y, z) = 0$).

By the ideas developed in Section 8.3, we see that this last is equivalent to requiring that the integral

$$\int_{t_1}^{t_2} [T + \lambda(t)G(x, y, z)]\, dt \tag{9.28}$$

be stationary with *no* side condition; here $\lambda(t)$ is an undetermined function of t (as in the theory of Lagrange multipliers). Euler's equations for the unconstrained variational problem (28) are

$$\begin{aligned} m\frac{d^2x}{dt^2} - \lambda G_x &= 0 \\ m\frac{d^2y}{dt^2} - \lambda G_y &= 0 \\ m\frac{d^2z}{dt^2} - \lambda G_z &= 0\,. \end{aligned}$$

When m and λ are eliminated from these equations, we find that

$$\frac{d^2x/dt^2}{G_x} = \frac{d^2y/dt^2}{G_y} = \frac{d^2z/dt^2}{G_z}\,.$$

Now the total energy $T + V = T$ of the particle is constant (this is intuitively clear, but can be established rigorously), so its speed is also constant. Therefore $s = kt$ for some constant k if the arc length s is measured from a suitable base point. This fact enables us to write our equations in the form

$$\frac{d^2x/ds^2}{G_x} = \frac{d^2y/ds^2}{G_y} = \frac{d^2z/ds^2}{G_z}\,.$$

These are precisely equation (27) at the end of Section 9.3.3. □

Problems for Review and Discovery

A. Drill Exercises

1. Find the extremals for the integral

$$I = \int_{x_1}^{x_2} f(x, y, y')\, dx$$

if the integrand is given by

(a) $(y')^2$
(b) $y^2 + (y')^2$
(c) $\dfrac{y}{1+(y')^2}$
(d) $\dfrac{y'}{1+y^2}$

2. Find the stationary function of

$$I = \int_0^2 [xy' + 2(y')^2]\, dx$$

which is determined by the boundary conditions $y(0) = 1$ and $y(2) = 0$.

3. Write down the integral that one must extremize in order to determine the geodesics on the surface (a paraboloid) $z = x^2 + y^2$ in space.

4. Use the method of Lagrange multipliers to determine the point on the surface $z = x^2 + y^2$ that is nearest to the point $(1, 1, 1)$.

B. Challenge Problems

1. For continuously differentiable functions $u : [a, b] \to \mathbb{R}^2$, consider the functional

$$E(u) \equiv \int_a^b \left|\frac{du}{dt}\right|^2 dt\,.$$

Calculate the Euler–Lagrange equations for this functional. Now define

$$L(u) \equiv \int_a^b \left|\frac{du}{dt}\right| dt\,.$$

Show that

$$L(u) \le \sqrt{(b-a)\cdot E(u)}\,.$$

Show that equality holds if and only if $|du/dt| \equiv$ constant.

2. Find all the functions of u which minimize the functional

$$J(u) \equiv \int_{-1}^{1} \left(1 - \frac{du}{dt}\right)^2 dt$$

with $u(-1) = 0 = u(1)$.

3. Consider all continuously differentiable curves $\gamma : \mathbb{R} \to$ (upper half plane). Write $\gamma(t) = (\gamma_1(t), \gamma_2(t))$. Define the functional

$$K(\gamma) \equiv \frac{1}{2} \cdot \int \frac{1}{[\gamma_2(t)]^2} \left|\frac{d\gamma}{dt}\right|^2 dt\,.$$

Calculate the Euler–Lagrange equations and determine all solutions.

4. For continuously differentiable curves $\gamma : \mathbb{R} \to$ (unit disc in $\mathbb{R}^2$), consider the functional

$$L(\gamma) \equiv \frac{1}{2} \int \frac{1}{(1-|\gamma(t)|^2)^2} \left|\frac{d\gamma}{dt}\right|^2 dt\,.$$

Calculate the Euler–Lagrange equations and find all solutions.

C. Problems for Discussion and Exploration

1. Consider a region R in the x-y plane bounded by a closed curve C. Let $z = z(x, y)$ be a function that is defined on R and assumes prescribed boundary values on C (but is otherwise arbitrary). This function can be thought of, for instance, as defining a variable surface fixed along its boundary in space. An integral of the form

$$I(z) = \iint\limits_R f(x, y, z, z_x, z_y)\, dxdy$$

will have values that depend on the choice of z, and we can consider the problem of finding a function z that gives a stationary value to this integral.

Follow the general line of reasoning presented in the text to derive Euler's equation

$$\frac{\partial}{\partial x}\left(\frac{\partial f}{\partial z_x}\right) + \frac{\partial}{\partial y}\left(\frac{\partial f}{\partial z_y}\right) - \frac{\partial f}{\partial z} = 0$$

for this problem.

2. Deduce the one-dimensional wave equation from Hamilton's principle by using the equation

$$\frac{\partial}{\partial x}\left(\frac{\partial f}{\partial z_x}\right) + \frac{\partial}{\partial y}\left(\frac{\partial f}{\partial z_y}\right) - \frac{\partial f}{\partial z} = 0$$

from Exercise 1.

3. Consider the collection $\mathcal{S}$ of continuously differentiable functions f on the interval $[-1, 1]$ with these properties:

(a) $f(x) \geq 0$ for $-1 \leq x \leq 1$

(b) $f(-1) = f(1) = 0$

(c) $f(0) \geq 1$

We pose the extremal problem of finding a function f in $\mathcal{S}$, which has the least area under it. Draw a picture to go with this problem. Explain why the problem has no solution. Which other variational problems of this sort have no solution? How can one tell in advance which problems have solutions and which do not?

4. For a twice continuously differentiable, positive $f : \mathbb{R} \to \mathbb{R}$, consider a surface of revolution in $\mathbb{R}^3$ having the form

$$S = \{(x, y, z) \in \mathbb{R}^3 : x^2 + y^2 = f(z)\}.$$

Can you say something about the geodesics on S? Are the curves $x = c$ or $y = c$ ever geodesics? Are the curves $z = c$ ever geodesics?

10

Systems of First-Order Equations

- The concept of a system of equations
- The solution of a system
- Linear systems
- Homogeneous linear systems
- Constant coefficients
- Nonlinear systems
- Predator-prey problems

10.1 Introductory Remarks

Systems of differential equations arise very naturally in many physical contexts. If $y_1, y_2, \ldots, y_n$ are functions of the variable x, then a system, for us, will have the form

$$\begin{aligned} y_1' &= f_1(x, y_1, \ldots, y_n) \\ y_2' &= f_2(x, y_1, \ldots, y_n) \\ &\cdots \\ y_n' &= f_n(x, y_1, \ldots, y_n) . \end{aligned} \tag{10.1}$$

In Section 4.5 we used a system of two second-order equations to describe the motion of coupled harmonic oscillators. In an example below we shall see how a system occurs in the context of dynamical systems having several degrees of freedom. In another context, we shall see a system of differential equations used to model a predator-prey problem in the study of population ecology.

From the mathematical point of view, systems of equations are useful in part because an nth order equation

$$y^{(n)} = f(x, y, y', \ldots, y^{(n-1)}) \tag{10.2}$$

can be regarded (after a suitable change of notation) as a system. To see this, we let

$$y_0 = y \, , \quad y_1 = y' \, , \quad \ldots \, , \quad y_{n-1} = y^{(n-1)} \, .$$

DOI: 10.1201/9781003214526-10

Then we have

$$\begin{aligned} y_0' &= y_1 \\ y_1' &= y_2 \\ y_2' &= y_3 \\ &\cdots \\ y_{n-1}' &= f(x, y_0, y_1, y_2, \ldots, y_{n-1}), \end{aligned}$$

and this system is equivalent to our original equation (2). In practice, it is sometimes possible to treat a system like this as a vector-valued, first-order differential equation, and to use techniques that we have studied in this book to learn about the (vector) solution.

For cultural reasons, and for general interest, we shall next turn to the n-body problem of classical mechanics. It, too, can be modeled by a system of ordinary differential equations. Imagine n particles with masses m_j and located at points (x_j, y_j, z_j) in three-dimensional space. Assume that these points exert a force on each other according to Newton's law of universal gravitation (which we shall formulate in a moment). If r_{ij} is the distance between m_i and m_j and if θ is the angle from the positive x-axis to the segment joining them (Figure 10.1), then the component of the force exerted on m_i by m_j is

$$\frac{Gm_i m_j}{r_{ij}^2}\cos\theta = \frac{Gm_i m_j (x_j - x_i)}{r_{ij}^3}.$$

Here G is a constant that depends on the force of gravity. Since the sum over j of all these components for $j \neq i$ equals $m_i(d^2 x_i/dt^2)$ (by Newton's second law), we obtain n second-order differential equations

$$\frac{d^2 x_i}{dt^2} = G \cdot \sum_{j \neq i} \frac{m_j (x_j - x_i)}{r_{ij}^3};$$

similarly,

$$\frac{d^2 y_i}{dt^2} = G \cdot \sum_{j \neq i} \frac{m_j (y_j - y_i)}{r_{ij}^3}$$

and

$$\frac{d^2 z_i}{dt^2} = G \cdot \sum_{j \neq i} \frac{m_j (z_j - z_i)}{r_{ij}^3}.$$

If we make the change of notation

$$v_{x_i} = \frac{dx_i}{dt}, \quad v_{y_i} = \frac{dy_i}{dt}, \quad v_{z_i} = \frac{dz_i}{dt},$$

then we can reduce our system of $3n$ second-order equations to $6n$ first-order equations with unknowns

$$x_1, v_{x_1}, x_2, v_{x_2}, \ldots, x_n, v_{x_n}, y_1, v_{y_1}, y_2, v_{y_2}, \ldots, y_n, v_{y_n}, z_1, v_{z_1}, z_2, v_{z_2}, \ldots, z_n, v_{z_n}.$$

We can also make the substitution

$$r_{ij}^3 = \left[(x_i - x_j)^2 + (y_i - y_j)^2 + (z_i - z_j)^2\right]^{3/2}.$$

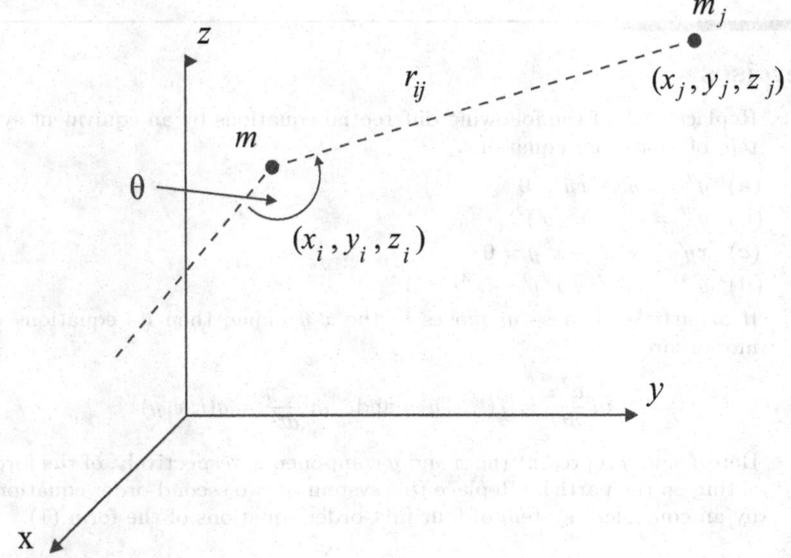

FIGURE 10.1
The force between two planets.

Then it can be proved that, if initial positions and velocities are specified for each of the n particles and if the particles do not collide (i.e., r_{ij} is never 0), then the subsequent position and velocity of each particle in the system are uniquely determined. This conclusion is at the heart of the once-popular philosophy of mechanistic determinism. The tenet is that the universe is nothing more than a giant machine whose future is inexorably fixed by its state at any given moment.[1]

This is the Newtonian model of the universe. It is completely deterministic. If $n = 2$, then the system was completely solved by Newton, giving rise to Kepler's Laws (Section 5.2). But for $n \geq 3$ there is a great deal that is not known. Of course this mathematical model can be taken to model the motions of the planets in our solar system. It is not known, for example, whether one of the planets (the Earth, let us say) will one day leave its orbit and go crashing into the sun. It is also not clear whether another planet will suddenly leave its orbit and go shooting out to infinity.

---

[1] The philosophy also led Sir James Jeans to define the universe as "a self-solving system of $6N$ simultaneous differential equations, where N is Eddington's number." The allusion is to Sir Arthur Eddington, who asserted (with more whimsy than precision) that

$$N = \frac{3}{2} \times 136 \times 2^{256}$$

is the total number of particles of matter in the universe. See [JEA] or [EDD].

Exercises

1. Replace each of the following differential equations by an equivalent system of first-order equations.
   (a) $y'' - xy' - xy = 0$
   (b) $y''' = y'' - x^2(y')^2$
   (c) $xy'' - x^2y' - x^3y = 0$
   (d) $y^{(4)} - xy''' + x^2y'' - x^3y = 1$
2. If a particle of mass m moves in the x-y plane, then its equations of motion are

$$m\,\frac{d^2x}{dt^2} = f(t, x, y) \quad \text{and} \quad m\,\frac{d^2y}{dt^2} = g(t, x, y)\,.$$

   Here f and g represent the x and y components, respectively, of the force acting on the particle. Replace this system of two second-order equations by an equivalent system of four first-order equations of the form (1).

10.2 Linear Systems

Our experience in this subject might lead us to believe that systems of *linear* equations will be the most tractable. That is indeed the case; we treat them in this section. By way of introduction, we shall concentrate on systems of two first-order equations in two unknown functions. Thus we have

$$\begin{cases} \dfrac{dx}{dt} &= F(t, x, y) \\ \dfrac{dy}{dt} &= G(t, x, y) \end{cases}$$

The brace is used here to stress that the equations are linked; the choice of t for the independent variable and of x and y for the dependent variables is traditional and will be borne out in the ensuing discussions.

In fact our system will have an even more special form because of linearity:

$$\begin{cases} \dfrac{dx}{dt} &= a_1(t)x + b_1(t)y + f_1(t) \\ \dfrac{dy}{dt} &= a_2(t)x + b_2(t)y + f_2(t)\,. \end{cases} \tag{10.3}$$

It will be convenient, and it is physically natural, for us to assume that the coefficient functions a_j, b_j, f_j, $j = 1, 2$, are continuous on a closed interval $[a, b]$ in the t-axis.

In the special case that $f_1 = f_2 \equiv 0$, we call the system *homogeneous.* Otherwise it is *nonhomogeneous.* A solution of this system is of course a *pair* of functions $(x(t), y(t))$ that satisfy both differential equations. We shall write

$$\begin{cases} x &= x(t) \\ y &= y(t)\,. \end{cases}$$

EXAMPLE 10.2.1 Verify that the system

$$\begin{cases} \dfrac{dx}{dt} &= 4x - y \\ \dfrac{dy}{dt} &= 2x + y \end{cases}$$

has

$$\begin{cases} x &= e^{3t} \\ y &= e^{3t} \end{cases}$$

and

$$\begin{cases} x &= e^{2t} \\ y &= 2e^{2t} \end{cases}$$

as solution sets.

Solution: We shall verify the first solution set, and leave the second for the reader.

Substituting $x = e^{3t}$, $y = e^{3t}$ into the first equation yields

$$\frac{d}{dt}e^{3t} = 4e^{3t} - e^{3t}$$

or

$$3e^{3t} = 3e^{3t}\,,$$

so that equation checks. For the second equation, we obtain

$$\frac{d}{dt}e^{3t} = 2e^{3t} + e^{3t}$$

or

$$3e^{3t} = 3e^{3t}\,,$$

so the second equation checks. □

We now give a sketch of the general theory of linear systems of first-order equations. Recall (Section 10.1) that any second-order, linear equation may be reduced to a first-order, linear system. Thus it will not be surprising that the theory we are about to describe is similar to the theory of second-order, linear equations.

We begin with a fundamental existence and uniqueness theorem.

Theorem 10.2.2 *Let $[a, b]$ be an interval and $t_0 \in [a, b]$. Suppose that a_j, b_j, f_j are continuous functions on $[a, b]$ for $j = 1, 2$. Let x_0 and y_0 be arbitrary numbers. Then there is one and only one solution to the system*

$$\begin{cases} \dfrac{dx}{dt} &= a_1(t)x + b_1(t)y + f_1(t) \\ \dfrac{dy}{dt} &= a_2(t)x + b_2(t)y + f_2(t) \end{cases}$$

satisfying $x(t_0) = x_0$, $y(t_0) = y_0$.

This theorem is nothing other than a vector-valued variant of Picard's fundamental result for first-order ordinary differential equations (Section 13.2).

We next discuss the structure of the solution of (3) that is obtained when $f_1(t) = f_2(t) \equiv 0$ (the so-called *homogeneous* situation). Thus we have

$$\begin{cases} \dfrac{dx}{dt} &= a_1(t)x + b_1(t)y \\ \dfrac{dy}{dt} &= a_2(t)x + b_2(t)y\,. \end{cases} \tag{10.4}$$

Of course the identically zero solution ($x(t) \equiv 0$, $y(t) \equiv 0$) is a solution of this homogeneous system. The next theorem—familiar in form—will be the key to constructing more useful solutions.

Theorem 10.2.3 *If the homogeneous system (4) has two solutions*

$$\begin{cases} x &= x_1(t) \\ y &= y_1(t) \end{cases} \quad \text{and} \quad \begin{cases} x &= x_2(t) \\ y &= y_2(t) \end{cases} \tag{10.5}$$

on $[a, b]$ then, for any constants c_1 and c_2,

$$\begin{aligned} x &= c_1x_1(t) + c_2x_2(t) \\ y &= c_1y_1(t) + c_2y_2(t) \end{aligned} \tag{10.6}$$

is also a solution on $[a, b]$.

Proof This is obtained by direct substitution of the solution into each of the equations. Details are left to the reader. □

Note, in the last theorem, that a new solution is obtained from the original two by multiplying the first by c_1 and the second by c_2 and then adding. We therefore call the newly created solution a *linear combination* of the given solutions (see the Appendix). Thus Theorem 10.2.3 simply says that a linear combination of two solutions of the linear system is also a solution of the system. As an instance, in Example 10.2.1, any pair of functions of the form

$$\left\{ \begin{array}{lcl} x & = & c_1 e^{3t} + c_2 e^{2t} \\ y & = & c_1 e^{3t} + c_2 2 e^{2t} \end{array} \right. \qquad (10.7)$$

is a solution of the given system.

The next obvious question to settle is whether the collection of all linear combinations of two independent solutions of the homogeneous system is in fact *all* the solutions (i.e., the *general solution*) of the system. By Theorem 10.2.2, we can generate all possible solutions provided we can arrange to satisfy all possible sets of initial conditions. This will now reduce to a simple and familiar algebra problem.

Demanding that, for some choice of c_1 and c_2, the solution

$$\left\{ \begin{array}{lcl} x & = & c_1 e^{3t} + c_2 e^{2t} \\ y & = & c_1 e^{3t} + c_2 e^{2t} \end{array} \right.$$

satisfy $x(t_0) = x_0$ and $y(t_0) = y_0$ amounts to specifying that

$$x_0 = c_1 x_1(t_0) + c_2 x_2(t_0)$$

and

$$y_0 = c_1 y_1(t_0) + c_2 y_2(t_0)\,.$$

This will be possible, for any choice of x_0 and y_0, provided that the determinant of the coefficients of the linear system is not zero. In other words, we require that

$$W(t) = \det \begin{pmatrix} x_1(t) & x_2(t) \\ y_1(t) & y_2(t) \end{pmatrix} = x_1(t) y_2(t) - y_1(t) x_2(t) \neq 0$$

on the interval $[a, b]$. This determinant is called the *Wronskian* of the two solutions.

Our discussion thus far establishes the following theorem:

Theorem 10.2.4 *If the two solutions (5) of the homogeneous system (4) have a nonvanishing Wronskian on the interval $[a, b]$, then (6) is the general solution of the system on this interval.*

Thus, in particular, (7) is the general solution of the system of differential equations in Example 10.2.1—for the Wronskian of the two solution sets is

$$W(t) = \det \begin{pmatrix} e^{3t} & e^{2t} \\ e^{3t} & 2e^{2t} \end{pmatrix} = e^{5t} ,$$

and this function of course never vanishes.

As we shall see later (see in particular Section 13.1), it is the case that either the Wronskian is identically zero or else it is never vanishing. For the record, we enunciate this property formally.

Theorem 10.2.5 *If $W(t)$ is the Wronskian of the two solutions of our homogeneous system (4), then either W is identically zero or else it is nowhere vanishing.*

Proof We calculate that

$$\begin{aligned} \frac{d}{dt} W(t) &= \frac{d}{dt} [x_1(t)y_2(t) - y_1(t)x_2(t)] \\ &= \frac{dx_1}{dt} y_2 + x_1 \frac{dy_2}{dt} - \frac{dy_1}{dt} x_2 - y_1 \frac{dx_2}{dt} \\ &= [a_1 x_1 + b_1 y_1] y_2 + x_1 [a_2 x_2 + b_2 y_2] \\ &\qquad -[a_2 x_1 + b_2 y_1] x_2 - y_1 [a_1 x_2 + b_1 y_2] \\ &= a_1 [x_1 y_2 - y_1 x_2] + b_2 [x_1 y_2 - y_1 x_2] \\ &= [a_1 + b_2] W . \end{aligned}$$

Thus the Wronskian W satisfies a familiar first-order, linear ordinary differential equation; we know immediately that the solution is

$$W(t) = C \cdot e^{\int [a_1(t) + b_2(t)]\, dt}$$

for some constant C. If $C \neq 0$ then the Wronskian never vanishes; if instead $C = 0$, then of course the Wronskian is identically zero. □

We now develop an alternative approach to the question of whether a given pair of solutions generates the general solution of a system. This new method is often more direct and more convenient.

The two solutions (5) are called *linearly dependent* on the interval $[a, b]$ if one ordered pair (x_1, y_1) is a constant multiple of the other pair (x_2, y_2). Thus they are *linearly dependent* if there is a constant k such that

$$\begin{aligned} x_1(t) &= k \cdot x_2(t) \\ y_1(t) &= k \cdot y_2(t) \end{aligned} \qquad \text{or} \qquad \begin{aligned} x_2(t) &= k \cdot x_1(t) \\ y_2(t) &= k \cdot y_1(t) \end{aligned}$$

for some constant k and for all $t \in [a, b]$. The solutions are *linearly independent* if neither is a constant multiple of the other in the sense just indicated. Clearly linear dependence is equivalent to the condition that there exist two constants c_1 and c_2, not both zero such that

$$\begin{aligned} c_1x_1(t) + c_2x_2(t) &= 0 \\ c_1y_1(t) + c_2y_2(t) &= 0 \end{aligned}$$

for all $t \in [a, b]$.

Theorem 10.2.6 *If the two solutions (5) of the homogeneous system (4) are linearly independent on the interval* $[a, b]$, *then (6) is the general solution of (4) on this interval.*

Proof The solutions are linearly independent if and only if the Wronskian is never zero, just as we have discussed. □

The interest of this new test is that one can usually determine by inspection whether two solutions are linearly independent.

Now it is time to return to the general case—of nonhomogeneous systems. We conclude our discussion with this result (and, again, note the analogy with second-order linear equations).

Theorem 10.2.7 *If the two solutions (5) of the homogeneous system (4) are linearly independent on* $[a, b]$ *and if*

$$\left\{ \begin{aligned} x &= x_p(t) \\ y &= y_p(t) \end{aligned} \right.$$

is any particular solution of the system (3) on this interval, then

$$\left\{ \begin{aligned} x &= c_1x_1(t) + c_2x_2(t) + x_p(t) \\ y &= c_1y_1(t) + c_2y_2(t) + y_p(t) \end{aligned} \right.$$

is the general solution of (3) on $[a, b]$.

Proof It suffices to show that if

$$\left\{ \begin{aligned} x &= x(t) \\ y &= y(t) \end{aligned} \right.$$

is an arbitrary solution of (3), then

$$\left\{ \begin{aligned} x &= x(t) - x_p(t) \\ y &= y(t) - y_p(t) \end{aligned} \right.$$

is a solution of (4). This is an exercise in substitution and elementary algebra; we leave the details to the reader. □

Although we would like to end this section with a dramatic example tying all the ideas together, this is in fact not feasible. In general it is quite difficult to find both a particular solution and the general solution to the associated homogeneous equations for a given system. We shall be able to treat the matter most effectively for systems with constant coefficients. We learn about that situation in the next section.

Exercises

1. Let the second-order linear equation

$$\frac{d^2x}{dt^2} + P(t)\frac{dx}{dt} + Q(t)x = 0 \tag{$*$}$$

be reduced to the system

$$\left\{\begin{array}{rcl} \dfrac{dx}{dt} & = & y \\ \dfrac{dy}{dt} & = & -Q(t)x - P(t)y\,. \end{array}\right. \tag{$**$}$$

If $x_1(t), x_2(t)$ are solutions of $(*)$ and if

$$\left\{\begin{array}{rcl} x & = & x_1(t) \\ y & = & y_1(t) \end{array}\right. \quad \text{and} \quad \left\{\begin{array}{rcl} x & = & x_2(t) \\ y & = & y_2(t) \end{array}\right.$$

are the corresponding solutions of $(**)$, then show that the Wronskian of $(*)$ in our earlier sense of Wronskian for a single equation is equal to the Wronskian of $(**)$ in the sense of the present section.

2. **(a)** Show that

$$\left\{\begin{array}{rcl} x & = & e^{4t} \\ y & = & e^{4t} \end{array}\right. \quad \text{and} \quad \left\{\begin{array}{rcl} x & = & e^{-2t} \\ y & = & -e^{-2t} \end{array}\right.$$

are solutions of the homogeneous system

$$\left\{\begin{array}{rcl} \dfrac{dx}{dt} & = & x + 3y \\ \dfrac{dy}{dt} & = & 3x + y\,. \end{array}\right.$$

(b) Show in two ways that the given solutions of the system in part **(a)** are linearly independent on every closed interval, and write the general solution of this system.

(c) Find the particular solution

$$\begin{cases} x = x(t) \\ y = y(t) \end{cases}$$

of this system for which $x(0) = 5$ and $y(0) = 1$.

3. **(a)** Show that

$$\begin{cases} x = 2e^{4t} \\ y = 3e^{4t} \end{cases} \quad \text{and} \quad \begin{cases} x = e^{-t} \\ y = -e^{-t} \end{cases}$$

are solutions of the homogeneous system

$$\begin{cases} \dfrac{dx}{dt} = x + 2y \\ \dfrac{dy}{dt} = 3x + 2y\,. \end{cases}$$

(b) Show in two different ways that the given solutions of the system in part **(a)** are linearly independent on every closed interval, and write the general solution of this system.

(c) Show that

$$\begin{cases} x = 3t - 2 \\ y = -2t + 3 \end{cases}$$

is a particular solution of the nonhomogeneous system

$$\begin{cases} \dfrac{dx}{dt} = x + 2y + t - 1 \\ \dfrac{dy}{dt} = 3x + 2y - 5t - 2\,. \end{cases}$$

Write the general solution of this system.

4. Obtain the given solutions of the homogeneous system in Exercise **3**

(a) by differentiating the first equation with respect to t and eliminating y;

(b) by differentiating the second equation with respect to t and eliminating x.

5. Use the method of Exercise **4** to find the general solution of the system

$$\begin{cases} \dfrac{dx}{dt} = x + y \\ \dfrac{dy}{dt} = y\,. \end{cases}$$

6. **(a)** Find the general solution of the system

$$\begin{cases} \dfrac{dx}{dt} = x \\ \dfrac{dy}{dt} = y\,. \end{cases}$$

(b) Show that any second-order equation obtained from the system in **(a)** is not equivalent to this system, in the sense that it has solutions that are not part of any solution of the system. Thus, although higher-order equations are equivalent to systems, the reverse is not true, and systems are definitely more general.

7. Give a complete proof of Theorem 10.2.3.

8. Finish the proof of Theorem 10.2.7.

10.3 Homogeneous Linear Systems with Constant Coefficients

It is now time for us to give a complete and explicit solution of the system

$$\left\{\begin{array}{rcl} \dfrac{dx}{dt} & = & a_1 x + b_1 y \\ \dfrac{dy}{dt} & = & a_2 x + b_2 y \,. \end{array}\right. \tag{10.8}$$

Here a_1, a_2, b_1, and b_2 are given constants. Sometimes a system of this type can be solved by differentiating one of the two equations, eliminating one of the dependent variables, and then solving the resulting second-order linear equation. In this section we propose an alternative method that is based on constructing a pair of linearly independent solutions directly from the given system.

Working by analogy with our studies of first-order linear equations, we now posit that our system has a solution of the form

$$\left\{\begin{array}{rcl} x & = & Ae^{mt} \\ y & = & Be^{mt} \,. \end{array}\right. \tag{10.9}$$

We substitute (9) into (8) and obtain

$$\begin{array}{rcl} Ame^{mt} & = & a_1 Ae^{mt} + b_1 Be^{mt} \\ Bme^{mt} & = & a_2 Ae^{mt} + b_2 Be^{mt} \,. \end{array}$$

(It is worth noting here that m is an eigenvalue for the differential operator—see the discussion in Section 13.1.) Dividing out the common factor of e^{mt} and rearranging yields the associated linear algebraic system

$$\begin{array}{rcl} (a_1 - m)A + b_1 B & = & 0 \\ a_2 A + (b_2 - m)B & = & 0 \end{array} \tag{10.10}$$

in the unknowns A and B.

Of course the system (10) has the trivial solution $A = B = 0$. This makes (9) the trivial solution of (8). We are of course seeking nontrivial solutions. The algebraic system (10) will have nontrivial solutions precisely when the determinant of the coefficients vanishes, i.e.,

$$\det \begin{pmatrix} a_1 - m & b_1 \\ a_2 & b_2 - m \end{pmatrix} = 0 .$$

Expanding the determinant, we find this quadratic expression for the unknown m:

$$m^2 - (a_1 + b_2)m + (a_1 b_2 - a_2 b_1) = 0 . \tag{10.11}$$

We call this the *associated equation* (or sometimes the *auxiliary equation*) for the original system (8).

Let m_1 and m_2 be the roots of equation (11). If we replace m by m_1 in (11), then we know that the resulting equations have a nontrivial solution set A_1, B_1 so that

$$\begin{cases} x = A_1 e^{m_1 t} \\ y = B_1 e^{m_1 t} \end{cases} \tag{10.12}$$

is a nontrivial solution of the original system (10). Proceeding similarly with m_2, we find another nontrivial solution

$$\begin{cases} x = A_2 e^{m_2 t} \\ y = B_2 e^{m_2 t} . \end{cases} \tag{10.13}$$

In order to be sure that we obtain two linearly independent solutions, and hence the general solution for (10), we must examine in detail each of the three possibilities for m_1 and m_2.

Distinct Real Roots. When m_1 and m_2 are distinct real numbers, then the solutions (12) and (13) are linearly independent. For, in fact, $e^{m_1 t}$ and $e^{m_2 t}$ are linearly independent. Thus

$$\begin{cases} x = c_1 A_1 e^{m_1 t} + c_2 A_2 e^{m_2 t} \\ y = c_1 B_1 e^{m_1 t} + c_2 B_2 e^{m_2 t} \end{cases}$$

is the general solution of (8).

EXAMPLE 10.3.1 Find the general solution of the system

$$\begin{cases} \dfrac{dx}{dt} = x + y \\ \dfrac{dy}{dt} = 4x - 2y . \end{cases}$$

Solution: The associated algebraic system is

$$\begin{aligned} (1 - m)A + B &= 0 \\ 4A + (-2 - m)B &= 0 . \end{aligned} \tag{10.14}$$

The auxiliary equation is then

$$m^2 + m - 6 = 0 \qquad \text{or} \qquad (m+3)(m-2) = 0\,,$$

so that $m_1 = -3$, $m_2 = 2$.

With m_1, (14) becomes

$$\begin{aligned} 4A + B &= 0 \\ 4A + B &= 0\,. \end{aligned}$$

Since these equations are identical, it is plain that the determinant of the coefficients is zero and there are nontrivial solutions.

A simple nontrivial solution of our system is $A = 1$, $B = -4$. Thus

$$\begin{cases} x = e^{-3t} \\ y = -4e^{-3t} \end{cases}$$

is a nontrivial solution of our original system of differential equations.

With m_2, (14) becomes

$$\begin{aligned} -A + B &= 0 \\ 4A - 4B &= 0\,. \end{aligned}$$

Plainly these equations are multiples of each other, and there are nontrivial solutions.

A simple nontrivial solution of our system is $A = 1$, $B = 1$. Thus

$$\begin{cases} x = e^{2t} \\ y = e^{2t} \end{cases}$$

is a nontrivial solution of our original system of differential equations.

Clearly the two solution sets that we have found are linearly independent. Thus

$$\begin{cases} x = c_1 e^{-3t} + c_2 e^{2t} \\ y = -4c_1 e^{-3t} + c_2 e^{2t} \end{cases}$$

is the general solution of our system. □

Distinct Complex Roots. In fact the only way that complex roots can occur as roots of a quadratic equation is as distinct conjugate roots $a \pm ib$, where a and b are real numbers and $b \neq 0$. In this case we expect the coefficients A and B to be complex numbers (which, for convenience, we shall call A_j^* and B_j^*), and we obtain the two linearly independent solutions

$$\begin{cases} x = A_1^* e^{(a+ib)t} \\ y = B_1^* e^{(a+ib)t} \end{cases} \quad \text{and} \quad \begin{cases} x = A_2^* e^{(a-ib)t} \\ y = B_2^* e^{(a-ib)t}\,. \end{cases} \tag{10.15}$$

However, these are complex-valued solutions. On physical grounds, we often want real-valued solutions; we therefore need a procedure for extracting such solutions.

We write $A_1^* = A_1 + iA_2$ and $B_1^* = B_1 + iB_2$, and we apply Euler's formula to the exponential. Thus the first indicated solution becomes

$$\left\{\begin{array}{rcl} x & = & (A_1 + iA_2)e^{at}(\cos bt + i \sin bt) \\ y & = & (B_1 + iB_2)e^{at}(\cos bt + i \sin bt) . \end{array}\right.$$

We may rewrite this as

$$\left\{\begin{array}{rcl} x & = & e^{at}\left[(A_1 \cos bt - A_2 \sin bt) + i(A_1 \sin bt + A_2 \cos bt)\right] \\ y & = & e^{at}\left[(B_1 \cos bt - B_2 \sin bt) + i(B_1 \sin bt + B_2 \cos bt)\right] . \end{array}\right.$$

From this information, just as in the case of single differential equations (Section 4.1), we deduce that there are two real-valued solutions to the system:

$$\left\{\begin{array}{rcl} x & = & e^{at}(A_1 \cos bt - A_2 \sin bt) \\ y & = & e^{at}(B_1 \cos bt - B_2 \sin bt) \end{array}\right. \tag{10.16}$$

and

$$\left\{\begin{array}{rcl} x & = & e^{at}(A_1 \sin bt + A_2 \cos bt) \\ y & = & e^{at}(B_1 \sin bt + B_2 \cos bt) . \end{array}\right. \tag{10.17}$$

One can use just algebra to see that these solutions are linearly independent (exercise for the reader). Thus the general solution to our linear system of ordinary differential equations is

$$\left\{\begin{array}{rcl} x & = & e^{at}\left[c_1(A_1 \cos bt - A_2 \sin bt) + c_2(A_1 \sin bt + A_2 \cos bt)\right] \\ y & = & e^{at}\left[c_1(B_1 \cos bt - B_2 \sin bt) + c_2(B_1 \sin bt + B_2 \cos bt)\right] . \end{array}\right.$$

Since this already gives us the general solution of our system, there is no need to consider the second of the two solutions given in (15). Just as in the case of a single differential equation of second order, our analysis of that second solution would give rise to the same general solution.

Repeated Real Roots. When $m_1 = m_2 = m$ then (12) and (13) are not linearly independent; in this case we have just the one solution

$$\left\{\begin{array}{rcl} x & = & Ae^{mt} \\ y & = & Be^{mt} . \end{array}\right.$$

Our experience with repeated roots of the auxiliary equation in the case of second-order linear equations with constant coefficients might lead us to guess that there is a second solution obtained by introducing into each of x and y a coefficient of t. In fact the present situation calls for something a bit more elaborate. We seek a second solution of the form

$$\left\{\begin{array}{rcl} x & = & (A_1 + A_2 t)e^{mt} \\ y & = & (B_1 + B_2 t)e^{mt} . \end{array}\right. \tag{10.18}$$

The general solution is then

$$\left\{\begin{array}{rcl} x & = & c_1 A e^{mt} + c_2 (A_1 + A_2 t) e^{mt} \\ y & = & c_1 B e^{mt} + c_2 (B_1 + B_2 t) e^{mt} . \end{array}\right. \tag{10.19}$$

The constants A_1, A_2, B_1, and B_2 are determined by substituting (18) into the original system of differential equations. Rather than endeavor to carry out this process in complete generality, we now illustrate the idea with a simple example.[2]

EXAMPLE 10.3.2 Find the general solution of the system

$$\left\{\begin{array}{rcl} \dfrac{dx}{dt} & = & 3x - 4y \\ \dfrac{dy}{dt} & = & x - y . \end{array}\right.$$

Solution: The associated linear algebraic system is

$$\begin{array}{rcl} (3-m)A - 4B & = & 0 \\ A + (-1-m)B & = & 0 . \end{array}$$

The auxiliary quadratic equation is then

$$m^2 - 2m + 1 = 0 \qquad \text{or} \qquad (m-1)^2 = 0 .$$

Thus $m_1 = m_2 = m = 1$.

With $m = 1$, the linear system becomes

$$\begin{array}{rcl} 2A - 4B & = & 0 \\ A - 2B & = & 0 . \end{array}$$

Of course $A = 2$, $B = 1$ is a solution, so we have

$$\left\{\begin{array}{rcl} x & = & 2e^t \\ y & = & e^t \end{array}\right.$$

as a nontrivial solution of the given system.

---

[2]There is an exception to the general discussion we have just presented that we ought to at least note. Namely, in case the coefficients of the system of ordinary differential equations satisfy $a_1 = b_2 = a$ and $a_2 = b_1 = 0$ then the associated quadratic equation is $m^2 - 2ma + a^2 = (m-a)^2 = 0$. Thus $m = a$ and the constants A and B are completely unrestricted (i.e., the putative equations that we usually solve for A and B reduce to a trivial tautology). In this case the general solution of our system of differential equations is just

$$\left\{\begin{array}{rcl} x & = & c_1 e^{mt} \\ y & = & c_2 e^{mt} . \end{array}\right.$$

What is going on here is that each differential equation can be solved independently; there is no interdependence. We call such a system *uncoupled*.

We now seek a second linearly independent solution of the form

$$\left\{\begin{array}{rcl} x & = & (A_1 + A_2 t)e^t \\ y & = & (B_1 + B_2 t)e^t \,. \end{array}\right. \tag{10.20}$$

When these expressions are substituted into our system of differential equations, we find that

$$\begin{array}{rcl} (A_1 + A_2 t + A_2)e^t & = & 3(A_1 + A_2 t)e^t - 4(B_1 + B_2 t)e^t \\ (B_1 + B_2 t + B_2)e^t & = & (A_1 + A_2 t)e^t - (B_1 + B_2 t)e^t \,. \end{array}$$

Using a little algebra, these can be reduced to

$$\begin{array}{rcl} (2A_2 - 4B_2)t + (2A_1 - A_2 - 4B_1) & = & 0 \\ (A_2 - 2B_2)t + (A_1 - 2B_1 - B_2) & = & 0 \,. \end{array}$$

Since these last are to be identities in the variable t, we can only conclude that

$$\begin{array}{ll} 2A_2 - 4B_2 = 0 & 2A_1 - A_2 - 4B_1 = 0 \\ A_2 - 2B_2 = 0 & A_1 - 2B_1 - B_2 = 0 \,. \end{array}$$

The two equations on the left have $A_2 = 2$, $B_2 = 1$ as a solution. With these values, the two equations on the right become

$$\begin{array}{rcl} 2A_1 - 4B_1 & = & 2 \\ A_1 - 2B_1 & = & 1 \,. \end{array}$$

Of course their solution is $A_1 = 1$, $B_1 = 0$. We now insert these numbers into (20) to obtain

$$\left\{\begin{array}{rcl} x & = & (1 + 2t)e^t \\ y & = & te^t \,. \end{array}\right.$$

This is our second solution.

Since it is clear from inspection that the two solutions we have found are linearly independent, we conclude that

$$\left\{\begin{array}{rcl} x & = & 2c_1 e^t + c_2(1 + 2t)e^t \\ y & = & c_1 e^t + c_2 te^t \end{array}\right.$$

is the general solution of our system of differential equations. □

Exercises

1. Use the methods treated in this section to find the general solution of each of the following systems.

(a) $\begin{cases} \dfrac{dx}{dt} = -3x + 4y \\ \dfrac{dy}{dt} = -2x + 3y \end{cases}$ (e) $\begin{cases} \dfrac{dx}{dt} = 2x \\ \dfrac{dy}{dt} = 3y \end{cases}$

(b) $\begin{cases} \dfrac{dx}{dt} = 4x - 2y \\ \dfrac{dy}{dt} = 5x + 2y \end{cases}$ (f) $\begin{cases} \dfrac{dx}{dt} = -4x - y \\ \dfrac{dy}{dt} = x - 2y \end{cases}$

(c) $\begin{cases} \dfrac{dx}{dt} = 5x + 4y \\ \dfrac{dy}{dt} = -x + y \end{cases}$ (g) $\begin{cases} \dfrac{dx}{dt} = 7x + 6y \\ \dfrac{dy}{dt} = 2x + 6y \end{cases}$

(d) $\begin{cases} \dfrac{dx}{dt} = 4x - 3y \\ \dfrac{dy}{dt} = 8x - 6y \end{cases}$ (h) $\begin{cases} \dfrac{dx}{dt} = x - 2y \\ \dfrac{dy}{dt} = 4x + 5y \end{cases}$

2. Show that the condition $a_2 b_1 > 0$ implies that the system (8) has two real-valued linearly independent solutions of the form (9). However, the converse is not true.

3. Show that the Wronskian of the two solutions (16) and (17) is

$$W(t) = (A_1 B_2 - A_2 B_1)e^{2at}.$$

Prove that $A_1 B_2 - A_2 B_1 \neq 0$.

4. Show that in formula (19) the constants A_2 and B_2 satisfy the same linear algebraic system as the constants A and B, and that consequently we may put $A_2 = A$ and $B_2 = B$ without any loss of generality.

5. Consider the nonhomogeneous linear system

$$\begin{cases} \dfrac{dx}{dt} = a_1(t)x + b_1(t)y + f_1(t) \\ \dfrac{dy}{dt} = a_2(t)x + b_2(t)y + f_2(t) \end{cases} \qquad (*)$$

and the corresponding homogeneous system

$$\begin{cases} \dfrac{dx}{dt} = a_1(t)x + b_1(t)y \\ \dfrac{dy}{dt} = a_2(t)x + b_2(t)y\,. \end{cases} \qquad (**)$$

(a) Suppose that

$$\begin{cases} x = x_1(t) \\ y = y_1(t) \end{cases} \quad \text{and} \quad \begin{cases} x = x_2(t) \\ y = y_2(t) \end{cases}$$

are linearly independent solutions of (**) so that

$$\begin{cases} x = c_1 x_1(t) + c_2 x_2(t) \\ y = c_1 y_1(t) + c_2 y_2(t) \end{cases}$$

is its general solution. Then show that

$$\begin{cases} x = v_1(t)x_1(t) + v_2(t)x_2(t) \\ y = v_1(t)y_1(t) + v_2(t)y_2(t) \end{cases}$$

is a particular solution of $(*)$ if the functions v_1 and v_2 satisfy the system

$$\begin{aligned} v_1'x_1 + v_2'x_2 &= f_1 \\ v_1'y_1 + v_2'y_2 &= f_2 \,. \end{aligned}$$

This technique for finding particular solutions of nonhomogeneous linear systems is called the *method of variation of parameters.*

(b) Apply the method described in part **(a)** to find a particular solution of the nonhomogeneous system

$$\begin{cases} \dfrac{dx}{dt} = x + y - 5t + 2 \\ \dfrac{dy}{dt} = 4x - 2y - 8t - 8 \,. \end{cases}$$

Note that the corresponding homogeneous system is solved in Example 10.3.1.

10.4 Nonlinear Systems: Volterra's Predator-Prey Equations

Imagine an island inhabited by foxes and rabbits. Foxes eat rabbits; rabbits, in turn, develop methods of evasion to avoid being eaten. The resulting interaction is a fascinating topic for study, and is amenable to differential equations.

To appreciate the nature of the dynamic between the foxes and the rabbits, let us describe some of the vectors at play. We take it that the foxes eat rabbits—that is their source of food—and the rabbits eat only clover. We assume that there is an endless supply of clover; the rabbits never run out of food. The big worry for rabbits is foxes.

When the rabbits are abundant, then the foxes flourish from eating rabbits and their population grows. When the foxes become too numerous and eat too many rabbits, then the rabbit population declines; as a result, the foxes enter a period of famine and their population begins to decline. As the foxes decrease in number, the rabbits become relatively safe and their population starts to increase again. This triggers a new increase in the fox population—as the foxes now have an increased source of food. As time goes on, we see an endlessly repeating cycle of interrelated increases and decreases in the populations of the two species. See Figure 10.2, in which the sizes of the populations (x for rabbits, y for foxes) are plotted against time.

Problems of the sort that we have described here have been studied, for many years, by both mathematicians and biologists. It is pleasing to see how the mathematical analysis confirms the intuitive perception of the situation as described above. In our analysis below, we shall follow the approach of Vito Volterra (1860–1940), who was one of the pioneers in this subject.

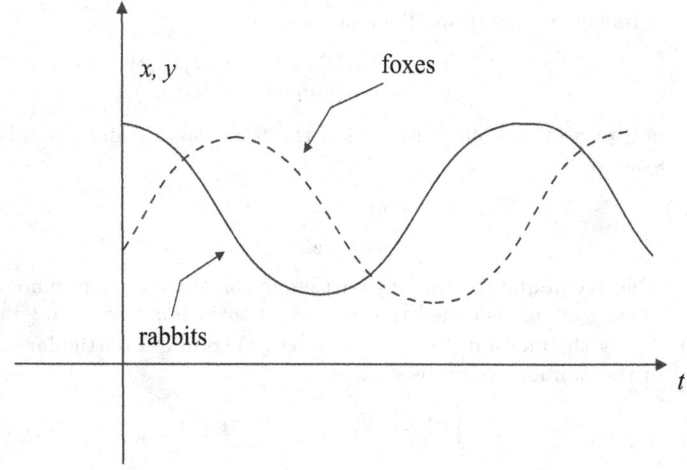

FIGURE 10.2
The rabbit and fox populations.

Math Nugget

Vito Volterra (1860–1940) was an eminent Italian mathematician whose interest in mathematics began at age 11 when he studied Legendre's *Geometry.* At the age of 13 he was already considering the three-body problem (see Section 5.2) and made some contributions by partitioning the time into small intervals over which the force could be considered to be constant.

Volterra's family was extremely impoverished (his father died when he was just two years old); but, after attending lectures at Florence, he was able to continue his studies in Pisa in 1878. There he studied under Betti, graduating Doctor of Physics in 1882. His thesis on hydrodynamics included some results of Stokes, which Volterra discovered later but independently.

Volterra became Professor of Mechanics at Pisa in 1883. After Betti's death, he occupied the Chair of Mathematical Physics. He was later appointed to the Chair of Mechanics at Turin, and then to the Chair of Mathematical Physics in Rome in 1900.

Along with work of Hilbert and Fredholm, Volterra's early studies of integral equations ushered in the full-scale development of linear analysis that dominated the first half of the twentieth century. In later life Volterra engaged in profound studies of the interplay of mathematics and biology. "Volterra's equation" is encountered in ecological studies of lynxes and hares in Canada and other original analyses.
During World War I, Volterra joined the Air Force. He made many journeys to France and England in order to promote scientific collaboration. After the war, he returned to the University of Rome; at that time his interests moved to mathematical biology. He studied the Verhulst equation and logistic curves. He also wrote about predator-prey problems.
Fascism overtook Italy in 1922, and Volterra fought valiantly against it in the Italian Parliament. In 1930, however, the Parliament was abolished; when Volterra refused to take an oath of allegiance to the Fascist Government in 1931 he was then forced to resign from the University of Rome. He spent the rest of his life living abroad, mostly in Paris (but also in Spain and other countries).
In 1938, Volterra was offered an honorary degree by the University of St. Andrews. But his doctor would not allow him to travel abroad to receive it. It is noteworthy that Volterra gave a total of four plenary lectures to various International Congresses of Mathematicians—more than any other scholar in history. The International Congress is *the* important, worldwide gathering of mathematicians that takes place every four years.

If x is the number of rabbits at time t, then the relation

$$\frac{dx}{dt} = ax\,, \qquad a > 0\,,$$

should hold, provided that the rabbits' food supply is unlimited and there are no foxes. This simply says that the rate of increase of the number of rabbits is proportional to the number present (just because the rabbits will keep reproducing).

It is natural to assume that the number of "encounters" between rabbits and foxes per unit of time is jointly proportional to x and y. If we furthermore make the plausible assumption that a certain proportion of those encounters results in a rabbit being eaten, then we have

$$\frac{dx}{dt} = ax - bxy\,, \qquad a, b > 0\,.$$

In the same way, we notice that in the absence of rabbits the foxes die out, and their increase depends on the number of encounters with rabbits. Thus the same logic leads to the companion differential equation

$$\frac{dy}{dt} = -cy + gxy\,, \qquad c, g > 0\,.$$

We have derived the following nonlinear system describing the interaction of the foxes and the rabbits:

$$\left\{\begin{array}{rcl} \dfrac{dx}{dt} & = & x(a - by) \\ \dfrac{dy}{dt} & = & -y(c - gx)\,. \end{array}\right. \tag{10.21}$$

Equations (21) are called *Volterra's predator-prey equations.* It is a basic and unavoidable fact that this system cannot be solved explicitly in terms of elementary functions. On the other hand, we can perform what is known as a *phase plane analysis* and learn a great deal about the behavior of $x(t)$ and $y(t)$.

To be more specific, instead of endeavoring to describe x as a function of t and y as a function of t, we instead think of

$$\left\{\begin{array}{rcl} x & = & x(t) \\ y & = & y(t) \end{array}\right.$$

as the parametric equations of a curve in the x-y plane. We shall be able to learn a great deal about the rectangular equations of this curve.

We begin by eliminating t in (21) and separating the variables. Thus

$$\begin{array}{rcl} \dfrac{dx}{x(a - by)} & = & dt \\ \dfrac{dy}{-y(c - gx)} & = & dt \end{array}$$

hence

$$\frac{dx}{x(a - by)} = \frac{dy}{-y(c - gx)}$$

or

$$\frac{(a - by)\,dy}{y} = -\frac{(c - gx)\,dx}{x}\,.$$

Integration now yields

$$a \ln y - by = -c \ln x + gx + C\,.$$

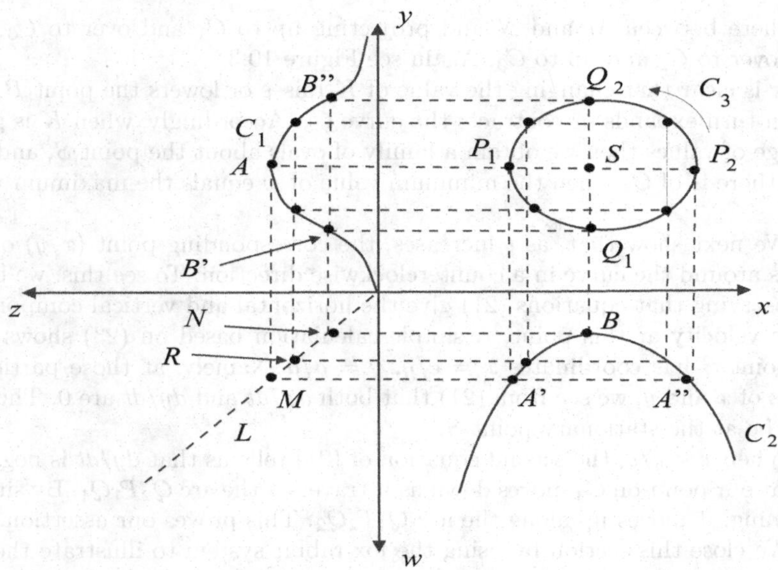

FIGURE 10.3
Plots of the graphs of C_1 and C_2.

In other words,

$$y^a e^{-by} = e^C x^{-c} e^{gx} \tag{10.22}$$

If we take it that $x(0) = x_0$ and $y(0) = y_0$ then we may solve this last equation for e^C and find that

$$e^C = x_0^c y_0^a e^{-gx_0 - by_0} .$$

It is convenient to let $e^C = K$.

In fact we cannot solve (22) for either x or y. But we can utilize an ingenious method of Volterra to find points on the curve. To proceed, we give the left-hand side of (22) the name of z and the right-hand side the name of w. Then we plot the graphs C_1 and C_2 of the functions

$$z = y^a e^{-by} \qquad \text{and} \qquad w = Kx^{-c}e^{gx} \tag{10.23}$$

as shown in Figure 10.3. Since $z = w$ (by (22)), we must in the third quadrant depict this relationship with the dotted line L. To the maximum value of z given by the point A on C_1, there corresponds one value of y and—via M on L and the corresponding points A' and A'' on C_2—two values of x, and these determine the bounds between which x may vary.

Similarly, the minimum value of w given by B on C_2 leads to N on L and hence to B' and B'' on C_1; these points determine the limiting values for y. In this way we find the points P_1, P_2, and Q_1, Q_2 on the desired curve C_3. Additional points are easily found by starting on L at a point R (let us say) anywhere between M and N and projecting up to C_1 and over to C_3, and then over to C_2 and up to C_3. Again see Figure 10.3.

It is clear that changing the value of K raises or lowers the point B, and this in turn expands or contracts the curve C_3. Accordingly, when K is given a range of values then we obtain a family of ovals about the point S, and this is all there is of C_3 when the minimum value of w equals the maximum value of z.

We next show that, as t increases, the corresponding point (x, y) on C_3 moves around the curve in a counterclockwise direction. To see this, we begin by observing that equations (21) give the horizontal and vertical components of the velocity at this point. A simple calculation based on (23) shows that the point S has coordinates $x = c/g$, $y = a/b$. Namely, at those particular values of x and y, we see from (21) that both dx/dt and dy/dt are 0. Thus we must be at the stationary point S.

When $x < c/g$, the second equation of (21) tells us that dy/dt is negative so that our point on C_3 moves down as it traverses the arc $Q_2P_1Q_1$. By similar reasoning, it moves up along the arc $Q_1P_2Q_2$. This proves our assertion.

We close this section by using the fox-rabbit system to illustrate the important method of *linearization.*[3] First note that, if the rabbit and fox populations are, respectively, constantly equal to

$$x = \frac{c}{g} \qquad \text{and} \qquad y = \frac{a}{b}, \tag{10.24}$$

then the system (21) is satisfied and we have $dx/dt \equiv 0$ and $dy/dt \equiv 0$. Thus there is no increase or decrease in either x or y. The populations (24) are called *equilibrium populations*; the populations x and y can maintain themselves indefinitely at these constant levels. This is the special case in which the minimum of w equals the maximum of z so that the oval C_3 reduces to the point S.

We now return to the general case and put

$$x = \frac{c}{g} + X \qquad \text{and} \qquad y = \frac{a}{b} + Y\,;$$

here we think of X and Y as the deviations of x and y from their equilibrium values. An easy calculation shows that, if we replace x and y in (21) with X and Y (which simply amounts to translating the point $(c/g, a/b)$ to the origin), then (21) becomes

$$\left\{\begin{array}{lcl} \dfrac{dX}{dt} & = & -\dfrac{bc}{g}Y - bXY \\ \dfrac{dY}{dt} & = & \dfrac{ag}{b}X + gXY\,. \end{array}\right. \tag{10.25}$$

The process of linearization now consists of assuming that, if X and Y

[3]The remainder of the book will devote a good deal of time and effort to the linearization technique.

are small, then the XY terms in (25) can be treated as negligible and hence discarded. This process results in (25) simplifying to the linear system (hence the name)

$$\begin{cases} \dfrac{dX}{dt} = -\dfrac{bc}{g}Y \\ \dfrac{dY}{dt} = \dfrac{ag}{b}X. \end{cases} \tag{10.26}$$

It is straightforward to solve (25) by the methods developed in this chapter. Easier still is to divide the left sides and right sides, thus eliminating dt, to obtain

$$\frac{dY}{dX} = -\frac{ag^2}{b^2c}\frac{X}{Y}.$$

The solution of this last equation is immediately seen to be

$$ag^2X^2 + b^2cY^2 = C^2.$$

This is a family of ellipses centered at the origin in the X-Y plane. Since ellipses are qualitatively similar to the ovals of Figure 10.3, we may hope that (26) is a reasonable approximation to (25).

One of the important themes that we have introduced in this chapter, that arose naturally in our study of systems, is that of nonlinearity. Nonlinear equations have none of the simple structure, nor any concept of "general solution," that the more familiar linear equations have. They are currently a matter of intense study.

In studying a system like (21), we have learned to direct our attention to the behavior of solutions near points in the x-y plane at which the right sides both vanish. We have seen why periodic solutions (i.e., those that yield simple closed curves like C_3 in Figure 10.3) are important and advantageous for our analysis. And we have given a brief hint of how it can be useful to study a nonlinear system by approximation with a linear system. In Chapter 12, we shall take a closer look at nonlinear theory, and we shall develop some of the present themes in greater detail.

Exercises

1. Eliminate y from the system (21) and obtain the nonlinear second-order equation satisfied by the function $x(t)$.
2. Show that $d^2y/dt^2 > 0$ whenever $dx/dt > 0$. What is the meaning of this result in terms of Figure 10.2?

Anatomy of an Application

SOLUTION OF SYSTEMS WITH MATRICES AND EXPONENTIALS

We now introduce an alternative notational device for studying linear systems with constant coefficients. It is in some ways more elegant than the technique presented in the text. It is also more efficient from a calculational point of view so that if one wanted to program a computer to solve systems of equations, then one would likely use this technique. The method that we shall present will depend on matrix theory and the elegant properties that exponential functions have vis à vis differential equations.

Consider a system

$$\left\{\begin{array}{rcl} \frac{dx}{dt} & = & \alpha x + \beta y \\ \frac{dy}{dt} & = & \gamma x + \delta y \,. \end{array}\right. \tag{10.27}$$

We define the matrix

$$A = \begin{pmatrix} \alpha & \beta \\ \gamma & \delta \end{pmatrix}$$

and the column vector

$$X(t) = \begin{pmatrix} x(t) \\ y(t) \end{pmatrix} .$$

Then we may rewrite the system (27) as

$$\frac{d}{dt}X = AX \,, \tag{10.28}$$

where it is understood that the right-hand side consists of the column vector X multiplied by the square matrix A. Of course equation (28) is reminiscent of first-order differential equations that we studied in Chapter 1. Those equations had exponential functions as solutions. With this thought in mind, we are tempted to guess that equation (28) has

$$\widetilde{X}(t) = e^{tA} \tag{10.29}$$

as its solution.

But what can this mean? How does one exponentiate a matrix?

Recall that the rigorous definition of the exponential function is with a power series:

$$e^x = \sum_{j=0}^{\infty} \frac{x^j}{j!} = 1 + x + \frac{x^2}{2!} + \frac{x^3}{3!} + \cdots .$$

What prevents us from putting a matrix in for x? Thus

$$e^{tA} = I + tA + \frac{t^2}{2!}A^2 + \frac{t^3}{3!}A^3 + \cdots .$$

A bit of interpretation is in order. First, I represents the identity matrix

$$I = \begin{pmatrix} 1 & 0 \\ 0 & 1 \end{pmatrix} .$$

Second, we multiply a matrix by a scalar by simply multiplying each matrix entry by that scalar.

But a more incisive comment is this. We are seeking a solution vector of the form

$$X(t) = \begin{pmatrix} x(t) \\ y(t) \end{pmatrix} ,$$

but the alleged solution $\widetilde{X}$ that we have found is a 2×2 matrix. What does this mean?

Let the columns of $\widetilde{X}$ be C_1 and C_2. Thus

$$\widetilde{X} = \begin{pmatrix} C_1 & C_2 \end{pmatrix} .$$

Since, formally,

$$\frac{d}{dt}\widetilde{X} = A\widetilde{X} ,$$

it follows from the way that we define matrix multiplication that

$$\frac{d}{dt}C_1 = AC_1$$

and

$$\frac{d}{dt}C_2 = AC_2 .$$

So C_1 and C_2 should be the two linearly independent solutions that we seek.

The best way to understand these abstract ideas is by way of some concrete examples.

EXAMPLE Consider the system

$$\begin{cases} \dfrac{dx}{dt} = 4x - y \\ \dfrac{dy}{dt} = 2x + y . \end{cases}$$

We define

$$A = \begin{pmatrix} 4 & -1 \\ 2 & 1 \end{pmatrix}$$

and

$$X(t) = \begin{pmatrix} x(t) \\ y(t) \end{pmatrix} .$$

Then our system is

$$\frac{d}{dt}X = AX .$$

Our idea is that the columns of e^{tA} should be the solutions to this system. So we need to calculate e^{tA}.

Now

$$\begin{aligned} e^{tA} &= I + tA + \frac{1}{2!}(tA)^2 + \frac{1}{3!}(tA)^3 + \frac{1}{4!}(tA)^4 + \cdots \\ &= \begin{pmatrix} 1 & 0 \\ 0 & 1 \end{pmatrix} + t\begin{pmatrix} 4 & -1 \\ 2 & 1 \end{pmatrix} + \frac{t^2}{2!}\begin{pmatrix} 14 & -5 \\ 10 & -1 \end{pmatrix} \\ &\quad + \frac{t^3}{3!}\begin{pmatrix} 46 & -19 \\ 38 & -11 \end{pmatrix} + \frac{t^4}{4!}\begin{pmatrix} 146 & -65 \\ 130 & -49 \end{pmatrix} + \cdots . \end{aligned}$$

Now we may actually sum up these matrices to obtain

$$\begin{aligned} \widetilde{X}(t) &= e^{tA} \\ &= \begin{pmatrix} \begin{matrix} 1 + 4t + 7t^2 + \frac{23}{3}t^3 \\ +\frac{73}{12}t^4 + \cdots \end{matrix} & \begin{matrix} 0 - t - \frac{5}{2}t^2 - \frac{19}{6}t^3 \\ -\frac{65}{24}t^4 + \cdots \end{matrix} \\ & \\ \begin{matrix} 0 + 2t + 5t^2 + \frac{19}{3}t^3 \\ +\frac{65}{12}t^4 + \cdots \end{matrix} & \begin{matrix} 1 + t - \frac{1}{2}t^2 - \frac{11}{6}t^3 \\ -\frac{49}{24}t^4 + \cdots \end{matrix} \end{pmatrix} . \end{aligned}$$

According to our discussion, the first column

$$C_1 = \begin{pmatrix} 1 + 4t + 7t^2 + \frac{23}{3}t^3 + \frac{73}{12}t^4 + \cdots \\ 0 + 2t + 5t^2 + \frac{19}{3}t^3 + \frac{65}{12}t^4 + \cdots \end{pmatrix}$$

is a solution to our system. That is to say,

$$\begin{aligned} x(t) &= 1 + 4t + 7t^2 + \frac{23}{3}t^3 + \frac{73}{12}t^4 + \cdots \\ y(t) &= 0 + 2t + 5t^2 + \frac{19}{3}t^3 + \frac{65}{12}t^4 + \cdots \end{aligned}$$

solves our system. And you can check this for yourself: differentiate $x(t)$ term-by-term and see that it equals $4x(t) - y(t)$; differentiate $y(t)$ term-by-term and see that it equals $2x(t) + y(t)$. Likewise,

$$C_2 = \begin{pmatrix} 0 - t - \frac{5}{2}t^2 - \frac{19}{6}t^3 - \frac{65}{24}t^4 + \cdots \\ 1 + t - \frac{t^2}{2} - \frac{11}{6}t^3 - \frac{49}{24}t^4 + \cdots \end{pmatrix}$$

is a solution to our system. That is to say,

$$\begin{aligned} x(t) &= 0 - t - \frac{5}{2}t^2 - \frac{19}{6}t^3 - \frac{65}{24}t^4 + \cdots \\ y(t) &= 1 + t - \frac{t^2}{2} - \frac{11}{6}t^3 - \frac{49}{24}t^4 + \cdots \end{aligned}$$

solves our system. Again, you may check this claim directly.

One drawback of the method that we are presenting here is that it is not always easy to recognize the solutions that we find in the language of familiar functions. But there are techniques for producing more familiar solutions. For example, suppose we consider

$$\begin{aligned} E_1 &= \frac{1}{3}C_1 + \frac{1}{3}C_2 \\ &= \begin{pmatrix} 1 + 3t + \frac{1}{2!}(3t)^2 + \frac{1}{3!}(3t)^3 + \frac{1}{4!}(3t)^4 \cdots \\ 1 + 3t + \frac{1}{2!}(3t)^2 + \frac{1}{3!}(3t)^3 + \frac{1}{4!}(3t)^4 \cdots \end{pmatrix} \\ &= \begin{pmatrix} e^{3t} \\ e^{3t} \end{pmatrix} \end{aligned}$$

and

$$\begin{aligned} E_2 &= \frac{1}{2}C_1 + C_2 \\ &= \begin{pmatrix} 1 + 2t + \frac{1}{2!}(2t)^2 + \frac{1}{3!}(2t)^3 + \frac{1}{4!}(2t)^4 + \cdots \\ 2 \cdot 1 + 2t + 2 \cdot \frac{1}{2!}(2t)^2 + 2 \cdot \frac{1}{3!}(2t)^3 + 2 \cdot \frac{1}{4!}(2t)^4 + \cdots \end{pmatrix} \\ &= \begin{pmatrix} e^{2t} \\ 2e^{2t} \end{pmatrix}. \end{aligned}$$

You see that we have now found two independent solutions to our system, and they coincide with the two solutions that we found by classical techniques in Example 10.2.1. ∎

The method that we presented in the text for solving systems consisted in finding the eigenvalues of the matrix A. When those eigenvalues were not distinct, or were complex, there were extra complications in finding the solution set of the system of differential equations. The method of matrices and exponentials that we are presenting now is immune to these difficulties. It works all the time. We illustrate with a final example.

EXAMPLE 10.4.1 Consider the system

$$\begin{cases} \dfrac{dx}{dt} = 3x - 4y \\ \dfrac{dy}{dt} = x - y\,. \end{cases}$$

We studied this system in Example 10.3.2 using classical methods. The eigenvalues of the system are repeated. Nonetheless, we can use the matrix/exponential method without any changes.

Set

$$A = \begin{pmatrix} 3 & -4 \\ 1 & -1 \end{pmatrix}.$$

We calculate that

$$\begin{aligned} e^{tA} &= I + tA + \frac{1}{2!}(tA)^2 + \frac{1}{3!}(tA)^3 + \frac{1}{4!}(tA)^4 + \cdots \\ &= \begin{pmatrix} 1 & 0 \\ 0 & 1 \end{pmatrix} + t\begin{pmatrix} 3 & -4 \\ 1 & -1 \end{pmatrix} + \frac{t^2}{2!}\begin{pmatrix} 5 & -8 \\ 2 & -3 \end{pmatrix} \\ &\quad + \frac{t^3}{3!}\begin{pmatrix} 7 & -12 \\ 3 & -5 \end{pmatrix} + \frac{t^4}{4!}\begin{pmatrix} 9 & -16 \\ 4 & -7 \end{pmatrix} + \cdots . \end{aligned}$$

Now we may actually sum up these matrices to obtain

$$\begin{aligned} \widetilde{X}(t) &= e^{tA} \\ &= \begin{pmatrix} \begin{array}{c} 1 + 3t + \frac{5}{2}t^2 \\ +\frac{7}{6}t^3 + \frac{9}{24}t^4 + \cdots \end{array} & \begin{array}{c} 0 - 4t - \frac{8}{2}t^2 \\ -\frac{12}{6}t^3 - \frac{16}{24}t^4 + \cdots \end{array} \\ & \\ \begin{array}{c} 0 + t + \frac{2}{2}t^2 \\ +\frac{3}{6}t^3 + \frac{4}{24}t^4 + \cdots \end{array} & \begin{array}{c} 1 - t - \frac{3}{2}t^2 \\ -\frac{5}{6}t^3 - \frac{7}{24}t^4 + \cdots \end{array} \end{pmatrix} . \end{aligned}$$

According to our discussion, the first column gives one solution to the system and the second column gives another. We leave it to you to find linear combinations of these two solutions that yield the two classical solutions that we found in Example 10.3.2:

$$\begin{aligned} x(t) &= 2e^t \\ y(t) &= e^t \end{aligned}$$

and

$$\begin{aligned} x(t) &= (1 + 2t)e^t \\ y(t) &= te^t . \end{aligned}$$

∎

Problems for Review and Discovery

A. Drill Exercises

1. Replace each of the following ordinary differential equations by an equivalent system of first-order equations.

(a) $y''' + x^2y'' - xy' + y = x$

(b) $y'' - [\sin x]y' + [\cos x]y = 0$

(c) $y^{(iv)} + x^2y''' - xy'' + y = 1$

(d) $xy'' - x^3y' + xy = \cos x$

2. In each of the following problems, show that the given solution set indeed satisfies the system of differential equations.

(a) $x(t) = \frac{1}{2}Ae^{-5t} + 2Be^{t}$, $y(t) = Ae^{-5t} + Be^{t}$

$$\begin{aligned} x'(t) &= 3x(t) - 4y(t) \\ y'(t) &= 4x(t) - 7y(t) \end{aligned}$$

(b) $x(t) = Ae^{3t} + Be^{-t}$, $y(t) = 2Ae^{3t} - 2Be^{-t}$

$$\begin{aligned} x'(t) &= x(t) + y(t) \\ y'(t) &= 4x(t) + y(t) \end{aligned}$$

(c) $x(t) = Ae^{-t} - \sqrt{2}Be^{-4t}$, $y(t) = \sqrt{2}Ae^{-t} + Be^{-t}$

$$\begin{aligned} x'(t) &= -3x(t) + \sqrt{2}y(t) \\ y'(t) &= \sqrt{2}x(t) - 2y(t) \end{aligned}$$

(d) $x(t) = Ae^{-t} + Be^{2t}$, $y(t) = -2Ae^{-t} - Be^{2t}$

$$\begin{aligned} x'(t) &= 5x(t) + 3y(t) \\ y'(t) &= -6x(t) - 4y(t) \end{aligned}$$

3. Solve each of the following systems of linear ordinary differential equations.

(a)
$$\begin{aligned} x'(t) &= 3x(t) + 2y(t) \\ y'(t) &= -2x(t) - y(t) \end{aligned}$$

(b)
$$\begin{aligned} x'(t) &= x(t) + y(t) \\ y'(t) &= -x(t) + y(t) \end{aligned}$$

(c)
$$\begin{aligned} x'(t) &= 3x(t) - 5y(t) \\ y'(t) &= -x(t) + 2y(t) \end{aligned}$$

(d)
$$\begin{aligned} x'(t) &= x(t) + 2y(t) \\ y'(t) &= -4x(t) + y(t) \end{aligned}$$

(e)
$$\begin{aligned} x'(t) &= 3x(t) + 2y(t) + z(t) \\ y'(t) &= -2x(t) - y(t) + 3z(t) \\ z'(t) &= x(t) + y(t) + z(t) \end{aligned}$$

(f)
$$\begin{aligned} x'(t) &= -x(t) + y(t) - z(t) \\ y'(t) &= 2x(t) - y(t) - 4z(t) \\ z'(t) &= 3x(t) - y(t) + z(t) \end{aligned}$$

4. Use the method of variation of parameters, introduced in Exercise 5 of Section 10.3, to solve each of these systems.

(a)
$$\begin{aligned} x'(t) &= x(t) + 2y(t) - 3t + 1 \\ y'(t) &= -x(t) + 2y(t) + 3t + 4 \end{aligned}$$

(b)
$$\begin{aligned} x'(t) &= -2x(t) + y(t) - t + 3 \\ y'(t) &= x(t) + 4y(t) + t - 2 \end{aligned}$$

(c)
$$\begin{aligned} x'(t) &= -4x(t) + y(t) - t + 3 \\ y'(t) &= -x(t) - 5y(t) + t + 1 \end{aligned}$$

B. Challenge Problems

1. Attempt to solve the system

$$\begin{aligned} x'(t) &= x(t)y(t) + 1\,, \quad x(0) = 2 \\ y'(t) &= -x(t) + y(t)\,, \quad y(0) = -1 \end{aligned}$$

by guessing that $x(t)$ has a power series expansion in t and $y(t)$ has a power series expansion in t. Substitute these series into the system and solve for the coefficients to obtain a pair of recursion relations.

2. Apply the method of Exercise 1 to the system

$$\begin{aligned} x'(t) &= ty(t) + 1\,, \quad x(0) = 1 \\ y'(t) &= -tx(t) + y(t)\,, \quad y(0) = 0 \end{aligned}$$

3. Solve the system

$$\begin{aligned} x'(t) &= x(t)y(t) + 1\,, \quad x(0) = 2 \\ y'(t) &= -x(t) + y(t)\,, \quad y(0) = -1 \end{aligned}$$

by taking the Laplace transform of both sides in each equation and thereby converting the problem to an algebraic one (this is the same system that we treated in Exercise 1).

4. Solve the system

$$\begin{aligned} x'(t) &= ty(t) + 1\,, \quad x(0) = 1 \\ y'(t) &= -tx(t) + y(t)\,, \quad y(0) = 0 \end{aligned}$$

by taking the Laplace transform of both sides in each equation and thereby converting the problem to an algebraic one (this is the same system that we treated in Exercise 2).

C. Problems for Discussion and Exploration

1. Two springs, two masses, and a dashpot are attached linearly on a horizontal surface without friction as depicted in Figure 10.4. Derive the system of differential equations for the displacements of the two points x_1 and x_2.

2. A pair of identical pendulums is coupled by a spring. Their motion is modeled by the pair of differential equations

$$\begin{aligned} mx_1'' &= -\frac{mg}{\ell}x_1 - k(x_1 - x_2) \\ mx_2'' &= -\frac{mg}{\ell}x_2 + k(x_1 - x_2) \end{aligned}$$

for small displacements of the pendulums. As usual, g is the gravitational constant. The number ℓ is the length of each pendulum. The number k is the spring constant. The number m is the mass of each pendulum. See Figure 10.5. Justify these equations as physical models for the systems.

Now solve the system and determine the natural (or *normal* frequencies for the system.

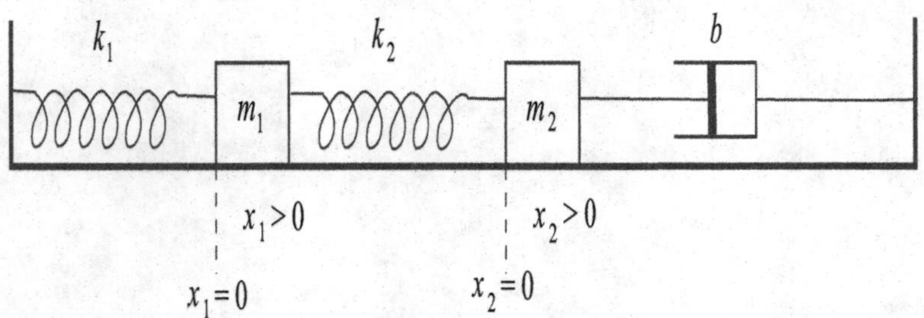

FIGURE 10.4
Two springs, two masses, and a dashpot.

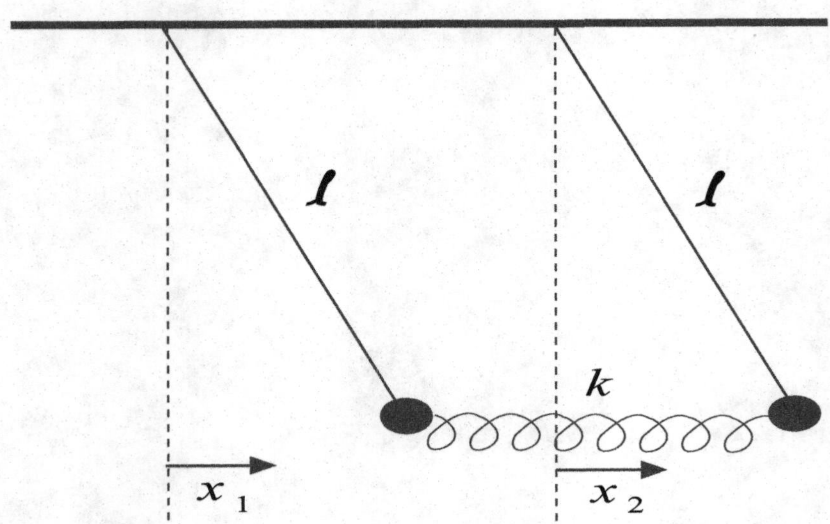

FIGURE 10.5
A pair of pendulums coupled by a spring.

11

Partial Differential Equations and Boundary Value Problems

- Boundary value problems
- Ideas from physics
- The wave equation
- The heat equation
- The Laplacian
- The Dirichlet problem
- The Poisson integral
- Sturm–Liouville problems

11.1 Introduction and Historical Remarks

In the middle of the eighteenth century, much attention was given to the problem of determining the mathematical laws governing the motion of a vibrating string with fixed endpoints at 0 and π (Figure 11.1). An elementary analysis of tension shows that, if $y(x,t)$ denotes the ordinate of the string at time t above the point x, then $y(x,t)$ satisfies the *wave equation*

$$\frac{\partial^2 y}{\partial t^2} = a^2 \frac{\partial^2 y}{\partial x^2}.$$

Here a is a parameter that depends on the tension of the string. A change of scale will allow us to assume that $a = 1$. (A bit later we shall actually provide a formal derivation of the wave equation. See also [KRA3] for a more thorough consideration of these matters.)

In 1747 d'Alembert showed that solutions of this equation have the form

$$y(x,t) = \frac{1}{2}\left(f(t+x) + g(t-x)\right), \tag{11.1}$$

where f and g are "any" functions of one variable. (The following technicality must be noted: the functions f and g are initially specified on the interval $[0,\pi]$. We extend f and g to $[-\pi,0]$ and to $[\pi,2\pi]$ by odd reflection. Continue f and g to the rest of the real line so that they are 2π-periodic.)

DOI: 10.1201/9781003214526-11

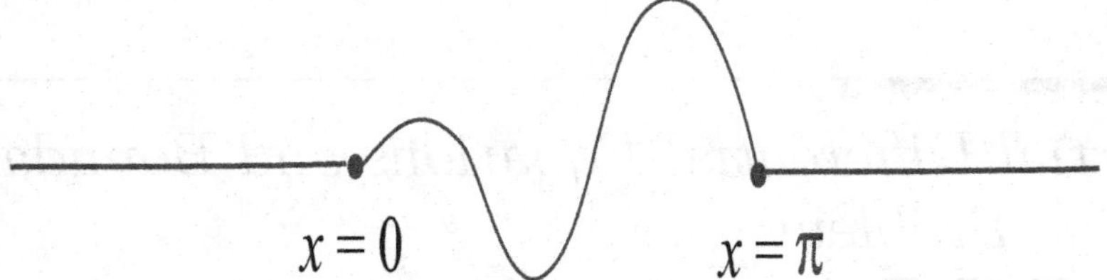

FIGURE 11.1
The wave equation.

In fact the wave equation, when placed in a "well-posed" setting, comes equipped with two initial conditions:

$$\begin{array}{lrcl} \text{(i)} & y(x,0) & = & \phi(x) \\ \text{(ii)} & \partial_t y(x,0) & = & \psi(x). \end{array}$$

These conditions mean **(i)** that the wave has an initial configuration that is the graph of the function ϕ and **(ii)** that the string is released with initial velocity ψ.

If (1) is to be a solution of this initial value problem then f and g must satisfy

$$\frac{1}{2}\left(f(x)+g(-x)\right)=\phi(x) \tag{11.2}$$

and

$$\frac{1}{2}\left(f'(x)+g'(-x)\right)=\psi(x). \tag{11.3}$$

Integration of (3) gives a formula for $f(x)-g(-x)$. That and (2) give a system that may be solved for f and g with elementary algebra.

The converse statement holds as well: for any functions f and g, a function y of the form (1) satisfies the wave equation (Exercise). The work of d'Alembert brought to the fore a controversy which had been implicit in the work of Daniel Bernoulli, Leonhard Euler, and others: what is a "function"? (We recommend the article [LUZ] for an authoritative discussion of the controversies that grew out of classical studies of the wave equation. See also [LAN].)

It is clear, for instance, in Euler's writings that he did not perceive a function to be an arbitrary "rule" that assigns points of the range to points of the domain; in particular, Euler did not think that a function could be specified in a fairly arbitrary fashion at different points of the domain. Once a function was specified on some small interval, Euler thought that it could only be extended in one way to a larger interval. Therefore, on physical grounds, Euler objected to d'Alembert's work. He claimed that the initial position of the vibrating string could be specified by several different functions pieced together continuously so that a single f could not generate the motion of the string.

Daniel Bernoulli solved the wave equation by a different method (separation of variables, which we treat below) and was able to show that there are infinitely many solutions of the wave equation having the form

$$\phi_j(x,t)=\sin jx\cos jt\,, \quad j\geq 1 \text{ an integer}\,.$$

Proceeding formally, he posited that all solutions of the wave equation satisfying $y(0,t)=y(\pi,t)=0$ and $\partial_t y(x,0)=0$ will have the form

$$y=\sum_{j=1}^{\infty} a_j \sin jx\cos jt.$$

Setting $t = 0$ indicates that the initial form of the string is $f(x) \equiv \sum_{j=1}^{\infty} a_j \sin jx$. In d'Alembert's language, the initial form of the string is $\frac{1}{2}(f(x) - f(-x))$, for we know that

$$0 \equiv y(0,t) = f(t) + g(t)$$

(because the endpoints of the string are held stationary), hence $g(t) = -f(t)$. If we suppose that d'Alembert's function is odd (as is $\sin jx$, each j), then the initial position is given by $f(x)$. Thus the problem of reconciling Bernoulli's solution to d'Alembert's reduces to the question of whether an "arbitrary" function f on $[0, \pi]$ may be written in the form $\sum_{j=1}^{\infty} a_j \sin jx$.

Since most mathematicians contemporary with Bernoulli believed that properties such as continuity, differentiability, and periodicity were preserved under (even infinite) addition, the consensus was that arbitrary f could *not* be represented as a (even infinite) trigonometric sum. The controversy extended over some years and was fueled by further discoveries (such as Lagrange's technique for interpolation by trigonometric polynomials) and more speculations.

In the 1820s, the problem of representation of an "arbitrary" function by trigonometric series was given a satisfactory answer as a result of two events. First, there is the sequence of papers by Joseph Fourier culminating with the tract [FOU]. Fourier gave a formal method of expanding an "arbitrary" function f into a trigonometric series. He computed some partial sums for some sample fs and verified that they gave very good approximations to f. Second, Dirichlet proved the first theorem giving sufficient (and very general) conditions for the Fourier series of a function f to converge pointwise to f. *Dirichlet was one of the first, in 1828, to formalize the notions of partial sum and convergence of a series*; his ideas certainly had antecedents in the work of Gauss and Cauchy.

For all practical purposes, these events mark the beginning of the mathematical theory of Fourier series (see [LAN]).

Math Nugget

The Bernoulli family was one of the foremost in all of the history of science. In three generations this remarkable Swiss family produced eight mathematicians, three of them outstanding. These in turn produced a swarm of descendants who distinguished themselves in many fields.

James Bernoulli (1654–1705) studied theology at the insistence of his father, but soon threw it over in favor of his love for science. He quickly learned the new "calculus" of Newton and Leibniz, became Professor of Mathematics at the University of Basel, and held that position until his death. James Bernoulli studied infinite series, special curves, and many other topics. He invented polar coordinates and introduced the Bernoulli numbers that appear in so many contexts in differential equations and special functions. In his book *Ars Conjectandi* he formulated what is now known as the law of large numbers (or Bernoulli's theorem). This is both an important philosophical and an important mathematical fact; it is still a source of study.

James's younger brother John (Johann) Bernoulli (1667–1748) also made a false start by first studying medicine and earning a doctor's degree at Basel in 1694 with a thesis on muscle contraction. He also became fascinated by calculus, mastered it quickly, and applied it to many problems in geometry, differential equations, and mechanics. In 1695 he was appointed Professor of Mathematics at Groningen in Holland. On James Bernoulli's death, John succeeded him in the chair at Basel. The Bernoulli brothers sometimes worked on the same problems; this was unfortunate in view of the family trait of touchiness and jealousy. On occasion their inherent friction flared up into nasty public feuds, more resembling barroom brawls than scientific debates. In particular, both James and John were solvers of the celebrated brachistochrone problem (along with Newton and Leibniz). They quarreled for years over the relative merits of their solutions (John's was the more elegant, James's the more general). John Bernoulli was particularly cantankerous in his personal affairs. He once threw his own son (Daniel) out of the house for winning a prize from the French Academy that he himself coveted.

Daniel Bernoulli (1700–1782) studied medicine like his father, and took a degree with a thesis on the action of the lungs. He soon yielded to his inborn talent and became Professor of Mathematics at St. Petersburg. In 1733 he returned to Basel and was, successively, professor of botany, anatomy, and physics. He won ten prizes from the French Academy (including the one that infuriated his father), and over the years published many works on physics, probability, calculus, and differential equations. His famous book *Hydrodynamica* discusses fluid mechanics and gives the earliest treatment of the kinetic theory of gases. Daniel Bernoulli was arguably the first mathematical physicist.

11.2 Eigenvalues, Eigenfunctions, and the Vibrating String

11.2.1 Boundary Value Problems

We wish to motivate the physics of the vibrating string. We begin this discussion by seeking a nontrivial solution y of the differential equation

$$y'' + \lambda y = 0 \tag{11.4}$$

subject to the conditions

$$y(0) = 0 \qquad \text{and} \qquad y(\pi) = 0\,. \tag{11.5}$$

Notice that this is a different situation from the one we have studied in earlier parts of the book. In Chapter 4 on second-order linear equations, we usually had *initial conditions* $y(x_0) = y_0$ and $y'(x_0) = y_1$. Now we have what are called *boundary conditions*: we specify one condition (in this instance the *value*) for the function at two different points. For instance, in the discussion of the vibrating string in the last section, we wanted our string to be pinned down at the two endpoints. These are typical boundary conditions coming from a physical problem.

The situation with boundary conditions is quite different from that for initial conditions. The latter is a sophisticated variation of the fundamental theorem of calculus. The former is rather more subtle. So let us begin to analyze.

First, if $\lambda < 0$, then Theorem 13.3.2 (the result that says that such a differential equation has a solution with only one zero) says that any solution of (4) has at most one zero. So it certainly cannot satisfy the boundary conditions (5). Alternatively, we could just solve the equation explicitly when $\lambda < 0$ and see that the independent solutions are a pair of exponentials, no linear combination of which can satisfy (5).

If $\lambda = 0$, then the general solution of (4) is the linear function $y = Ax + B$. Such a function cannot vanish at two points unless it is identically zero.

So the only interesting case is $\lambda > 0$. In this situation, the general solution of (4) is

$$y = A\sin\sqrt{\lambda}x + B\cos\sqrt{\lambda}x\,.$$

Since $y(0) = 0$, this in fact reduces to

$$y = A\sin\sqrt{\lambda}x\,.$$

In order for $y(\pi) = 0$, we must have $\sqrt{\lambda}\pi = n\pi$ for some positive integer n, thus $\lambda = n^2$. These values of λ are termed the *eigenvalues* of the problem, and the corresponding solutions

$$\sin x\,, \quad \sin 2x\,, \quad \sin 3x\ \ldots$$

are called the *eigenfunctions* of the problem (4), (5).

We note these immediate properties of the eigenvalues and eigenfunctions for our problem:

(i) If ϕ is an eigenfunction for eigenvalue λ, then so is $c \cdot \phi$ for any constant c.

(ii) The eigenvalues $1, 4, 9, \ldots$ form an increasing sequence that approaches $+\infty$.

(iii) The nth eigenfunction $\sin nx$ vanishes at the endpoints $0, \pi$ (as we originally mandated) and has exactly $n - 1$ zeros in the interval $(0, \pi)$.

11.2.2 Derivation of the Wave Equation

Now let us re-examine the vibrating string from the last section and see how eigenfunctions and eigenvalues arise naturally in a physical problem. We consider a flexible

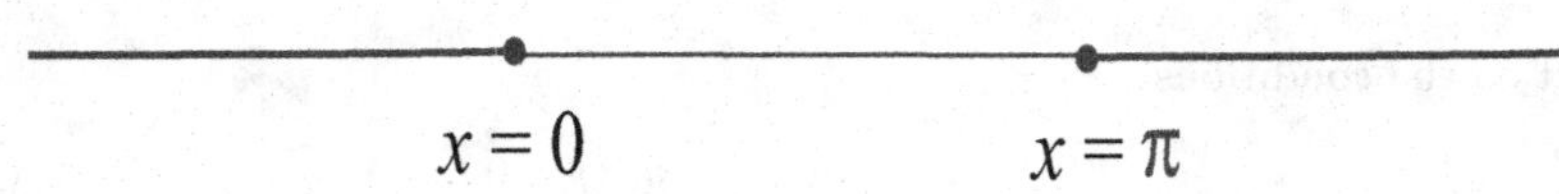

FIGURE 11.2
The string in relaxed position.

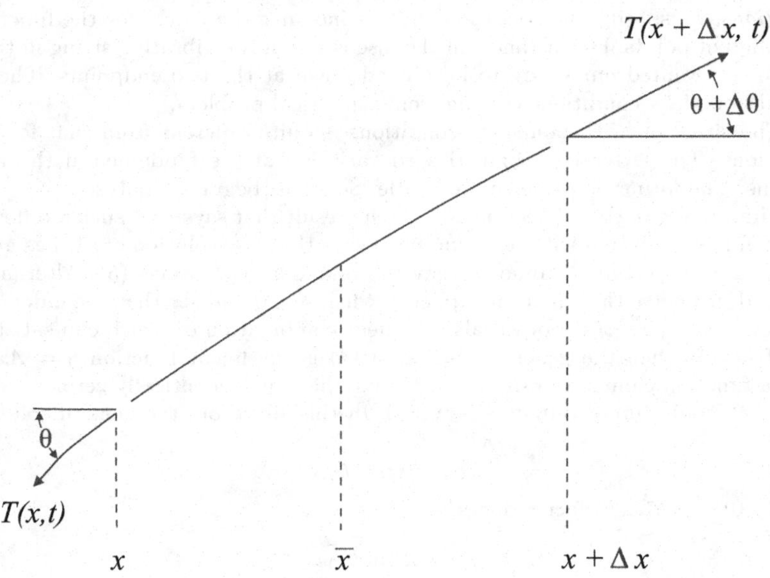

FIGURE 11.3
An element of the plucked string.

string with negligible weight that is fixed at its ends at the points $(0, 0)$ and $(\pi, 0)$. The curve is deformed into an initial position $y = f(x)$ in the x-y plane and then released.

Our analysis will ignore damping effects, such as air resistance. We assume that, in its relaxed position, the string is as in Figure 11.2. The string is plucked in the vertical direction, and is thus set in motion in a vertical plane. We will be supposing that the oscillation has small amplitude.

We focus attention on an "element" Δx of the string (Figure 11.3) that lies between x and $x + \Delta x$. We adopt the usual physical conceit of assuming that the displacement (motion) of this string element is *small*, so that there is only a slight error in supposing that the motion of each point of the string element is strictly vertical. We let the tension of the string, at the point x at time t, be denoted by $T(x, t)$. Note that T acts only in the tangential direction (i.e., along the string). We denote the mass density of the string by ρ.

Since *there is no horizontal component of acceleration*, we see that

$$T(x + \Delta x, t) \cdot \cos(\theta + \Delta\theta) - T(x, t) \cdot \cos(\theta) = 0. \tag{11.6}$$

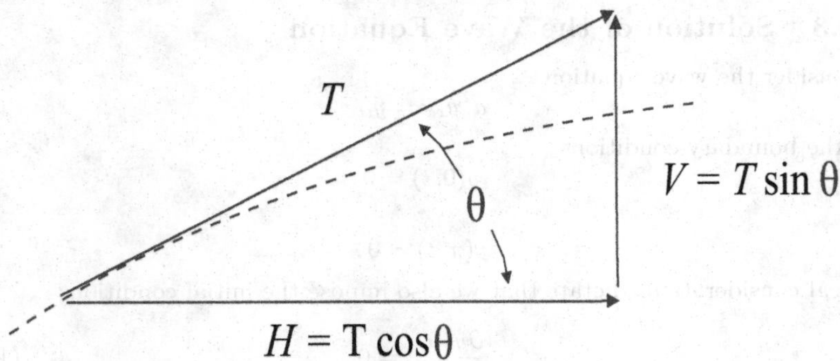

FIGURE 11.4
The horizontal component of the tension.

(Refer to Figure 11.4: The expression $T(\star)\cdot\cos(\star)$ denotes $H(\star)$, the horizontal component of the tension.) Thus equation (6) says that H is independent of x.

Now we look at the vertical component of force (acceleration):

$$T(x+\Delta x,t)\cdot\sin(\theta+\Delta\theta)-T(x,t)\cdot\sin(\theta)=\rho\cdot\Delta x\cdot y_{tt}(\overline{x},t). \tag{11.7}$$

Here $\overline{x}$ is the mass center of the string element and we are applying Newton's second law—that the external force is the mass of the string element times the acceleration of its center of mass. We use subscripts to denote derivatives. We denote the vertical component of $T(\star)$ by $V(\star)$. Thus equation (7) can be written as

$$\frac{V(x+\Delta x,t)-V(x,t)}{\Delta x}=\rho\cdot y_{tt}(x,t).$$

Letting $\Delta x\to 0$ yields

$$V_x(x,t)=\rho\cdot y_{tt}(x,t). \tag{11.8}$$

We would like to express equation (8) entirely in terms of y, so we notice that

$$V(x,t)=H(t)\tan\theta=H(t)\cdot y_x(x,t).$$

(We have used the fact that the derivative in x is the slope of the tangent line, which is $\tan\theta$.) Substituting this expression for V into (8) yields

$$(Hy_x)_x=\rho\cdot y_{tt}.$$

But H is independent of x, so this last line simplifies to

$$H\cdot y_{xx}=\rho\cdot y_{tt}.$$

For small displacements of the string, θ is nearly zero, so $H=T\cos\theta$ is nearly T. We are most interested in the case where T is constant. And of course ρ is constant. Thus we finally write our equation as

$$\frac{T}{\rho}y_{xx}=y_{tt}.$$

It is traditional to denote the constant T/ρ on the left by a^2. We finally arrive at the *wave equation*

$$a^2y_{xx}=y_{tt}.$$

11.2.3 Solution of the Wave Equation

We consider the wave equation

$$a^2 y_{xx} = y_{tt} \tag{11.9}$$

with the boundary conditions

$$y(0,t) = 0$$

and

$$y(\pi,t) = 0 \,.$$

Physical considerations dictate that we also impose the initial conditions

$$\left.\frac{\partial y}{\partial t}\right|_{t=0} = 0 \tag{11.10}$$

(indicating that the initial velocity of the string is 0) and

$$y(x,0) = f(x) \tag{11.11}$$

(indicating that the initial configuration of the string is the graph of the function f).

We solve the wave equation using a classical technique known as "separation of variables." For convenience, we assume that the constant $a = 1$. We guess a solution of the form $y(x,t) = u(x) \cdot v(t)$. Putting this guess into the differential equation

$$y_{xx} = y_{tt}$$

gives

$$u''(x)v(t) = u(x)v''(t) \,.$$

We may obviously separate variables, in the sense that we may write

$$\frac{u''(x)}{u(x)} = \frac{v''(t)}{v(t)} \,.$$

The left-hand side depends only on x while the right-hand side depends only on t. The only way this can be true is if

$$\frac{u''(x)}{u(x)} = \lambda = \frac{v''(t)}{v(t)}$$

for some constant λ. But this gives rise to two second-order linear, ordinary differential equations that we can solve explicitly:

$$u'' = \lambda \cdot u \tag{11.12}$$

$$v'' = \lambda \cdot v \,. \tag{11.13}$$

Observe that this is the *same* constant λ in both of these equations. Now, as we have already discussed, we want the initial configuration of the string to pass through the points $(0,0)$ and $(\pi,0)$. We can achieve these conditions by solving (12) with $u(0) = 0$ and $u(\pi) = 0$. But of course this is the eigenvalue problem that we treated at the beginning of the section. The problem has a nontrivial solution if and only if $\lambda = -n^2$ for some positive integer n, and the corresponding eigenfunction is

$$u_n(x) = \sin nx \,.$$

For this same λ, the general solution of (10) is

$$v(t) = A\sin nt + B\cos nt\,.$$

If we impose the requirement that $v'(0) = 0$, so that (10) is satisfied, then $A = 0$ and we find the solution

$$v(t) = B\cos nt\,.$$

This means that the solution we have found of our differential equation with boundary and initial conditions is

$$y_n(x,t) = \sin nx\cos nt\,. \tag{11.14}$$

And in fact any finite sum with coefficients (or *linear combination*) of these solutions will also be a solution:

$$y = \alpha_1\sin x\cos t + \alpha_2\sin 2x\cos 2t + \cdots \alpha_k\sin kx\cos kt\,.$$

Ignoring the rather delicate issue of convergence (which was discussed a bit in Section 7.2), we may claim that any *infinite* linear combination of the solutions (11) will also be a solution:

$$y = \sum_{j=1}^{\infty} b_j\sin jx\cos jt\,. \tag{11.15}$$

Now we must examine the initial condition (11). The mandate $y(x,0) = f(x)$ translates to

$$\sum_{j=1}^{\infty} b_j\sin jx = y(x,0) = f(x) \tag{11.16}$$

or

$$\sum_{j=1}^{\infty} b_j u_j(x) = y(x,0) = f(x)\,. \tag{11.17}$$

Thus we demand that f have a valid Fourier series expansion. We know from our studies in Chapter 7 that such an expansion is correct for a rather broad class of functions f. Thus the wave equation is solvable in considerable generality.

Now fix $m \neq n$. We know that our eigenfunctions u_j satisfy

$$u_m'' = -m^2 u_m \qquad\text{and}\qquad u_n'' = -n^2 u_n\,.$$

Multiply the first equation by u_n and the second by u_m and subtract. The result is

$$u_n u_m'' - u_m u_n'' = (n^2 - m^2)u_n u_m$$

or

$$[u_n u_m' - u_m u_n']' = (n^2 - m^2)u_n u_m\,.$$

We integrate both sides of this last equation from 0 to π and use the fact that $u_j(0) = u_j(\pi) = 0$ for every j. The result is

$$0 = [u_n u_m' - u_m u_n']\Big|_0^{\pi} = (n^2 - m^2)\int_0^{\pi} u_m(x)u_n(x)\,dx\,.$$

Thus

$$\int_0^{\pi}\sin mx\sin nx\,dx = 0 \qquad \text{for } n \neq m \tag{11.18}$$

or

$$\int_0^{\pi} u_m(x)u_n(x)\,dx = 0 \qquad \text{for } n \neq m\,. \tag{11.19}$$

Of course this is a standard fact from calculus. But now we understand it as an orthogonality condition (see Sections 7.5, 11.5), and we see how the condition arises naturally from the differential equation. A little later, we shall fit this phenomenon into the general context of Sturm–Liouville problems.

In view of the orthogonality condition (19), it is natural to integrate both sides of (17) against $u_k(x)$. The result is

$$\begin{aligned}\int_0^{\pi} f(x)\cdot u_k(x)\,dx &= \int_0^{\pi}\left(\sum_{j=0}^{\infty} b_j u_j(x)\right)\cdot u_k(x)\,dx\\ &= \sum_{j=0}^{\infty} b_j \int_0^{\pi} u_j(x)u_k(x)\,dx\\ &= \frac{\pi}{2} b_k\,.\end{aligned}$$

The b_k are the Fourier coefficients that we studied in Chapter 7. Using these coefficients, we have *Bernoulli's solution* (15) of the wave equation.

Exercises

1. Find the eigenvalues λ_n and the eigenfunctions y_n for the equation $y'' + \lambda y = 0$ in each of the following instances.

(a) $y(0) = 0\,,\quad y(\pi/2) = 0$
(b) $y(0) = 0\,,\quad y(2\pi) = 0$
(c) $y(0) = 0\,,\quad y(1) = 0$
(d) $y(0) = 0\,,\quad y(L) = 0$ for $L > 0$
(e) $y(-L) = 0\,,\quad y(L) = 0$ for $L > 0$
(f) $y(a) = 0\,,\quad y(b) = 0$ for $a < b$

Solve the following two exercises without worrying about convergence of series or differentiability of functions.

2. If $y = F(x)$ is an arbitrary function, then $y = F(x + at)$ represents a wave of fixed shape that moves to the left along the x-axis with velocity a (Figure 11.5).

Similarly, if $y = G(x)$ is another arbitrary function, then $y = G(x - at)$ is a wave moving to the right, and the most general one-dimensional wave with velocity a is

$$y(x,t) = F(x + at) + G(x - at)\,. \tag{$*$}$$

(a) Show that $(*)$ satisfies the wave equation (9).

(b) It is easy to see that the constant a in equation (9) has the dimensions of velocity. Also, it is intuitively clear that if a stretched string is disturbed, then the waves will move in both directions away from the source of the disturbance. These considerations suggest introducing

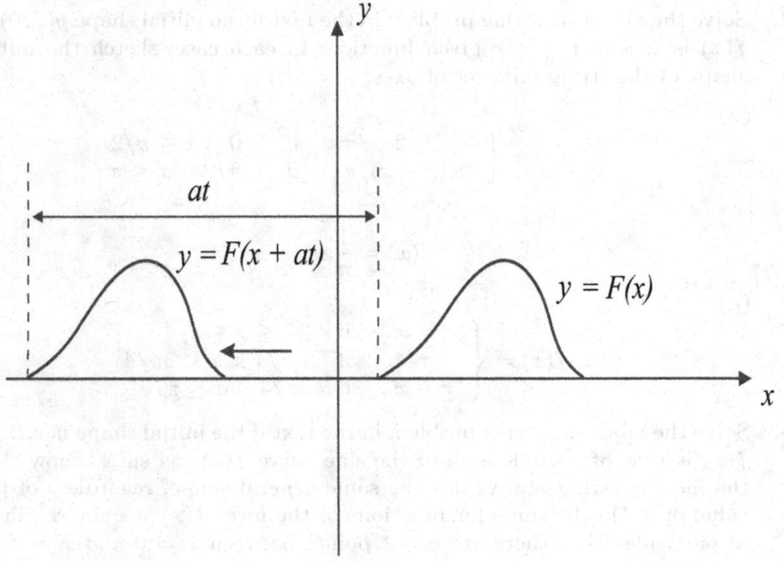

FIGURE 11.5
Wave of fixed shape moving to the left.

the new variables $\alpha = x + at$, $\beta = x - at$. Show that with these independent variables, equation (9) becomes

$$\frac{\partial^2 y}{\partial \alpha \partial \beta} = 0 .$$

From this derive $(*)$ by integration. Formula $(*)$ is called *d'Alembert's solution* of the wave equation. It was also obtained, slightly later and independently, by Euler.

3. Consider an infinite string stretched taut on the x-axis from $-\infty$ to $+\infty$. Let the string be drawn aside into a curve $y = f(x)$ and released, and assume that its subsequent motion is described by the wave equation (9).

(a) Use $(*)$ in Exercise **2** to show that the string's displacement is given by *d'Alembert's formula*

$$y(x,t) = \frac{1}{2}[f(x+at) + f(x-at)] . \qquad (**)$$

Hint: Remember the initial conditions (10) and (11).

(b) Assume further that the string remains motionless at the points $x = 0$ and $x = \pi$ (such points are called *nodes*) so that $y(0,t) = y(\pi,t) = 0$, and use $(**)$ to show that f is an odd function that is periodic with period 2π (that is, $f(-x) = -f(x)$ and $f(x + 2\pi) = f(x)$).

(c) Show that since f is odd and periodic with period 2π then f necessarily vanishes at 0 and π.

(d) Show that Bernoulli's solution (15) of the wave equation can be written in the form $(**)$. *Hint:* $2 \sin nx \cos nat = \sin[n(x+at)] + \sin[n(x-at)]$.

4. Solve the vibrating string problem in the text if the initial shape $y(x, 0) = f(x)$ is specified by the given function. In each case, sketch the initial shape of the string on a set of axes.

(a)

$$f(x) = \begin{cases} 2cx/\pi & \text{if} \quad 0 \le x \le \pi/2 \\ 2c(\pi - x)/\pi & \text{if} \quad \pi/2 \le x \le \pi \end{cases}$$

(b)

$$f(x) = \frac{1}{\pi} x(\pi - x)$$

(c)

$$f(x) = \begin{cases} x & \text{if} \quad 0 \le x \le \pi/4 \\ \pi/4 & \text{if} \quad \pi/4 < x < 3\pi/4 \\ \pi - x & \text{if} \quad 3\pi/4 \le x \le \pi \end{cases}$$

5. Solve the vibrating string problem in the text if the initial shape $y(x, 0) = f(x)$ is that of a single arch of the sine curve $f(x) = c \sin x$. Show that the moving string always has the same general shape, regardless of the value of c. Do the same for functions of the form $f(x) = c \sin nx$. Show in particular that there are $n - 1$ points between $x = 0$ and $x = \pi$ at which the string remains motionless; these points are called *nodes*, and these solutions are called *standing waves.* Draw sketches to illustrate the movement of the standing waves.

6. The problem of the *struck string* is that of solving the wave equation with the boundary conditions

$$y(0, t) = 0, \quad y(\pi, t) = 0$$

and the initial conditions

$$\left.\frac{\partial y}{\partial t}\right|_{t=0} = g(x) \qquad \text{and} \qquad y(x, 0) = 0\,.$$

(These initial conditions mean that the string is initially in the equilibrium position, and has an initial velocity $g(x)$ at the point x as a result of being struck.) By separating variables and proceeding formally, obtain the solution

$$y(x, t) = \sum_{j=1}^{\infty} c_j \sin jx \sin jat\,,$$

where

$$c_j = \frac{2}{\pi j a} \int_0^{\pi} g(x) \sin jx\, dx\,.$$

11.3 The Heat Equation

Fourier's Point of View

In [FOU], Fourier considered variants of the following basic question. Let there be given an insulated, homogeneous rod of length π with initial temperature at each $x \in [0, \pi]$ given by a function $f(x)$ (Figure 11.6). Assume

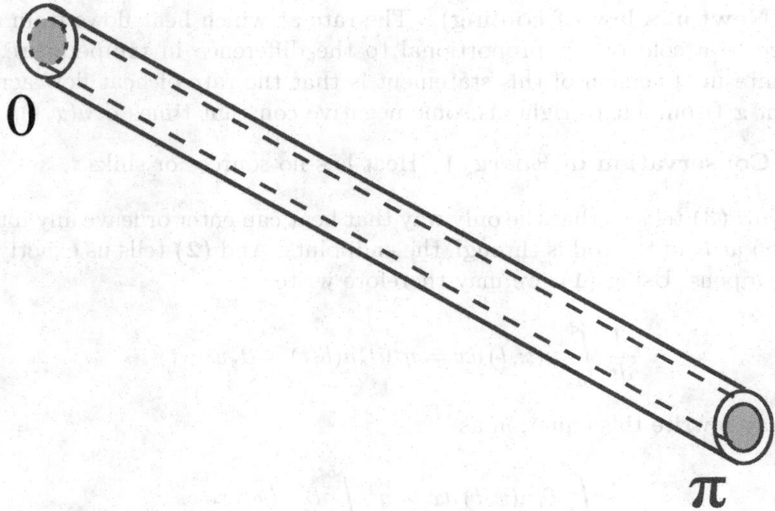

FIGURE 11.6
The insulated rod.

that the endpoints are held at temperature 0, and that the temperature of each cross-section is constant. The problem is to describe the temperature $u(x, t)$ of the point x in the rod at time t. Fourier perceived the fundamental importance of this problem as follows:

> Primary causes are unknown to us; but are subject to simple and constant laws, which may be discovered by observation, the study of them being the object of natural philosophy.
>
> Heat, like gravity, penetrates every substance of the universe, its rays occupying all parts of space. The object of our work is to set forth the mathematical laws which this element obeys. The theory of heat will hereafter form one of the most important branches of general physics.
>
> . . . .
>
> I have deduced these laws from prolonged study and attentive comparison of the facts known up to this time; all these facts I have observed afresh in the course of several years with the most exact instruments that have hitherto been used.

Let us now describe the manner in which Fourier solved his problem. First, it is required to write a differential equation which u satisfies. We shall derive such an equation using three physical principles:

(1) The density of heat energy is proportional to the temperature u; hence, the amount of heat energy in any interval $[a, b]$ of the rod is proportional to $\int_a^b u(x, t)\, dx$.

(2) (Newton's law of cooling) The rate at which heat flows from a hot place to a cold one is proportional to the difference in temperature. The infinitesimal version of this statement is that the rate of heat flow across a point x (from left to right) is some negative constant times $\partial_x u(x,t)$.

(3) (Conservation of Energy) Heat has no sources or sinks.

Now **(3)** tells us that the only way that heat can enter or leave any interval portion $[a,b]$ of the rod is through the endpoints. And **(2)** tells us exactly how this happens. Using **(1)**, we may therefore write

$$\frac{d}{dt}\int_a^b u(x,t)\,dx = \eta^2[\partial_x u(b,t) - \partial_x u(a,t)].$$

We may rewrite this equation as

$$\int_a^b \partial_t u(x,t)\,dx = \eta^2 \int_a^b \partial_x^2 u(x,t)\,dx.$$

Differentiating in b, we find that

$$\partial_t u = \eta^2 \partial_x^2 u, \tag{11.20}$$

and that is the heat equation.

Math Nugget

The English biologist J. B. S. Haldane (1892–1964) had this remark about the one-dimensional heat equation: "In scientific thought we adopt the simplest theory which will explain all the facts under consideration and enable us to predict new facts of the same kind. The catch in this criterion lies in the word 'simplest.' It is really an aesthetic canon such as we find implicit in our criticism of poetry or painting. The layman finds such a law as

$$a^2\frac{\partial^2 w}{\partial x^2} = \frac{\partial w}{\partial t}$$

much less simple than 'it oozes,' of which it is the mathematical statement. The physicist reverses this judgment, and his statement is certainly the more fruitful of the two, so far as prediction is concerned. It is, however, a statement about something very unfamiliar to the plain man, namely, the rate of change of a rate of change."

Suppose for simplicity that the constant of proportionality η^2 equals 1. Fourier guessed that equation (20) has a solution of the form $u(x,t) = \alpha(x)\beta(t)$. Substituting this guess into the equation yields

$$\alpha(x)\beta'(t) = \alpha''(x)\beta(t)$$

or

$$\frac{\beta'(t)}{\beta(t)} = \frac{\alpha''(x)}{\alpha(x)}.$$

Since the left side is independent of x and the right side is independent of t, it follows that there is a constant K such that

$$\frac{\beta'(t)}{\beta(t)} = K = \frac{\alpha''(x)}{\alpha(x)}$$

or

$$\begin{aligned} \beta'(t) &= K\beta(t) \\ \alpha''(x) &= K\alpha(x). \end{aligned}$$

We conclude that $\beta(t) = Ce^{Kt}$. The nature of β, and hence of α, thus depends on the sign of K. But physical considerations tell us that the temperature will dissipate as time goes on, so we conclude that $K \leq 0$. Therefore $\alpha(x) = \cos\sqrt{-K}x$ and $\alpha(x) = \sin\sqrt{-K}x$ are solutions of the differential equation for α. The initial conditions $u(0,t) = u(\pi,t) = 0$ (since the ends of the rod are held at constant temperature 0) eliminate the first of these solutions and force $K = -j^2$, j an integer. Thus Fourier found the solutions

$$u_j(x,t) = e^{-j^2 t}\sin jx\ , \quad j \in \mathbb{N}$$

of the heat equation. By linearity, any finite linear combination

$$u(x,t) = \sum_j b_j e^{-j^2 t}\sin jx \tag{11.21}$$

of these solutions is also a solution. It is plausible to extend this assertion to infinite linear combinations. Using the initial condition $u(x,0) = f(x)$ again raises the question of whether "any" function $f(x)$ on $[0,\pi]$ can be written as a (infinite) linear combination of the functions $\sin jx$.

Fourier's solution to this last problem (of the sine functions spanning essentially everything) is roughly as follows: Suppose f is a function that is so representable:

$$f(x) = \sum_j b_j \sin jx. \tag{11.22}$$

Setting $x = 0$ gives

$$f(0) = 0.$$

Differentiating both sides of (22) and setting $x = 0$ gives

$$f'(0) = \sum_{j=1}^{\infty} j b_j. \tag{11.23}$$

Successive differentiation of (22), and evaluation at 0, gives

$$f^{(k)}(0) = \sum_{j=1}^{\infty} j^k b_j (-1)^{\lfloor k/2 \rfloor}$$

for k odd (by oddness of f, the even derivatives must be 0 at 0). Here $\lfloor \quad \rfloor$ denotes the greatest integer function. Thus Fourier devised a system of infinitely many equations in the infinitely many unknowns $\{b_j\}$. He proceeded to solve this system by truncating it to an $N \times N$ system (the first N equations restricted to the first N unknowns), solving that truncated system, and then letting N tend to ∞. Suffice it to say that Fourier's arguments contained many dubious steps (see [FOU] and [LAN]).

The upshot of Fourier's intricate and lengthy calculations was that

$$b_j = \frac{2}{\pi} \int_0^{\pi} f(x) \sin jx \, dx. \tag{11.24}$$

By modern standards, Fourier's reasoning was specious; for he began by assuming that f possessed an expansion in terms of sine functions. The formula (24) hinges on that supposition, together with steps in which one compensated division by zero with a later division by ∞. Nonetheless, Fourier's methods give an actual *procedure* for endeavoring to expand any given f in a series of sine functions.

Fourier's abstract arguments constitute the first part of his book. The bulk, and remainder, of the book consists of separate chapters in which the expansions for particular functions are computed.

EXAMPLE 11.3.1 Suppose that the thin rod in the setup of the heat equation is first immersed in boiling water so that its temperature is uniformly 100°C. Then imagine that it is removed from the water at time $t = 0$ with its ends immediately put into ice so that these ends are kept at temperature 0°C. Find the temperature $u = u(x,t)$ under these circumstances.

Solution: The initial temperature distribution is given by the constant function

$$f(x) = 100 \, , \quad 0 < x < \pi \, .$$

The two boundary conditions, and the other initial condition, are as usual. Thus our job is simply this: to find the sine series expansion of this function

f. We calculate that

$$\begin{aligned} b_j &= \frac{2}{\pi}\int_0^\pi 100\sin jx\,dx \\ &= -\frac{200}{\pi}\frac{\cos jx}{j}\bigg|_0^\pi \\ &= -\frac{200}{\pi}\left[\frac{(-1)^j}{j} - \frac{1}{j}\right] \\ &= \begin{cases} 0 & \text{if} \quad j = 2\ell \text{ is even} \\ \dfrac{400}{\pi j} & \text{if} \quad j = 2\ell - 1 \text{ is odd}. \end{cases} \end{aligned}$$

Thus

$$f(x) = \frac{400}{\pi}\left(\sin x + \frac{\sin 3x}{3} + \frac{\sin 5x}{5} + \cdots\right).$$

Now, referring to formula (21) from our general discussion of the heat equation, we know that

$$u(x,t) = \frac{400}{\pi}\left(e^{-t}\sin x + \frac{1}{3}e^{-9t}\sin 3x + \frac{1}{5}e^{-25t}\sin 5x + \cdots\right). \qquad \square$$

EXAMPLE 11.3.2 Find the steady-state temperature of the thin rod from our analysis of the heat equation if the fixed temperatures at the ends $x = 0$ and $x = \pi$ are w_1 and w_2 respectively.

Solution: The phrase "steady state" means that $\partial u/\partial t = 0$, so that the heat equation reduces to $\partial^2 u/\partial x^2 = 0$ or $d^2u/dx^2 = 0$. The general solution is then $u = Ax + B$. The values of these two constants A and B are forced by the two boundary conditions.

In fact a little high school algebra tells us that

$$u = w_1 + \frac{1}{\pi}(w_2 - w_1)x. \qquad \square$$

The steady-state version of the three-dimensional heat equation

$$a^2\left(\frac{\partial^2 u}{\partial x^2} + \frac{\partial^2 u}{\partial y^2} + \frac{\partial^2 u}{\partial z^2}\right) = \frac{\partial u}{\partial t}$$

is

$$\frac{\partial^2 u}{\partial x^2} + \frac{\partial^2 u}{\partial y^2} + \frac{\partial^2 u}{\partial z^2} = 0.$$

This last is called *Laplace's equation.* The study of this equation and its solutions and subsolutions and their applications is a deep and rich branch of mathematics called *potential theory.* There are applications to heat, to gravitation, to electromagnetics, and to many other parts of physics. The equation plays a central role in the theory of partial differential equations, and is also an integral part of complex variable theory.

Exercises

1. Solve the boundary value problem

$$\begin{aligned} a^2 \frac{\partial^2 w}{\partial x^2} &= \frac{\partial w}{\partial t} \\ w(x,0) &= f(x) \\ w(0,t) &= 0 \\ w(\pi,t) &= 0 \end{aligned}$$

if the last three conditions—the boundary conditions—are changed to

$$\begin{aligned} w(x,0) &= f(x) \\ w(0,t) &= w_1 \\ w(\pi,t) &= w_2\,. \end{aligned}$$

Hint: Write $w(x,t) = W(x,t) + g(x)$, where $g(x)$ is the function that we produced in Example 7.3.2.

2. Suppose that the lateral surface of the thin rod that we analyzed in the text is not insulated, but in fact radiates heat into the surrounding air. If Newton's law of cooling (that a body cools at a rate proportional to the difference of its temperature with the temperature of the surrounding air) is assumed to apply, then show that the one-dimensional heat equation becomes

$$a^2 \frac{\partial^2 w}{\partial x^2} = \frac{\partial w}{\partial t} + c(w - w_0)$$

where c is a positive constant and w_0 is the temperature of the surrounding air.

3. In Exercise **2**, find $w(x,t)$ if the ends of the rod are kept at 0°C, $w_0 = 0^\circ$C, and the initial temperature distribution on the rod is $f(x)$.

4. In Example 7.3.1, suppose that the ends of the rod are insulated instead of being kept fixed at 0°C. What are the new boundary conditions? Find the temperature $w(x,t)$ in this case by using just common sense—and *not* calculating.

5. Solve the problem of finding $w(x,t)$ for the rod with insulated ends at $x = 0$ and $x = \pi$ (see the preceding exercise) if the initial temperature distribution is given by $w(x,0) = f(x)$.

6. The two-dimensional heat equation is

$$a^2 \left(\frac{\partial^2 w}{\partial x^2} + \frac{\partial^2 w}{\partial y^2} \right) = \frac{\partial w}{\partial t}\,.$$

Use the method of separation of variables to find a steady-state solution of this equation in the infinite half-strip of the x-y plane bounded by the lines $x = 0$, $x = \pi$, and $y = 0$ if the following boundary conditions are satisfied:

$$\begin{aligned} w(0,y,t) &= 0 & \qquad w(\pi,y,t) &= 0 \\ w(x,0,0) &= f(x) & \qquad \lim_{y\to+\infty} w(x,y,t) &= 0\,. \end{aligned}$$

7. Derive the three-dimensional heat equation

$$a^2\left(\frac{\partial^2 w}{\partial x^2}+\frac{\partial^2 w}{\partial y^2}+\frac{\partial^2 w}{\partial z^2}\right)=\frac{\partial w}{\partial t}$$

by adapting the reasoning in the text to the case of a small box with edges Δx, Δy, Δz contained in a region R in x-y-z space where the temperature function $w(x,y,z,t)$ is sought. *Hint:* Consider the flow of heat through two opposite faces of the box, first perpendicular to the x-axis, then perpendicular to the y-axis, and finally perpendicular to the z-axis.

11.4 The Dirichlet Problem for a Disc

We now study the two-dimensional Laplace equation, which is

$$\Delta w = \frac{\partial^2 w}{\partial x^2}+\frac{\partial^2 w}{\partial y^2}=0 .$$

Common notations for the Laplace operator are Δ and ∇^2.

It will be useful for us to write this equation in polar coordinates. To do so, recall that

$$r^2 = x^2+y^2 \ , \quad x = r\cos\theta \ , \quad y = r\sin\theta .$$

Thus

$$\begin{aligned}\frac{\partial}{\partial r} &= \frac{\partial x}{\partial r}\frac{\partial}{\partial x}+\frac{\partial y}{\partial r}\frac{\partial}{\partial y} = \cos\theta\frac{\partial}{\partial x}+\sin\theta\frac{\partial}{\partial y}\\ \frac{\partial}{\partial \theta} &= \frac{\partial x}{\partial \theta}\frac{\partial}{\partial x}+\frac{\partial y}{\partial \theta}\frac{\partial}{\partial y} = -r\sin\theta\frac{\partial}{\partial x}+r\cos\theta\frac{\partial}{\partial y} .\end{aligned}$$

We may solve these two equations for the unknowns $\partial/\partial x$ and $\partial/\partial y$. The result is

$$\frac{\partial}{\partial x} = \cos\theta\frac{\partial}{\partial r}-\frac{\sin\theta}{r}\frac{\partial}{\partial\theta} \quad\text{and}\quad \frac{\partial}{\partial y} = \sin\theta\frac{\partial}{\partial r}+\frac{\cos\theta}{r}\frac{\partial}{\partial\theta} .$$

A tedious calculation now reveals that

$$\begin{aligned}\Delta = \frac{\partial^2}{\partial x^2}+\frac{\partial^2}{\partial y^2} &= \left(\cos\theta\frac{\partial}{\partial r}-\frac{\sin\theta}{r}\frac{\partial}{\partial\theta}\right)\left(\cos\theta\frac{\partial}{\partial r}-\frac{\sin\theta}{r}\frac{\partial}{\partial\theta}\right)\\ &\quad+\left(\sin\theta\frac{\partial}{\partial r}+\frac{\cos\theta}{r}\frac{\partial}{\partial\theta}\right)\left(\sin\theta\frac{\partial}{\partial r}+\frac{\cos\theta}{r}\frac{\partial}{\partial\theta}\right)\\ &= \frac{\partial^2}{\partial r^2}+\frac{1}{r}\frac{\partial}{\partial r}+\frac{1}{r^2}\frac{\partial^2}{\partial\theta^2} . \qquad (11.25)\end{aligned}$$

Let us fall back once again on the separation of variables method. We shall seek a solution $w = w(r,\theta) = u(r)\cdot v(\theta)$ of the Laplace equation. Using the polar form (25) of the Laplacian, we find that this leads to the equation

$$u''(r)\cdot v(\theta) + \frac{1}{r}u'(r)\cdot v(\theta) + \frac{1}{r^2}u(r)\cdot v''(\theta) = 0 .$$

Thus

$$\frac{r^2u''(r) + ru'(r)}{u(r)} = -\frac{v''(\theta)}{v(\theta)} .$$

Since the left-hand side depends only on r, and the right-hand side only on θ, both sides must be constant. Denote the common constant value by λ.

Then we have

$$v'' + \lambda v = 0 \tag{11.26}$$

and

$$r^2u'' + ru' - \lambda u = 0 . \tag{11.27}$$

If we demand that v be continuous and periodic, then we must have (because of equation (26)) that $\lambda > 0$ and in fact that $\lambda = n^2$ for some nonnegative integer n (so that we end up with solutions $v = \sin n\theta$ and $v = \cos n\theta$). We have studied this situation in detail in Section 11.2. For $n = 0$ the only suitable solution is $v \equiv$ constant and for $n > 0$ the general solution (with $\lambda = n^2$) is

$$v = A\cos n\theta + B\sin n\theta .$$

We set $\lambda = n^2$ in equation (27) and obtain[1]

$$r^2u'' + ru' - n^2u = 0 .$$

We may solve this equation by guessing $u(r) = r^m$. Plugging that guess into the differential equation, we obtain

$$r^m(m^2 - n^2) = 0 .$$

Now there are two cases:

(i) If $n = 0$, then we obtain the repeated root $m = 0, 0$. Now we proceed by analogy with our study of second-order linear equations with constant coefficients, and hypothesize a second solution of the form $u(r) = \ln r$. This works, so we obtain the general solution

$$u(r) = A + B\ln r .$$

(ii) If $n > 0$ then $m = \pm n$, and the general solution is

$$u(r) = Ar^n + Br^{-n} .$$

---

[1] This is Euler's equidimensional equation. The change of variables $r = e^z$ transforms this equation to a linear equation with constant coefficients, and that can in turn be solved with our standard techniques.

We are most interested in solutions u that are continuous at the origin, so we take $B = 0$ in all cases. The resulting solutions are

$$\begin{aligned} n &= 0\,, & w &= a_0/2 \text{ a constant} \\ n &= 1\,, & w &= r(a_1 \cos\theta + b_1 \sin\theta) \\ n &= 2\,, & w &= r^2(a_2 \cos 2\theta + b_2 \sin 2\theta) \\ n &= 3\,, & w &= r^3(a_3 \cos 3\theta + b_3 \sin 3\theta) \\ & & & \ldots \end{aligned}$$

Of course any finite sum of solutions of Laplace's equation is also a solution. The same is true for infinite sums. Thus we are led to consider

$$w = w(r, \theta) = \frac{1}{2}a_0 + \sum_{j=1}^{\infty} r^j (a_j \cos j\theta + b_j \sin j\theta)\,.$$

On a formal level, letting $r \to 1^-$ in this last expression gives

$$\frac{1}{2}a_0 + \sum_{j=1}^{\infty}(a_j \cos j\theta + b_j \sin j\theta)\,.$$

We draw all these ideas together with the following physical rubric. Consider a thin aluminum disc of radius 1, and imagine applying a heat distribution to the boundary of that disc. In polar coordinates, this distribution is specified by a function $f(\theta)$. We seek to understand the steady-state heat distribution on the entire disc. So we seek a function $w(r, \theta)$, continuous on the closure of the disc, which agrees with f on the boundary and which represents the steady-state distribution of heat inside. Some physical analysis shows that such a function w is the solution of the boundary value problem

$$\begin{aligned} \triangle w &= 0 \\ w\big|_{\partial D} &= f\,. \end{aligned}$$

Here we use the notation ∂D to denote the boundary of D.

According to the calculations we performed prior to this last paragraph, a natural approach to this problem is to expand the given function f in its Fourier series:

$$f(\theta) = \frac{1}{2}a_0 + \sum_{j=1}^{\infty}(a_j \cos j\theta + b_j \sin j\theta)$$

and then posit that the w we seek is

$$w(r, \theta) = \frac{1}{2}a_0 + \sum_{j=1}^{\infty} r^j (a_j \cos j\theta + b_j \sin j\theta)\,.$$

This process is known as solving *the Dirichlet problem on the disc with boundary data f*.

EXAMPLE 11.4.1 Follow the paradigm just sketched to solve the Dirichlet problem on the disc with $f(\theta) = 1$ on the top half of the boundary and $f(\theta) = -1$ on the bottom half of the boundary.

Solution: It is straightforward to calculate that the Fourier series (sine series) expansion for this f is

$$f(\theta) = \frac{4}{\pi}\left(\sin\theta + \frac{\sin 3\theta}{3} + \frac{\sin 5\theta}{5} + \cdots\right).$$

The solution of the Dirichlet problem is therefore

$$w(r,\theta) = \frac{4}{\pi}\left(r\sin\theta + \frac{r^3\sin 3\theta}{3} + \frac{r^5\sin 5\theta}{5} + \cdots\right). \qquad \square$$

11.4.1 The Poisson Integral

We have presented a formal procedure with series for solving the Dirichlet problem. But in fact it is possible to produce a closed formula (i.e., an integral formula) for this solution. We now make the construction explicit.

Referring back to our sine series expansion for f, and the resulting expansion for the solution of the Dirichlet problem, we recall that

$$a_j = \frac{1}{\pi}\int_{-\pi}^{\pi} f(\phi)\cos j\phi\,d\phi \quad \text{and} \quad b_j = \frac{1}{\pi}\int_{-\pi}^{\pi} f(\phi)\sin j\phi\,d\phi\,.$$

Thus

$$\begin{aligned} w(r,\theta) &= \frac{1}{2}a_0 + \sum_{j=1}^{\infty} r^j\left(\frac{1}{\pi}\int_{-\pi}^{\pi} f(\phi)\cos j\phi\,d\phi\cos j\theta\right. \\ &\qquad \left. + \frac{1}{\pi}\int_{-\pi}^{\pi} f(\phi)\sin j\phi\,d\phi\sin j\theta\right). \end{aligned}$$

This, in turn, equals

$$\begin{aligned} &\frac{1}{2}a_0 + \frac{1}{\pi}\sum_{j=1}^{\infty}\int_{-\pi}^{\pi} f(\phi)r^j(\cos j\phi\cos j\theta + \sin j\phi\sin j\theta)d\phi \\ &= \frac{1}{2}a_0 + \frac{1}{\pi}\sum_{j=1}^{\infty}\int_{-\pi}^{\pi} f(\phi)r^j\left(\cos j(\theta-\phi)d\phi\right). \end{aligned}$$

We finally simplify our expression to

$$w(r,\theta) = \frac{1}{\pi}\int_{-\pi}^{\pi} f(\phi)\left(\frac{1}{2} + \sum_{j=1}^{\infty} r^j\cos j(\theta-\phi)\right)d\phi\,.$$

It behooves us, therefore, to calculate the expression inside the large parentheses. For simplicity, we let $\alpha = \theta - \phi$ and then we let

$$z = re^{i\alpha} = r(\cos\alpha + i\sin\alpha)\,.$$

Likewise

$$z^n = r^n e^{in\alpha} = r^n(\cos n\alpha + i\sin n\alpha)\,.$$

In what follows, if $z = x + iy$, then we let $\operatorname{Re} z = x$ denote the *real part* of z and $\operatorname{Im} z = y$ denote the *imaginary part* of z. Also $\overline{z} = x - iy$ is the *conjugate* of z.

Then

$$\begin{aligned}
\frac{1}{2} + \sum_{j=1}^{\infty} r^j \cos j\alpha &= \operatorname{Re}\left(\frac{1}{2} + \sum_{j=1}^{\infty} z^j\right) \\
&= \operatorname{Re}\left(-\frac{1}{2} + \sum_{j=0}^{\infty} z^j\right) \\
&= \operatorname{Re}\left(-\frac{1}{2} + \frac{1}{1-z}\right) \\
&= \operatorname{Re}\left(\frac{1+z}{2(1-z)}\right) \\
&= \operatorname{Re}\left(\frac{(1+z)(1-\overline{z})}{2|1-z|^2}\right) \\
&= \frac{1-|z|^2}{2|1-z|^2} \\
&= \frac{1-r^2}{2(1-2r\cos\alpha + r^2)}\,.
\end{aligned}$$

Putting the result of this calculation into our original formula for w we finally obtain the Poisson integral formula:

$$w(r,\theta) = \frac{1}{2\pi}\int_{-\pi}^{\pi} \frac{1-r^2}{1-2r\cos(\theta-\phi)+r^2} f(\phi)\,d\phi\,.$$

Observe what this formula does for us: It expresses the solution of the Dirichlet problem with boundary data f as an explicit integral of a universal expression (called a *kernel*) against that data function f.

There is a great deal of information about w and its relation to f contained in this formula. As just one simple instance, we note that when r is set equal to 0 then we obtain

$$w(0,\theta) = \frac{1}{2\pi}\int_{-\pi}^{\pi} f(\phi)\,d\phi\,.$$

This says that the value of the steady-state heat distribution at the origin is just the average value of f around the circular boundary.

Math Nugget

Siméon Denis Poisson (1781–1840) was an eminent French mathematician and physicist. He succeeded Fourier in 1806 as Professor at the École Polytechnique. In physics, Poission's equation describes the variation of the potential inside a continuous distribution of mass or in an electric charge. Poisson made important theoretical contributions to the study of elasticity, magnetism, heat, and capillary action. In pure mathematics, the Poisson summation formula is a major tool in analytic number theory, and the Poisson integral pointed the way to many significant developments in Fourier analysis. In addition, Poisson worked extensively in probability theory. It was he who identified and named the "law of large numbers;" and the Poisson distribution—or "law of small numbers"—has fundamental applications in all parts of statistics and probability.
According to Abel, Poisson was a short, plump man. His family tried to encourage him in many directions, from being a doctor to being a lawyer, this last on the theory that perhaps he was fit for nothing better. But at last he found his place as a scientist and produced over 300 works in a relatively short lifetime. "La vie, c'est le travail (Life is work)," said Poisson—and he had good reason to know.

EXAMPLE 11.4.2 Consider an initial heat distribution on the boundary of the unit disc which is given by a "point mass." That is to say, there is a "charge of heat" at the point $(1, 0)$ of total mass 1 and with value 0 elsewhere on ∂D. What will be the steady-state heat distribution on the entire disc?

Solution: Think of the point mass as the limit of functions that take the value N on a tiny interval of length $1/N$. Convince yourself that the Poisson integral of such a function tends to the Poisson kernel itself. So the steady-state heat distribution in this case is given by the Poisson kernel. This shows, in particular, that the Poisson kernel is a harmonic function. □

Exercises

1. Solve the Dirichlet problem for the unit disc when the boundary function $f(\theta)$ is defined by

(a) $f(\theta) = \cos\theta/2, \quad -\pi \le \theta \le \pi$

(b) $f(\theta) = \theta, \quad -\pi < \theta \le \pi$

(c) $f(\theta) = \begin{cases} 0 & \text{if} \quad -\pi \le \theta < 0 \\ \sin\theta & \text{if} \quad 0 \le \theta \le \pi \end{cases}$

(d) $f(\theta) = \begin{cases} 0 & \text{if} \quad -\pi \le \theta < 0 \\ 1 & \text{if} \quad 0 \le \theta \le \pi \end{cases}$

(e) $f(\theta) = \theta^2/4, \quad -\pi \le \theta \le \pi$

2. Show that the Dirichlet problem for the disc $\{(x, y) : x^2 + y^2 \le R^2\}$, where $f(\theta)$ is the boundary function, has the solution

$$w(r, \theta) = \frac{1}{2}a_0 + \sum_{j=1}^{\infty} \left(\frac{r}{R}\right)^j (a_j \cos j\theta + b_j \sin j\theta)$$

where a_j and b_j are the Fourier coefficients of f. Show also that the Poisson integral formula for this more general disc setting is

$$w(r, \theta) = \frac{1}{2\pi} \int_{-\pi}^{\pi} \frac{R^2 - r^2}{R^2 - 2Rr\cos(\theta - \phi) + r^2} f(\phi)\, d\phi\,.$$

(**Hint:** Do not solve this problem from first principles. Rather, do a change of variables to reduce this new problem to the already-understood situation on the unit disc.)

3. Let w be a harmonic function in a planar region, and let C be any circle entirely contained (along with its interior) in this region. Prove that the value of w at the center of C is the average of its values on the circumference.

4. If $w = F(x, y) = \mathcal{F}(r, \theta)$, with $x = r\cos\theta$ and $y = r\sin\theta$, then show that

$$\begin{aligned} \frac{\partial^2 w}{\partial x^2} + \frac{\partial^2 w}{\partial y^2} &= \frac{1}{r}\left\{ \frac{\partial}{\partial r}\left(r\frac{\partial w}{\partial r}\right) + \frac{1}{r}\frac{\partial^2 w}{\partial \theta^2} \right\} \\ &= \frac{\partial^2 w}{\partial r^2} + \frac{1}{r}\frac{\partial w}{\partial r} + \frac{1}{r^2}\frac{\partial^2 w}{\partial \theta^2}\,. \end{aligned}$$

Hint: We can calculate that

$$\frac{\partial w}{\partial r} = \frac{\partial w}{\partial x}\cos\theta + \frac{\partial w}{\partial y}\sin\theta \quad \text{and} \quad \frac{\partial w}{\partial \theta} = \frac{\partial w}{\partial x}(-r\sin\theta) + \frac{\partial w}{\partial y}(r\cos\theta)\,.$$

Similarly, compute $\dfrac{\partial}{\partial r}\left(r\dfrac{\partial w}{\partial r}\right)$ and $\dfrac{\partial^2 w}{\partial \theta^2}$.

5. Use your symbol manipulation software, such as `Maple` or `Mathematica`, to calculate the Poisson integral of the given function on $[-\pi, \pi]$.

(a) $f(\theta) = \ln^2\theta$

(b) $f(\theta) = \theta^3 \cdot \cos\theta$

(c) $f(\theta) = e^\theta \cdot \sin\theta$

(d) $f(\theta) = e^\theta \cdot \ln\theta$

11.5 Sturm—Liouville Problems

We wish to place the idea of eigenvalues and eigenfunctions into a broader context. This setting is the fairly extensive and far-reaching subject of Sturm–Liouville problems.

Recall that a sequence y_j of functions such that

$$\int_a^b y_m(x) y_n(x)\, dx = 0 \qquad \text{for } m \neq n$$

is said to be an *orthogonal system* on the interval $[a, b]$. If

$$\int_a^b y_j^2(x)\, dx = 1$$

for each j then we call this an *orthonormal system* or *orthonormal sequence.* It turns out (and we have seen several instances of this phenomenon) that the sequence of eigenfunctions associated with a wide variety of boundary value problems enjoys the orthogonality property.

Now consider a differential equation of the form

$$\frac{d}{dx}\left(p(x)\frac{dy}{dx}\right) + [\lambda q(x) + r(x)]y = 0\,; \tag{11.28}$$

we shall be interested in solutions valid on an interval $[a, b]$. We know that, under suitable conditions on the coefficients, a solution of equation (28) that takes a prescribed value and a prescribed derivative value at a fixed point $x_0 \in [a, b]$ will be uniquely determined. In other circumstances, we may wish to prescribe the values of y at two distinct points, say at a and at b. We now begin to examine the conditions under which such a *boundary value problem* has a nontrivial solution.

EXAMPLE 11.5.1 Consider equation (28) with $p(x) \equiv q(x) \equiv 1$ and $r(x) \equiv 0$. Then the differential equation becomes

$$y'' + \lambda y = 0\,.$$

We take the domain interval to be $[0, \pi]$ and the boundary conditions to be

$$y(0) = 0, \quad y(\pi) = 0\,.$$

What are the eigenvalues and eigenfunctions for this problem?

Solution: Of course we completely analyzed this problem in Section 11.2. But now, as motivation for the work in this section, we review. We know that, in order for this boundary value problem to have a solution, the parameter λ can only assume the values $\lambda_n = n^2$, $n = 1, 2, 3, \ldots$. The corresponding solutions to the differential equation are $y_n(x) = \sin nx$. We call λ_n the *eigenvalues* for the problem and y_n the *eigenfunctions* (or sometimes the *eigenvectors*) for the problem. □

It will turn out—and this is the basis for the Sturm–Liouville theory—that, if $p, q > 0$ on $[a, b]$, then equation (28) will have a solvable boundary value problem—for a certain discrete set of values of λ—with data specified at points a and b. These special values of λ will of course be the eigenvalues for the boundary value problem. They are real numbers that we shall arrange in their natural order

$$\lambda_1 < \lambda_2 < \cdots < \lambda_n < \cdots ,$$

and we shall learn that $\lambda_j \to +\infty$. The corresponding eigenfunctions will then be ordered as $y_1, y_2, \ldots$.

Now let us examine possible orthogonality properties for the eigenfunctions of the boundary value problem for equation (28). Consider the differential equation (28) with two different eigenvalues λ_m and λ_n and y_m and y_n the corresponding eigenfunctions:

$$\frac{d}{dx}\left(p(x)\frac{dy_m}{dx}\right) + [\lambda_m q(x) + r(x)]y_m = 0$$

and

$$\frac{d}{dx}\left(p(x)\frac{dy_n}{dx}\right) + [\lambda_n q(x) + r(x)]y_n = 0 .$$

We convert to the more convenient prime notation for derivatives, multiply the first equation by y_n and the second by y_m, and subtract. The result is

$$y_n(py'_m)' - y_m(py'_n)' + (\lambda_m - \lambda_n)qy_my_n = 0 .$$

We move the first two terms to the right-hand side of the equation and integrate from a to b. Hence

$$\begin{aligned}(\lambda_m - \lambda_n)\int_a^b qy_my_n\,dx &= \int_a^b y_m(py'_n)'\,dx - \int_a^b y_n(py'_m)'\,dx \\ &\overset{\text{(parts)}}{=} \left[y_m(py'_n)\right]_a^b - \int_a^b y'_m(py'_n)\,dx \\ &\qquad -\left[y_n(py'_m)\right]_a^b + \int_a^b y'_n(py'_m)\,dx \\ &= p(b)[y_m(b)y'_n(b) - y_n(b)y'_m(b)] \\ &\qquad -p(a)[y_m(a)y'_n(a) - y_n(a)y'_m(a)] .\end{aligned}$$

Notice that the two integrals have canceled.

Let us denote by $W(x)$ the Wronskian determinant of the two solutions y_m, y_n. Thus

$$W(x) = y_m(x)y_n'(x) - y_n(x)y_m'(x) .$$

Then our last equation can be written in the more compact form

$$(\lambda_m - \lambda_n)\int_a^b q y_m y_n \, dx = p(b)W(b) - p(a)W(a) .$$

Notice that things have turned out so nicely, and certain terms have canceled, just because of the special form of the original differential equation.

We want the right-hand side of this last equation to vanish. This will certainly be the case if we require the familiar boundary condition

$$y(a) = 0 \qquad \text{and} \qquad y(b) = 0$$

or instead we require that

$$y'(a) = 0 \qquad \text{and} \qquad y'(b) = 0 .$$

Either of these will guarantee that the Wronskian vanishes, and therefore

$$\int_a^b y_m \cdot y_n \cdot q \, dx = 0 .$$

This is called an *orthogonality condition with weight q*.

With such a condition in place, we can consider representing an arbitrary function f as a linear combination of the y_j:

$$f(x) = a_1y_1(x) + a_2y_2(x) + \cdots + a_jy_j(x) + \cdots . \tag{11.29}$$

We may determine the coefficients a_j by multiplying both sides of this equation by $y_k \cdot q$ and integrating from a to b. Thus

$$\begin{aligned}
\int_a^b f(x)y_k(x)q(x)\,dx &= \int_a^b \Big(a_1y_1(x) + a_2y_2(x) + \cdots \\
&\qquad + a_jy_j(x) + \cdots\Big) y_k(x)q(x)\,dx \\
&= \sum_j a_j \int_a^b y_j(x)y_k(x)q(x)\,dx \\
&= a_k \int_a^b y_k^2(x)q(x)\,dx .
\end{aligned}$$

Thus

$$a_k = \frac{\int_a^b f(x)y_k(x)q(x)\,dx}{\int_a^b y_k^2(x)q(x)\,dx} .$$

There is an important question that now must be asked. Namely, are there *enough* of the eigenfunctions y_j so that virtually any function f can be expanded as in (29)? For instance, the functions $y_1(x) = \sin x, y_3(x) = \sin 3x, y_7(x) = \sin 7x$ are orthogonal on $[-\pi, \pi]$, and for any function f one can calculate coefficients a_1, a_3, a_7. But there is no hope that a large class of functions f can be spanned by just y_1, y_3, y_7. We need to know that our y_j's "fill out the space." The study of this question is beyond the scope of the present text, as it involves ideas from Hilbert space (see [RUD]). Our intention here has been merely to acquaint the reader with some of the language of Sturm–Liouville problems.

Exercises

1. If an equation of the form

$$P(x)y'' + Q(x)y' + R(x)y = 0$$

is not exact, it can often be made exact by multiplying through by a suitable integrating factor $\mu(x)$ (actually, there is always an integrating factor—but you may have trouble finding it). The function $\mu(x)$ must satisfy the condition that the equation

$$\mu(x)P(x)y'' + \mu(x)Q(x)y' + \mu(x)R(x)y = 0$$

can be expressed in the form

$$[\mu(x)P(x)y']' + [S(x)y]' = 0$$

for some appropriate function S. Show that this μ must be the solution of the *adjoint equation*

$$P(x)\mu''(x) + [2P'(x) - Q(x)]\,\mu'(x) + [P''(x) - Q'(x) + R(x)]\,\mu(x) = 0\,.$$

Often the adjoint equation is just as difficult to solve as the original differential equation. But not always. Find the adjoint equation in each of the following instances.

(a) **Legendre's equation:** $(1 - x^2)y'' - 2xy' + p(p+1)y = 0$

(b) **Bessel's equation:** $x^2y'' + xy' + (x^2 - p^2)y = 0$

(c) **Hermite's equation:** $y'' - 2xy' + 2py = 0$

(d) **Laguerre's equation:** $xy'' + (1 - x)y' + py = 0$

2. Consider the Euler equidimensional equation,

$$x^2y'' + xy' - n^2y = 0\,,$$

which we have seen before. Here n is a positive integer. Find the values of n for which this equation is exact, and for these values find the general solution by the method suggested in Exercise 1.

3. Consider the differential equation $u'' + \lambda u = 0$ with the endpoint conditions $u(0) = 0$ and $u(\pi) = u'(\pi) = 0$. Show that there is an infinite sequence of eigenfunctions with distinct eigenvalues. Identify the eigenvalues explicitly.

4. Refer to Exercise 1 for terminology. Solve the equation

$$y'' - \left(2x + \frac{3}{x}\right)y' - 4y = 0$$

by finding a simple solution of the adjoint equation by inspection.

5. Show that the adjoint of the adjoint of the equation $P(x)y'' + Q(x)y' + R(x)y = 0$ is just the original equation.

6. The equation $P(x)y'' + Q(x)y' + R(x)y = 0$ is called *self-adjoint* if its adjoint is just the same equation (after a possible change of notation).

(a) Show that this equation is self-adjoint if and only if $P'(x) = Q(x)$. In this case the equation becomes

$$P(x)y'' + P'(x)y' + R(y) = 0$$

or

$$[P(x)y']' + R(x)y = 0\,.$$

This is the standard form for a self-adjoint equation.

(b) Which of the equations in Exercise 1 are self-adjoint?

7. Show that any equation $P(x)y'' + Q(x)y' + R(x)y = 0$ can be made self-adjoint by multiplying through by

$$\frac{1}{P} \cdot e^{\int (Q/P)\,dx}\,.$$

8. Using Exercise 7 when appropriate, put each equation in Exercise 1 into the standard self-adjoint form described in Exercise 6.

Math Nugget

Charles Hermite (1822–1901) was one of the most eminent French mathematicians of the nineteenth century. He was particularly noted for the elegance, indeed the artistry, of his work. As a student, he courted disaster by neglecting his routine assignments in order to study the classic masters of mathematics. Although he nearly failed his examinations, he became a first-rate and highly creative mathematician while still in his early twenties. In 1870 he was appointed to a professorship at the Sorbonne, where he trained a whole generation of important French mathematicians; these include Picard, Borel, and Poincaré.

The unusual character of Hermite's mind is suggested by the following remark of Poincaré: "Talk with M. Hermite. He never evokes a concrete image, yet you soon perceive that the most abstract entities are to him like living creatures." He disliked geometry, but was strongly attracted to number theory and analysis; his favorite subject was elliptic functions, where these two subjects interact in remarkable ways.

Several of Hermite's purely mathematical discoveries had unexpected applications many years later to mathematical physics. For example, the Hermite forms and matrices which he invented in connection with certain problems of number theory turned out to be crucial for Heisenberg's 1925 formulation of quantum mechanics. Also Hermite polynomials and Hermite functions are useful in solving Schrödinger's wave equation.

Historical Note

Fourier

Jean Baptiste Joseph Fourier (1768–1830) was a mathematical physicist of some note. He was an acolyte of Napoleon Bonaparte and accompanied the fiery leader to Egypt in 1798. On his return, Fourier became the prefect of the district of Isère in southeastern France; in that post he built the first real road from Grenoble to Turin. He also became the friend and mentor of the boy Champollion, who later was to be the first to decipher the Rosetta Stone.

During these years he worked on the theory of the conduction of heat. Euler, Bernoulli, d'Alembert, and many others had studied the heat equation and made conjectures on the nature of its solutions. The most central issues hinged on the problem of whether an "arbitrary function" could be represented as a sum of sines and cosines. In those days, nobody was very sure what a function was and the notion of convergence of a series had not yet been defined, so the debate was largely metaphysical.

Fourier actually came up with a formula for producing the coefficients of a cosine or sine series of any given function. He presented it in the first chapter of his book *The Analytic Theory of Heat.* Fourier's ideas were controversial, and he had a difficult time getting the treatise published. In fact he only managed to do so when he became the Secretary of the French National Academy of Sciences and published the book himself.

The series that Fourier studied, and in effect put on the map, are now named after him. The subject area has had a profound influence on mathematics as a whole. Riemann's theory of the integral—the one that is used in most every calculus book—was developed specifically in order to study certain questions of the convergence of Fourier series. Cantor's theory of sets was cooked up primarily to address issues of sets of convergence for Fourier series. Many of the modern ideas in functional analysis—the uniform boundedness principle, for example—grew out of questions of the convergence of Fourier series. Dirichlet invented the modern rigorous notion of "function" as part of his study of Fourier series. As we have indicated in this chapter, Fourier analysis is a powerful tool in the study of partial differential equations (and ordinary differential equations as well).

Fourier's name has become universally known in modern analytical science. His ideas have been profound and influential. Harmonic analysis is the modern generalization of Fourier analysis, and wavelets are the latest implementation of these ideas.

Historical Note

Dirichlet

Peter Gustav Lejeune Dirichlet (1805–1859) was a German mathematician who was deeply influenced by the works of the Parisians—Cauchy, Fourier, Legendre, and many others. He was strongly influenced by Gauss's *Disquisitiones Arithmeticae*. This was quite a profound but impenetrable work, and Dirichlet was not satisfied until he had worked through the ideas himself in detail. He was not only the first to understand Gauss's famous book, but also the first to explain it to others.

In later life Dirichlet became a friend and disciple of Gauss, and also a friend and advisor to Riemann. In 1855, after lecturing in Berlin for many years, he succeeded Gauss in the professorship at Göttingen.

In 1829 Dirichlet achieved two milestones. One is that he gave a rigorous definition of the convergence of series. The other is that he gave the definition, that we use today, of a function. In particular, he freed the idea of function from any dependence on formulas or laws or mathematical operations. He applied both these ideas to the study of the convergence of Fourier series, and gave the first rigorously proved convergence criterion.

Between 1837 and 1839, Dirichlet developed some very remarkable applications of mathematical analysis to number theory. In particular, he proved that there are infinitely many primes in any arithmetical progression of the form $a + bn$ with a and b relatively prime. His studies of absolutely convergent series also appeared in 1837. Dirichlet's important convergence test for series was not published until after his death.

Dirichlet also engaged in studies of mathematical physics. These led, in part, to the important *Dirichlet principle* in potential theory. This idea establishes the existence of certain extremal harmonic functions. It was important historically, because it was the key to finally obtaining a rigorous proof of the Riemann mapping theorem. It is still used today in partial differential equations, the calculus of variations, differential geometry, and mathematical physics.

Dirichlet is remembered today for the Dirichlet problem, for his results in number theory (the useful "pigeonhole principle" was originally called the "Dirichletscher Schubfachschluss" or "Dirichlet's drawer-shutting principle"). He is one of the important mathematicians of the nineteenth century.

Anatomy of an Application

Some Ideas from Quantum Mechanics

Sturm–Liouville problems arise in many parts of mathematical physics—both elementary and advanced. One of these is quantum mechanics. Here we see how the study of matter on a very small scale can be effected with some of the ideas that we have been learning. We derive our exposition from the lovely book [KBO].

We think of a system as being specified by a *state function* $\psi(\mathbf{r}, t)$. Here $\mathbf{r}$ represents a position vector for a point in space (of dimension one, two, or three), and t is time. We think of ψ as a probability distribution, so it is appropriate to assume that

$$1 = \langle \psi, \psi \rangle = \iint_{\mathbb{R}^3} |\psi(\mathbf{r}, t)|^2 \, d\mathbf{r} \, .$$

That is, for each fixed time t, ψ has total mass 1.

One of the basic tenets of quantum mechanics is that, in each system, there is a linear operator H such that

$$i\hbar \frac{\partial \psi}{\partial t} = H\psi \, .$$

Here $\hbar$ is a constant specified by the condition that

$$2\pi\hbar \approx 6.62 \cdot 19^{-34} \, J \, s^{-1}$$

is Planck's constant (the unit J is a *joule*). The actual value of the constant $\hbar$ is of no interest for the present discussion. The operator H has the nice (Hermitian) property that

$$\langle Hy_1, y_2 \rangle = \langle y_1, Hy_2 \rangle \, .$$

Finally—and this is one of the key ideas of von Neumann's model for quantum mechanics—to each observable property of the system there corresponds a linear, Hermitian operator A. Moreover, any measurement of the observable property gives rise to an eigenvalue of A. As an example, as you learn in your physics course, the operator that corresponds to momentum is $-i\hbar\nabla$ and the operator that corresponds to energy is $i\hbar\partial/\partial t$.

Now that we have dispensed with the background, let us examine a specific system and see how a Sturm–Liouville problem comes into play. Consider a particle of mass m moving in a potential field $V(\mathbf{r}, t)$. Then, if p is the momentum of the particle, we have that

$$E = \frac{p^2}{2m} + V(\mathbf{r}, t) \, . \tag{11.30}$$

Observe that the first expression on the right is the kinetic energy and the second expression is the potential energy. Now we quantize this classical relation by substituting in appropriate operators for the potential and the energy. We then obtain

$$i\hbar \frac{\partial \psi}{\partial t} = -\frac{\hbar^2}{2m} \nabla^2 \psi + V(\mathbf{r}, t)\psi \, . \tag{11.31}$$

This important identity is known as *Schrödinger's equation.* It controls the evolution of the wave function.

To simplify matters, let us suppose that the potential V does not depend on time t. We may then seek to solve (31) by separation of variables. Let us write $\psi(\mathbf{r}, t) = \alpha(\mathbf{r})T(t)$. Substituting into the differential equation, we find that

$$i\hbar\alpha(\mathbf{r}) \frac{dT}{dt} = -\frac{\hbar^2}{2m} \nabla^2 \alpha \cdot T + V(\mathbf{r})\alpha(\mathbf{r})T$$

or

$$i\hbar \frac{(dT/dt)(t)}{T(t)} = -\frac{\hbar^2}{2m\alpha(\mathbf{r})} \nabla^2 \alpha(\mathbf{r}) + V(\mathbf{r}) \, . \tag{11.32}$$

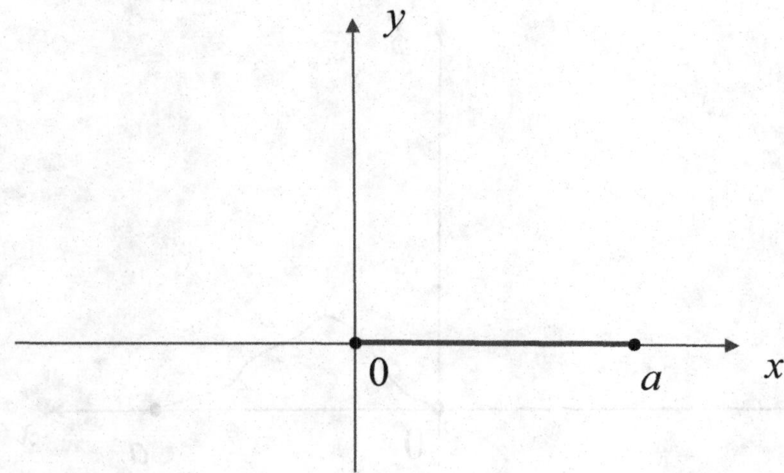

FIGURE 11.7
Particle trapped in a region of zero potential.

Observe that the left-hand side depends only on t, and the right-hand side only on $\mathbf{r}$. Thus both are equal to some constant μ.

In particular,

$$i\hbar \frac{dT}{dt} = \mu T .$$

So we may write

$$i\hbar \frac{\partial \psi/\partial t}{\psi} = i\hbar \frac{dT/dt}{T} = \mu$$

hence

$$i\hbar \frac{\partial \psi}{\partial t} = \mu \psi .$$

Thus we see that μ is the energy of the particle in our system.

Looking at the right-hand side of (32), we now see that

$$-\frac{\hbar^2}{2m}\nabla^2 \alpha + (V(\mathbf{r}) - \mu)\alpha = 0 . \tag{11.33}$$

This is the *time-independent Schrödinger equation.* It will turn out, contrary to the philosophy of classical physics, that the energy of this one-particle system must be one of the eigenvalues of the boundary value problem that we shall construct from equation (33). The energy is said to be *quantized.*

To consider a simple instance of the ideas we have been discussing, we examine a particle of mass m that is trapped in a region of zero potential by the infinite potentials at $x = 0$ and $x = a$. See Figure 11.7. Let us consider the possible energies that such a particle can have.

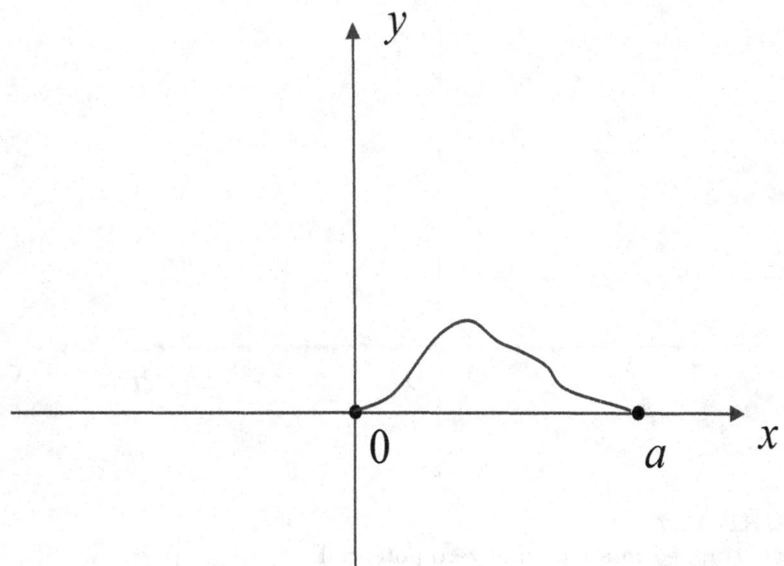

FIGURE 11.8
Graph of the continuous probability distribution ψ.

Thinking of ψ as a (continuous) probability distribution, we see that $\psi = 0$ outside the interval $(0, a)$, since the probability is zero that the particle will be found outside the interval. Thus the graph of ψ is as in Figure 11.8.

Thus our eigenvalue problem simplifies to

$$-\frac{\hbar^2}{2m}\frac{\partial^2\alpha}{\partial x^2} = \mu\alpha\,,$$

subject to the boundary conditions $\alpha(0) = \alpha(a) = 0$. Of course this is a familiar problem, and we solved it (except for a change of notation) in Example 11.3.1. Observe that the role of μ in that example is now being played by $2m\mu/h^2$ and the role of the function y is now played by α. Thus we find that the allowed energies in our system are simply $\hbar^2/2m$ times the eigenvalues that we found for the problem in Example 11.3.1, or $\mu_n = \hbar^2 n^2\pi^2/[2ma^2]$.

Problems for Review and Discovery

A. Drill Exercises

1. Find the eigenvalues λ_n and eigenfunctions y_n for the equation $y'' + \lambda y = 0$ in each of the following cases.

   (a) $y(-2) = 0\,, \quad y(2) = 0$
   (b) $y(0) = 0\,, \quad y(3) = 0$
   (c) $y(1) = 0\,, \quad y(4) = 0$
   (d) $y(-3) = 0\,, \quad y(0) = 0$

2. Solve the vibrating string problem in Section 6.2 if the initial shape $y(x, 0) = f(x)$ is specified by the function $f(x) = x + |x|$. Sketch the initial shape of the string on a set of axes.

3. Solve the Dirichlet problem for the unit disc when the boundary function $f(\theta)$ is defined by

   (a) $f(\theta) = \sin\theta/2\,, \quad -\pi \le \theta \le \pi$
   (b) $f(\theta) = \theta + |\theta|\,, \quad -\pi \le \theta \le \pi$
   (c) $f(\theta) = \theta^2\,, \quad -\pi \le \theta \le \pi$

4. Find the solution to the Dirichlet problem on the unit disc with boundary data

   (a) $f(\theta) = |\theta|$
   (b) $g(\theta) = \sin^2\theta$
   (c) $h(\theta) = \cos\theta/2$
   (d) $f(\theta) = \theta/2$

5. Find a solution to this Dirichlet problem for a half-annulus:

$$\begin{aligned}
\frac{\partial^2 u}{\partial r^2} + \frac{1}{r}\frac{\partial u}{\partial r} + \frac{1}{r^2}\frac{\partial^2 u}{\partial \theta^2} &= 0\,,\ 1 < r < 2\,,\ 0 < \theta < \pi \\
u(r, 0) &= \sin \pi r\,,\ 1 \le r \le 2 \\
u(r, \pi) &= 0\,,\ 1 \le r \le 2 \\
u(1, \theta) &= u(2, \theta) = 0\,,\ 0 \le \theta \le \pi\,.
\end{aligned}$$

B. Challenge Problems

1. Is it possible for a harmonic function to vanish on an entire line segment? Give an example.

2. Use methods introduced in this chapter to find a solution of the boundary value problem

$$\begin{aligned}
\frac{\partial u}{\partial t} &= 4\frac{\partial^2 u}{\partial x^2}\,,\ 0 < x < \pi\,,\ t > 0 \\
\frac{\partial u}{\partial x}(0, t) &= \frac{\partial u}{\partial x}(\pi, t) = 0\,,\ t > 0 \\
u(x, 0) &= 3x\,,\ 0 < x < \pi\,.
\end{aligned}$$

3. Use methods introduced in this chapter to find a solution of the boundary value problem

$$\begin{aligned}
\frac{\partial u}{\partial t} &= \frac{\partial^2 u}{\partial x^2}\,,\ 0 < x < 1\,,\ t > 0 \\
u(0, t) &= 0\,,\ u(1, t) + \frac{\partial u}{\partial x}(1, t) = 0\,,\ t > 0\,. \\
u(x, 0) &= f(x)\,,\ 0 < x < 1
\end{aligned}$$

4. Use methods introduced in this chapter to find a solution of the boundary value problem

$$\begin{aligned}\frac{\partial u}{\partial t} &= 2\frac{\partial^2 u}{\partial x^2} + 4 \ , \ 0 < x < 1 \ , \ t > 0 \\ u(0,t) &= u(1,t) = 1 \ , \ t > 0 \\ u(x,0) &= 1 \ , \ 0 < x < 1 \ .\end{aligned}$$

5. Use methods introduced in this chapter to find a solution of the boundary value problem

$$\begin{aligned}\frac{\partial^2 u}{\partial t^2} &= 9\frac{\partial^2 u}{\partial x^2} \ , \ 0 < x < 1 \ , \ t > 0 \\ u(0,t) &= u(1,t) = 0 \ , \ t > 0 \\ u(x,0) &= 1 - \cos^2 \pi x \ , \ 0 < x < 1 \\ \frac{\partial u}{\partial t}(x,0) &= 1 - \sin x \ , \ 0 < x < 1 \ .\end{aligned}$$

6. Use methods introduced in this chapter to find a solution of the boundary value problem

$$\begin{aligned}\frac{\partial^2 u}{\partial t^2} &= \frac{\partial^2 u}{\partial x^2} \ , \ 0 < x < 2 \ , \ t > 0 \\ u(0,t) &= u(2,t) = 0 \ , \ t > 0 \\ u(x,0) &= x(2-x) \ , \ 0 < x < 2 \\ \frac{\partial u}{\partial t}(x,0) &= \cos 4\pi x \ , \ 0 < x < 2 \ .\end{aligned}$$

C. Problems for Discussion and Exploration

1. Let w be a harmonic function in a planar region, and let D be any disc entirely contained in this region. Prove that the value of w at the center of D is the average of the values of w on the D.

2. Let w be a real-valued, twice continuously differentiable function on planar region U. Suppose that both w and w^2 are harmonic. Prove that w must be constant.

3. It is a fact (not obvious) that if u is harmonic on a connected region U in the plane and if $(x_0, y_0) \in U$ then u has a convergent power series about (x_0, y_0) of the form

$$u(x,y) = \sum_{j,k} a_{j,k}(x-x_0)^j(y-y_0)^k$$

$$\text{for } |(x-x_0)^2 + (y-y_0)^2| < \epsilon, \text{ some small } \epsilon > 0 \ .$$

Use this information to show that if u vanishes on some disc in U, then u is identically 0 on all of U.

4. Use methods introduced in this chapter to find a solution of the boundary value problem

$$\begin{aligned}
\frac{\partial u}{\partial t} &= \frac{\partial^2 u}{\partial x^2} + \frac{\partial^2 u}{\partial y^2} \ , \ 0 < x < 1 \ , 0 < y < 1 \ , \ t > 0 \\
\frac{\partial u}{\partial x}(0, y, t) &= \frac{\partial u}{\partial x}(1, y, t) = 0 \ , \ 0 < y < 1 \ , \ t > 0 \\
u(x, 0, t) &= u(x, 1, t) = 0 \ , \ 0 < x < 1 \ , \ t > 0 \\
u(x, y, 0) &= f(x, y) \ , 0 < x < 1 \ , \ 0 < y < 1 \, .
\end{aligned}$$

5. A vibrating circular membrane, or drum, of radius 1 with edges held fixed in the plane and with displacement $u(r, t)$ (r is radius and t is time) is given. That is to say, the displacement of any point of the drum depends only on the distance from the center of the drum and the elapsed time. This situation is described by the boundary value problem

$$\begin{aligned}
\frac{\partial^2 u}{\partial t^2} &= \alpha^2 \left(\frac{\partial^2 u}{\partial r^2} + \frac{1}{r}\frac{\partial u}{\partial r} \right) \ , \ 0 < r < 1, \ t > 0 \\
u(1, t) &= 0 \ , \ t > 0 \\
u(r, t) & \quad \text{remains bounded as } r \to 0^+ \\
u(r, 0) &= f(r) \ , \ 0 < r < 1 \\
\frac{\partial u}{\partial t}(r, 0) &= g(r) \ , \ 0 < r < 1 \, .
\end{aligned}$$

Here f is the initial displacement and g is the initial velocity. Use the method of separation of variables, as introduced in this chapter, to find a solution of this boundary value problem. (*Hint:* Bessel functions will be involved.)

12

The Nonlinear Theory

- The nature of a nonlinear problem
- Examples of nonlinear problems
- Critical points
- Stability
- Linear systems
- Lyapunov's direct method
- Nonlinear systems
- Nonlinear mechanics
- Periodic solutions
- Poincaré–Bendixson theory

Both the historical and the textbook development of the subject matter of this book might lead one to think that ordinary differential equations is primarily about *linear* equations. Historically speaking this has certainly been the case.

It is definitely true that the linear equations are the most tractable, they are the ones that we understand best, and they are the ones for which there is the most complete theory. But in fact the world is largely nonlinear. We have traditionally shied away from nonlinear equations just because they are so difficult, and because their solutions can rarely be written down in a closed formula.

But there is still much that can be said about nonlinear ordinary differential equations. Especially if we are willing to accept *qualitative* information rather than closed formulas, there is a considerable amount that one can learn using even elementary techniques. In the present chapter our intention is to acquaint the reader with some of the basic ideas of the nonlinear theory, and to present a number of important examples.

12.1 Some Motivating Examples

If x is the angle of deviation at time t of an undamped pendulum of length a whose bob has mass m, then the equation of motion is

$$\frac{d^2x}{dt^2} + \frac{g}{a}\sin x = 0\,. \tag{12.1}$$

DOI: 10.1201/9781003214526-12

Because of the presence of the term $\sin x$, this equation is definitely nonlinear. If there is present a damping force proportional to the velocity, then the equation becomes

$$\frac{d^2x}{dt^2} + \frac{c}{m}\frac{dx}{dt} + \frac{g}{a}\sin x = 0 \,. \tag{12.2}$$

It is common in elementary studies to replace $\sin x$ by x, which is a reasonable approximation for small oscillations (since $\sin x = x - x^3/3! + \cdots$). But this substitution is a gross distortion when x is large.

A second example comes from the theory of the vacuum tube, which leads to the famous *van der Pol equation:*

$$\frac{d^2x}{dt^2} + \mu(x^2 - 1)\frac{dx}{dt} + x = 0 \,. \tag{12.3}$$

This time the nonlinearity comes from the presence of the term x^2 (which is multiplied times dx/dt).

12.2 Specializing Down

In the present chapter we shall concentrate our attention on nonlinear equations of the form

$$\frac{d^2x}{dt^2} = f\left(x, \frac{dx}{dt}\right) \,. \tag{12.4}$$

If we imagine a simple dynamical system consisting of a particle of unit mass moving on the x-axis, and if $f(x, dx/dt)$ is the force acting on it, then (4) is the resulting equation of motion. This equation includes examples (1), (2), and (3) as special cases.

In equation (4), it is common to call the values of x (position) and dx/dt (velocity), that is, the quantities which at each instant characterize the state of the system, the *phases* of the system; the plane determined by these two variables is called the *phase plane.* If we introduce the substitution $y = dx/dt$, then equation (4) may be rewritten as

$$\left\{\begin{array}{rcl} \dfrac{dx}{dt} & = & y \\ \dfrac{dy}{dt} & = & f(x,y) \,. \end{array}\right. \tag{12.5}$$

A great deal can be learned about the solution of (4) by studying the system (5). If we think of t as a parameter, then a solution of (5) is a pair of functions $x(t)$, $y(t)$; these in turn define a curve in the x-y plane—i.e., the phase plane mentioned a moment ago. We shall learn in this chapter a great deal about the geometry of curves in the phase plane.

More generally, it is useful for us to study systems of the form

$$\left\{\begin{array}{rcl} \dfrac{dx}{dt} & = & F(x,y) \\ \dfrac{dy}{dt} & = & G(x,y) \,. \end{array}\right. \tag{12.6}$$

Here F and G are assumed to be continuously differentiable: each is continuous and has continuous first partial derivatives in the entire plane. A system of this kind, in which the independent variable t does not appear in the functions on the right, is called an *autonomous system.* We now examine the solutions of such a system.

It follows—at least in the linear case—from our standard existence result for systems (Theorem 10.2.2) that, if t_0 is any number and (x_0, y_0) is any point in the phase plane, then there is a unique solution

$$\begin{cases} x &= x(t) \\ y &= y(t) \end{cases} \tag{12.7}$$

of (6) such that $x(t_0) = x_0$ and $y(t_0) = y_0$. If the resulting $x(t)$ and $y(t)$ are not both constant functions, then (7) defines a curve in the phase plane which we call a *path* of the system (the terms *trajectory* and *characteristic* are also commonly used).

It is a trivial observation that, if (7) is a solution of (6), then

$$\begin{cases} x &= x(t+c) \\ y &= y(t+c) \end{cases} \tag{12.8}$$

is also a solution for any real constant c. Thus any path may be reparameterized by translation. Picard's existence and uniqueness result (Theorem 13.2.1) implies that any path through the point (x_0, y_0) must correspond to a solution of the form (8). As a result of these considerations, at most one path passes through each point of the phase plane. Furthermore, the direction induced by increasing t along a given path is the same for every parameterization. A path is therefore a *directed curve*, and its orientation is intrinsic. In our figures we shall use arrows to indicate the direction in which each path is traced out as t increases.

Since there is a path through *every point* (x_0, y_0), it follows that the entire phase plane is filled up by paths (i.e., no point (x_0, y_0) is missed). Moreover, the paths do not cross or intersect each other (again, by the uniqueness of solutions). The only exception to this last statement, where the uniqueness theory breaks down, is at points (x_0, y_0) where both F and G vanish: $F(x_0, y_0) = 0$ and $G(x_0, y_0) = 0$. These points are called *critical points*; at such a point our uniqueness theorem guarantees only that the unique solution is the constant solution $x \equiv x_0$, $y \equiv y_0$. A constant solution does not define a path (according to our definition of "path"), and thus no path passes through a critical point. We shall assume, in the problems that we study below, that all critical points are isolated. This means that there are only finitely many critical points, and they do not accumulate at any point of the phase plane.

With a view to finding a physical interpretation for critical points, let us consider the special autonomous system (5) arising from the dynamical systems equation (4). In this circumstance a critical point is a point $(x_0, 0)$ at which $y = 0$ and $f(x_0, 0) = 0$. That is to say, such a point corresponds to a state of the particle's motion in which both the velocity dx/dt and the acceleration $dy/dt = dx^2/dt^2$ vanish. Thus the particle is at rest with no force acting on it; it is therefore in a state of equilibrium. Obviously the states of equilibrium of a physical system are among its most important attributes, and this observation accounts in part for our interest in critical points. There will also be important geometrical reasons for focusing on these points.

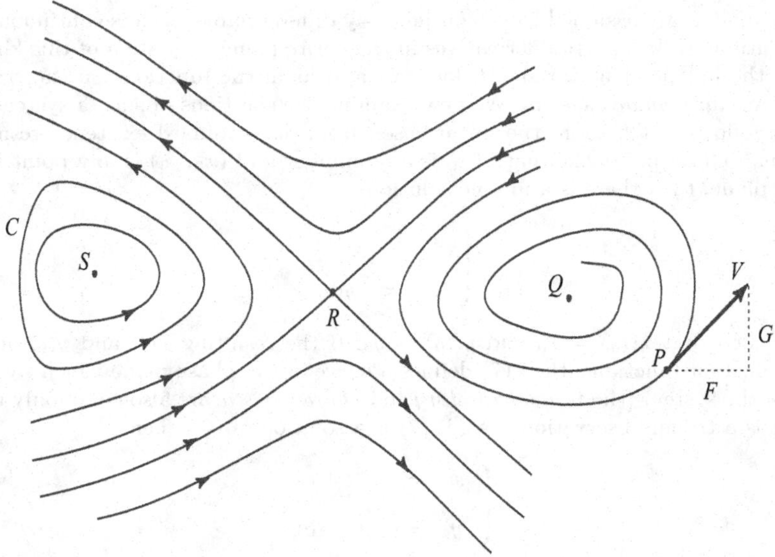

FIGURE 12.1
Paths and critical points.

An autonomous system of the form (6) does not always arise from a dynamical equation of the form (4). In this case, therefore, what physical significance can be assigned to the paths and critical points? As an aid to our thoughts, let us consider Figure 12.1 and the two-dimensional vector field defined by

$$\mathbf{V}(x,y) = F(x,y)\mathbf{i} + G(x,y)\mathbf{j}\,.$$

At a point $P = (x, y)$, the vector field has horizontal component $F(x, y)$ and vertical component $G(x, y)$. Since $dx/dt = F$ and $dy/dt = G$, this vector is tangent to the path at P and points in the direction of increasing t. If we think of t as time, then the vector $\mathbf{V}$ can be interpreted as the velocity vector of a particle moving along the path.

An alternative physical interpretation of this mathematical model is that we have not one, but many, particles in motion. Each path is the trail of a moving particle, preceded and followed by many others on the same path. Every path has particles and every particle has a path. This situation can be described as a two-dimensional *fluid motion*. Since the system (6) is autonomous, which means that the vector $\mathbf{V}(x, y)$ does not change with time, it follows that the fluid motion is *stationary*. The paths are the trajectories of the moving particles, and the critical points Q, R, and S (shown in the figure) are points of zero velocity where the particles are at rest (i.e. fixed points of the fluid motion). This particular physical interpretation will be especially useful in our studies for the rest of the chapter.

The most striking features of the fluid motion illustrated in Figure 12.1 are these:

(a) the critical points;

(b) the geometric configuration of the paths near the critical points;

(c) the stability or instability of critical points; that is, whether a particle near such a point remains near the critical point or wanders off into another part of the plane;

(d) closed paths (like C in the figure), which correspond to periodic solutions.

These features are essential components of what we call the *phase portrait* of the system (6). Here the phase portrait consists of a graphical depiction of the paths—or at least a representative collection of the paths (since if we were to try to draw *all* the paths, then we would end up with a solid black diagram). Our goal, given that we are generally unable to explicitly solve nonlinear equations and systems, is to derive as much information as possible about the phase portrait from the analytic properties of F and G.

As an instance of the type of reasoning we hope to develop, observe that if $x(t)$ is a periodic solution of the dynamical equation (4), then its derivative $y(t) = dx/dt$ is also periodic. Therefore the corresponding path in the system is closed. Conversely, any closed path of (5) corresponds to a periodic solution of (4). As a concrete instance of these ideas, the van der Pol equation (3)—which cannot be solved in closed form—can nevertheless be shown to have a unique periodic solution (if $\mu > 0$) by showing that its equivalent autonomous system has a unique closed path.

Exercises

1. Derive equation (2) by applying Newton's second law of motion to the bob of the pendulum.
2. Let (x_0, y_0) be a point in the phase plane. If $x_1(t)$, $y_1(t)$ and $x_2(t)$, $y_2(t)$ are solutions of (6) such that $x_1(t_1) = x_0$, $y_1(t_1) = y_0$ and $x_2(t_2) = x_0$, $y_2(t_2) = y_0$ for suitable t_1, t_2, then show that there exists a constant c such that
$$x_1(t+c) = x_2(t) \quad \text{and} \quad y_1(t+c) = y_2(t)\,.$$
3. Describe the relation between the phase portraits of the systems
$$\left\{\begin{aligned} \frac{dx}{dt} &= F(x,y) \\ \frac{dy}{dt} &= G(x,y) \end{aligned}\right. \quad \text{and} \quad \left\{\begin{aligned} \frac{dx}{dt} &= -F(x,y) \\ \frac{dy}{dt} &= -G(x,y)\,. \end{aligned}\right.$$
4. Sketch and describe the phase portrait of each of the following systems.

$$\text{(a)} \left\{\begin{aligned} \frac{dx}{dt} &= 0 \\ \frac{dy}{dt} &= 0 \end{aligned}\right. \qquad \text{(c)} \left\{\begin{aligned} \frac{dx}{dt} &= 1 \\ \frac{dy}{dt} &= 2 \end{aligned}\right.$$

$$\text{(b)} \left\{\begin{aligned} \frac{dx}{dt} &= x \\ \frac{dy}{dt} &= 0 \end{aligned}\right. \qquad \text{(d)} \left\{\begin{aligned} \frac{dx}{dt} &= -x \\ \frac{dy}{dt} &= -y \end{aligned}\right.$$

5. The critical points and paths of equation (4) are by definition those of the equivalent system (5). Use this rubric to find the critical points of equations (1), (2), and (3).
6. Find the critical points of

(a) $\frac{d^2x}{dt^2} + \frac{dx}{dt} - (x^3 + x^2 - 2x) = 0$

(b) $\left\{ \begin{aligned} \frac{dx}{dt} &= y^2 - 5x + 6 \\ \frac{dy}{dt} &= x - y \end{aligned} \right.$

7. Find all solutions of the nonautonomous system

$$\left\{ \begin{aligned} \frac{dx}{dt} &= x \\ \frac{dy}{dt} &= x + e^t \end{aligned} \right.$$

and sketch (in the x-y plane) some of the curves defined by these solutions.

12.3 Types of Critical Points: Stability

Consider an autonomous system

$$\left\{ \begin{aligned} \frac{dx}{dt} &= F(x, y) \\ \frac{dy}{dt} &= G(x, y) . \end{aligned} \right. \tag{12.9}$$

We assume, as usual, that the functions F and G are continuously differentiable. The critical points of the system (9) can be found by solving the system

$$\begin{aligned} F(x, y) &= 0 \\ G(x, y) &= 0 . \end{aligned}$$

There are four basic types of critical points that arise in this manner; the present section is devoted to describing those types. We do so by describing the behavior of nearby paths. First we need two supporting definitions.

Let (x_0, y_0) be an isolated critical point of (9). If $C = [x(t), y(t)]$ is a path of (9), then we say that

$$C \text{ approaches } (x_0, y_0) \text{ as } t \to +\infty$$

if

$$\lim_{t\to+\infty} x(t) = x_0 \quad \text{and} \quad \lim_{t\to+\infty} y(t) = y_0 \,. \tag{12.10}$$

Geometrically, this means that if $P = (x, y)$ is a point that traces out the curve C according to $x = x(t)$, $y = y(t)$, then $P \to (x_0, y_0)$ as $t \to \infty$. If it is also true that

$$\lim_{t\to\infty} \frac{y(t) - y_0}{x(t) - x_0} \tag{12.11}$$

exists, or if this quotient becomes positively or negatively infinite as $t \to \infty$, then we say that C *enters* the critical point (x_0, y_0) as $t \to \infty$. The quotient in (11) is the slope of the line joining (x_0, y_0) and the point $P = (x(t), y(t))$ so that the additional requirement for "entering" means that this line approaches a definite limiting slope as $t \to \infty$. Of course we could also, in this discussion, consider limits as $t \to -\infty$. These are properties of the path C, and do not depend on the particular parameterization that we choose.

Although it is occasionally possible to find explicit solutions of the system (1), it is unrealistic to hope that this will happen in any generality. What we usually do instead is to eliminate the variable t and thereby create the equation

$$\frac{dy}{dx} = \frac{G(x, y)}{F(x, y)} \,. \tag{12.12}$$

This first-order equation expresses the slope of the tangent to the path of (1) that passes through the point (x, y), provided that the functions F and G are not both zero at this point. If both functions *do* vanish at a given point, then that point is a critical point; no path will pass through it. The paths of equation (1) therefore coincide with the one-parameter family of integral curves of (12); we may often actually solve for this family using the methods of Chapter 1. Note, however, that while the paths of (1) are directed curves (having an orientation), the integral curves of (12) have no direction associated with them. We shall use the examples below to illustrate all the techniques for finding paths.

We now give geometric descriptions of the four types of critical points. We shall, in each instance, assume for convenience that the critical point is the origin $O = (0, 0)$.

Nodes. A critical point like that depicted in Figure 12.2 is called a *node*. Such a critical point is approached and also is entered by each path as $t \to \pm\infty$. For the node shown in the figure, there are four half-line paths—these are $\overrightarrow{AO}$, $\overrightarrow{BO}$, $\overrightarrow{CO}$, and $\overrightarrow{DO}$. Together with the origin, these make up the lines $\overleftrightarrow{AB}$ and $\overleftrightarrow{CD}$.

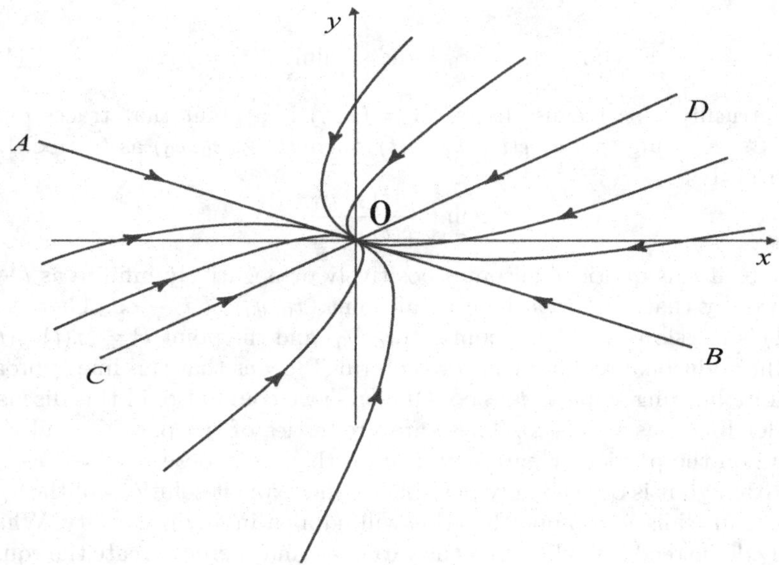

FIGURE 12.2
A node.

EXAMPLE 12.3.1 Consider the system

$$\begin{cases} \dfrac{dx}{dt} &= x \\ \dfrac{dy}{dt} &= -x + 2y \,. \end{cases} \tag{12.13}$$

Clearly the origin is the only critical point. The general solution of the system can be found quickly and easily by the methods of Section 10.2. It is

$$\begin{cases} x &= c_1 e^t \\ y &= c_1 e^t + c_2 e^{2t} \,. \end{cases} \tag{12.14}$$

When $c_1 = 0$, we have $x = 0$ and $y = c_2 e^{2t}$. In this case the path (Figure 12.3) is the positive y-axis when $c_2 > 0$ and the negative y-axis when $c_2 < 0$. Each path approaches and enters the origin as $t \to -\infty$. If instead $c_2 = 0$ then $x = c_1 e^t$ and $y = c_1 e^t$. This path is the half-line $y = x$, $x > 0$, when $c_1 > 0$, and is the half-line $y = x$, $x < 0$ when $c_1 < 0$. Again, both paths approach and enter the origin as $t \to -\infty$.

When both c_1 and c_2 are not zero then the paths lie on the parabolas $y = x + (c_2/c_1^2)x^2$, which go through the origin with slope 1. It should be clearly understood that each of these paths consists of part of a parabola—the part with $x > 0$ if $c_1 > 0$ and the part with $x < 0$ if $c_1 < 0$. Each of these paths also approaches and enters the origin as $t \to -\infty$; this assertion can be seen immediately from (14).

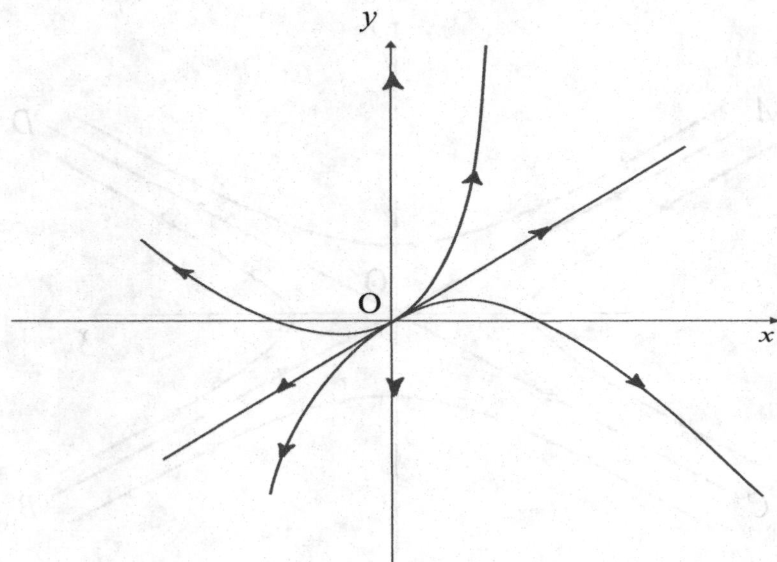

FIGURE 12.3
Example 12.3.1.

We proceed directly from (13) to the differential equation

$$\frac{dy}{dx} = \frac{-x + 2y}{x}, \tag{12.15}$$

giving the slope of the tangent to the path through (x, y) (provided that $(x, y) \neq (0, 0)$). Then, on solving (15) as a homogeneous equation, we find that $y = x + cx^2$.

Observe that this procedure yields the curves on which the paths lie (except those on the y-axis), but gives no information on the *direction* of the curves.

This discussion—and Figure 12.3—makes clear that the critical point $(0, 0)$ is a node. ∎

Saddle Points. A critical point like that in Figure 12.4 is called a *saddle point.* It is approached and entered by two half-line paths AO and BO as $t \to +\infty$, and these two paths of course lie on the line $\overleftrightarrow{AB}$. It is also approached and entered by two half-line paths CO and DO at $t \to -\infty$, and these two paths lie on another line $\overleftrightarrow{CD}$. The four half-lines determine four regions, and each contains a family of hyperbola-like paths. These paths do not approach O as $t \to \pm\infty$, but instead are asymptotic to one or another of the half-line paths as $t \to \pm\infty$.

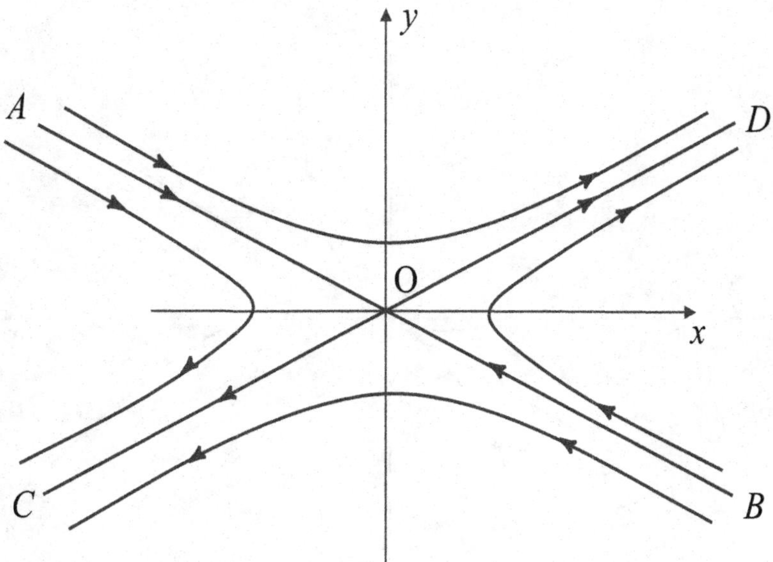

FIGURE 12.4
A saddle point.

Centers. A *center* (sometimes called a *vortex*) is a critical point that is surrounded by a family of closed paths. It is not approached by any path either as $t \to +\infty$ or as $t \to -\infty$.

EXAMPLE 12.3.2 The system

$$\left\{ \begin{array}{rcl} \dfrac{dx}{dt} & = & -y \\ \dfrac{dy}{dt} & = & x \end{array} \right. \tag{12.16}$$

has the origin as its only critical point. The general solution of the system is

$$\left\{ \begin{array}{rcl} x & = & -c_1 \sin t + c_2 \cos t \\ y & = & c_1 \cos t + c_2 \sin t \,. \end{array} \right. \tag{12.17}$$

The solution that satisfies $x(0) = 1$, $y(0) = 0$ is clearly

$$\left\{ \begin{array}{rcl} x & = & \cos t \\ y & = & \sin t \end{array} \right. \tag{12.18}$$

while the solution that satisfies $x(0) = 0$ and $y(0) = -1$ is

$$\left\{ \begin{array}{rcl} x & = & \sin t = \cos \left(t - \frac{\pi}{2}\right) \\ y & = & -\cos t = \sin \left(t - \frac{\pi}{2}\right) \,. \end{array} \right. \tag{12.19}$$

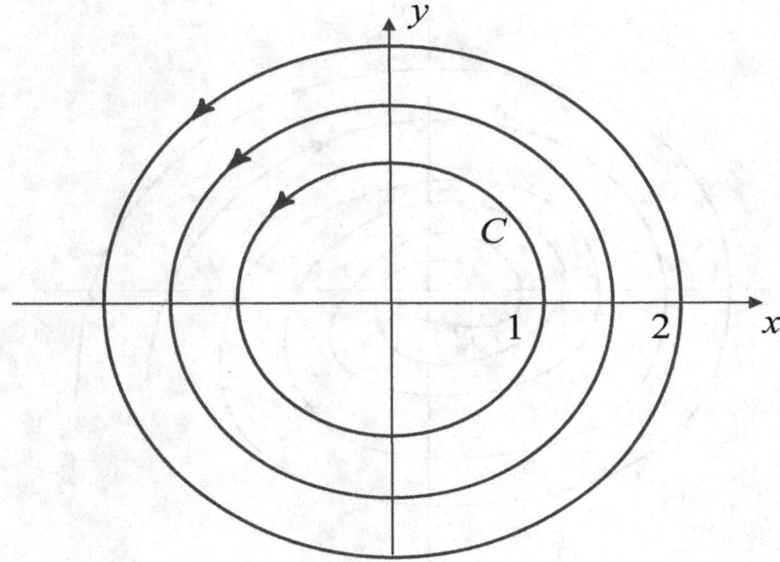

FIGURE 12.5
A center.

These two different solutions define the same path C (Figure 12.5), which is evidently the circle $x^2 + y^2 = 1$. Both (18) and (19) show that this path is traced out in the counterclockwise direction. If we eliminate t between the equations of the system, then we get

$$\frac{dy}{dx} = -\frac{x}{y} .$$

The general solution of this equation is $x^2 + y^2 = c^2$, yielding all the paths (a concentric family of circles) but without directions. Obviously, the critical point $(0, 0)$ is a center. ∎

Spirals. A critical point like the one shown in Figure 12.6 is a *spiral* (sometimes also called a *focus*). Such a point is characterized by the fact that the paths approach it along spiral-like trajectories that wind around it an infinite number of times as $t \to \pm\infty$. Note, in particular, that the paths approach O but they do not enter it. That is to say, each path approaching O does not approach any definite direction.

EXAMPLE 12.3.3 Let a be an arbitrary constant. Then the system

$$\begin{cases} \dfrac{dx}{dt} &= \quad ax - y \\ \dfrac{dy}{dt} &= \quad x + ay \end{cases} \tag{12.20}$$

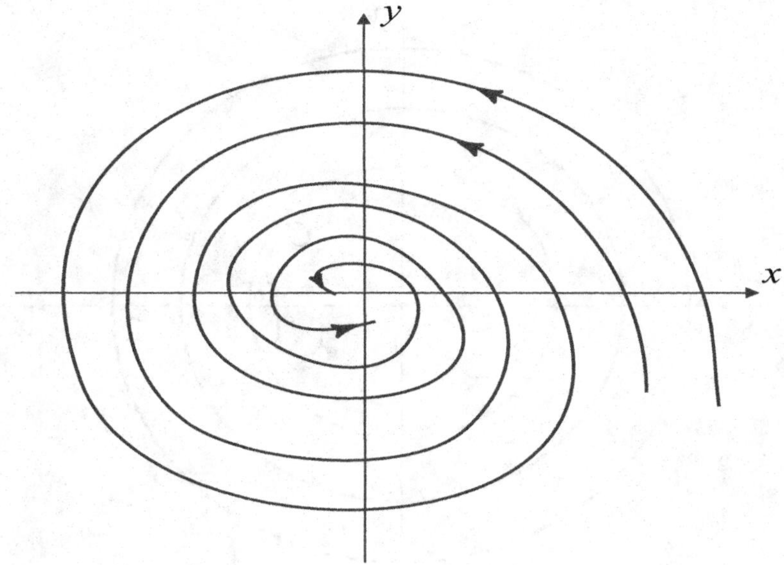

FIGURE 12.6
A spiral.

has the origin as its only critical point. The differential equation

$$\frac{dy}{dx} = \frac{x + ay}{ax - y} \tag{12.21}$$

for the paths is most easily solved by introducing polar coordinates:

$$\begin{aligned} x &= r\cos\theta \\ y &= r\sin\theta\,. \end{aligned}$$

Since

$$r^2 = x^2 + y^2 \quad \text{and} \quad \theta = \arctan\frac{y}{x}\,,$$

we see that

$$r\frac{dr}{dx} = x + y\frac{dy}{dx} \quad \text{and} \quad r^2\frac{d\theta}{dx} = x\frac{dy}{dx} - y\,.$$

With the aid of these equations, (21) can be written in the form

$$\frac{dr}{d\theta} = ar\,.$$

As a result,

$$r = ce^{a\theta} \tag{12.22}$$

is the polar equation for the paths. The two possible spiral configurations are

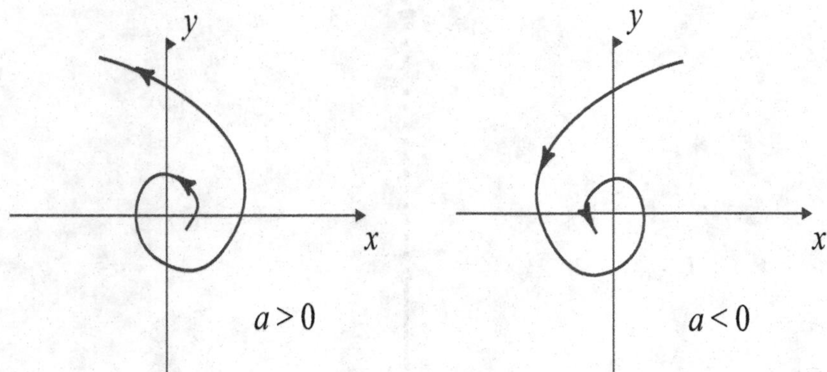

FIGURE 12.7
Two possible spiral configurations.

shown in Figure 12.7. The direction in which these paths are traversed can be derived from the fact that $dx/dt = -y$ when $x = 0$. In the left-most figure, we see that when y is positive, then dx/dt is negative so the orientation is to the left. In the right-most figure, we see that when y is negative, then dx/dt is positive so the orientation is to the right.

If $a = 0$ then (20) collapses to (16) and (22) becomes $r = c$, which is the polar equation of the family of concentric circles $x^2 + y^2 = c^2$. This example thus generalizes Example 12.3.2, and since the center shown in Figure 12.5 stands on the borderline between the spirals of Figure 12.7, we often think of a center as a borderline case for spirals. We shall encounter additional borderline cases in the next section. ∎

We now introduce the concept of *stability* as it applies to the critical points of the system (1). One of the most important questions in the study of a physical system is that of its steady states. A steady state has little real significance unless it has a reasonable degree of permanence, that is, unless it is stable. We begin with an intuitive discussion.

As a simple example, consider the pendulum of Figure 12.8. There are two steady states possible here: **(i)** when the bob is at rest at the highest point and **(ii)** when the bob is at rest at the lowest point. The first of these states is clearly unstable, for a slight disturbance will cause the bob to fall clear to the bottom. The second state is stable, because the bob will always return immediately from a slight perturbation. We now recall that a steady state of a simple physical system corresponds to an equilibrium point (or critical point) in the phase plane. These considerations suggest that a small disturbance at an unstable equilibrium point leads to a larger and larger departure from this point, while the opposite is true at a stable equilibrium point.

We now formulate these intuitive ideas in a more precise and rigorous fashion. Consider an isolated critical point of the system (1). Assume for

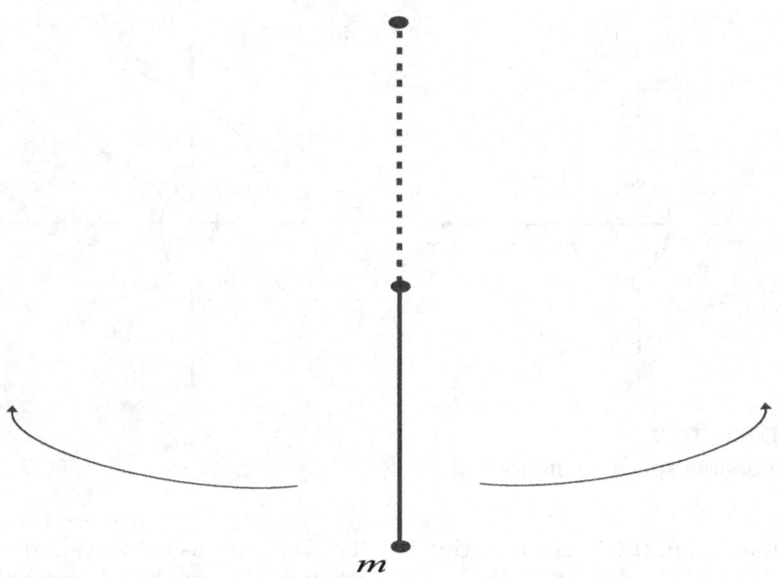

FIGURE 12.8
A pendulum.

simplicity that this point is located at the origin $O = (0, 0)$. This critical point is said to be *stable* if, for each positive number R, there is a positive number $0 < r \leq R$ such that every path which is inside the circle $x^2 + y^2 = r^2$ at some time $t = t_0$ remains inside the circle $x^2 + y^2 = R^2$ for all $t > t_0$. In other words, a critical point is stable if each path that gets sufficiently close to the point stays close to the point (Figure 12.9). Furthermore, our critical point is said to be *asymptotically stable* if it is stable and if, in addition, there is a circle $x^2 + y^2 = r_0^2$ such that every path which is inside this circle for some $t = t_0$ approaches the origin as $t \to +\infty$. Finally, if the critical point is not stable, then it is unstable.

As examples of these ideas, we observe that the node in Figure 12.3, the saddle point in Figure 12.4, and the spiral in the left-hand part of Figure 12.7 are all unstable, while the center in Figure 12.5 is stable but not asymptotically stable. The node in Figure 12.2, the spiral in Figure 12.6, and the spiral on the right in Figure 12.7 are all asymptotically stable.

Exercises

1. For each of these nonlinear systems, do the following:

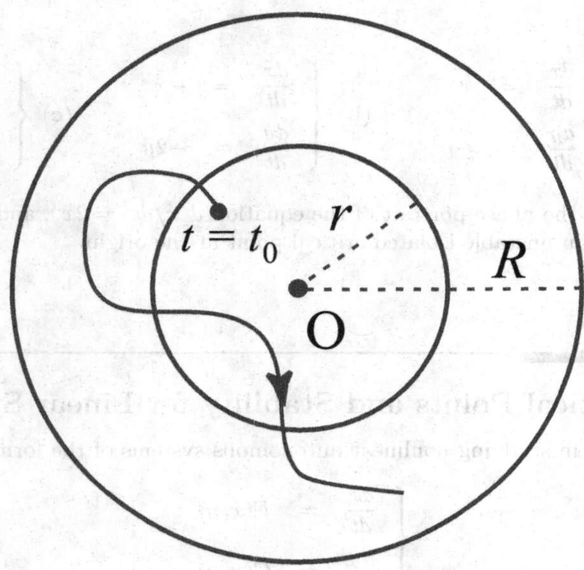

FIGURE 12.9
A stable critical point.

(i) Find the critical points;
(ii) Find the differential equation of the paths;
(iii) Solve this equation to find the paths;
(iv) Sketch a few of the paths and show the direction of increasing t.

(a) $\begin{cases} \dfrac{dx}{dt} = y(x^2+1) \\ \dfrac{dy}{dt} = 2xy^2 \end{cases}$ (c) $\begin{cases} \dfrac{dx}{dt} = e^y \\ \dfrac{dy}{dt} = e^y \cos x \end{cases}$

(b) $\begin{cases} \dfrac{dx}{dt} = y(x^2+1) \\ \dfrac{dy}{dt} = -x(x^2+1) \end{cases}$ (d) $\begin{cases} \dfrac{dx}{dt} = -x \\ \dfrac{dy}{dt} = 2x^2y^2 \end{cases}$

2. Each of the following linear systems has the origin as an isolated critical point. For each system,

(i) Find the general solution;
(ii) Find the differential equation of the paths;
(iii) Solve the equation found in part **(ii)** and sketch a few of the paths, showing the direction of increasing t;
(iv) Discuss the stability of the critical points.

(a) $\begin{cases} \dfrac{dx}{dt} &= x \\ \dfrac{dy}{dt} &= -y \end{cases}$ (b) $\begin{cases} \dfrac{dx}{dt} &= -x \\ \dfrac{dy}{dt} &= -2y \end{cases}$ (c) $\begin{cases} \dfrac{dx}{dt} &= 4y \\ \dfrac{dy}{dt} &= -x \end{cases}$

3. Sketch the phase portrait of the equation $d^2x/dt^2 = 2x^3$, and show that it has an unstable isolated critical point at the origin.

12.4 Critical Points and Stability for Linear Systems

Our main goal in studying nonlinear autonomous systems of the form

$$\begin{cases} \dfrac{dx}{dt} &= F(x,y) \\ \dfrac{dy}{dt} &= G(x,y) \end{cases}$$

is to classify the critical points of such a system with respect to their nature and stability. We shall see that, under suitable conditions, this problem can be solved for a given nonlinear system by studying a related linear system. In preparation for that strategy, we devote the present section to a complete analysis of the critical points of linear autonomous systems.

Consider the system

$$\begin{cases} \dfrac{dx}{dt} &= a_1x + b_1y \\ \dfrac{dy}{dt} &= a_2x + b_2y\,. \end{cases} \tag{12.23}$$

Such a system has the origin $O = (0,0)$ as an obvious critical point. We shall assume throughout this section that

$$\det \begin{pmatrix} a_1 & b_1 \\ a_2 & b_2 \end{pmatrix} \neq 0\,; \tag{12.24}$$

thus $(0,0)$ is the only critical point. It was proved in Section 12.2 that (23) has a nontrivial solution of the form

$$\begin{cases} x &= Ae^{mt} \\ y &= Be^{mt} \end{cases}$$

whenever m is a root of the auxiliary quadratic equation

$$m^2 - (a_1 + b_1)m + (a_1b_2 - a_2b_1) = 0\,. \tag{12.25}$$

Observe that condition (24) obviously implies that 0 cannot be a root of equation (25).

Let m_1 and m_2 be the roots of (25). We shall prove that the properties of the critical point $(0,0)$ of the system (23) are determined by the roots m_1, m_2. There will be three cases, according to whether **(i)** m_1, m_2 are real and distinct or **(ii)** m_1, m_2 are real and equal or **(iii)** m_1, m_2 are complex conjugate. Unfortunately, the situation is a bit more complicated than this, and some of these cases will have to be subdivided. There are a total of five cases. They are these:

Major Cases:

Case A. The roots m_1, m_2 are real, distinct, and of the same sign. This gives rise to a *node.*

Case B. The roots m_1, m_2 are real, distinct, and of opposite signs. This gives rise to a *saddle point.*

Case C. The roots m_1, m_2 are conjugate complex but not pure imaginary. This gives rise to a *spiral.*

Borderline Cases:

Case D. The roots m_1, m_2 are real and equal. This gives rise to a *node.*

Case E. The roots m_1, m_2 are pure imaginary. This gives rise to a *center.*

The reasoning behind the distinction between the major cases and the borderline cases will become clearer as the analysis develops. For the moment, we remark that the borderline cases are of less significance just because the circumstances under which they arise are unimportant in most physical applications. We now turn to an explication of these five cases.

Analysis of Case A. If the roots m_1, m_2 of (3) are real, distinct, and of the same sign, then the critical point $(0,0)$ is a node.

To see this, let us begin by assuming that both m_1, m_2 are negative. We may assume that $m_1 < m_2 < 0$. By Section 12.2, the general solution of (1) in this case is

$$\left\{\begin{array}{lcl} x & = & c_1A_1e^{m_1t} + c_2A_2e^{m_2t} \\ y & = & c_1B_1e^{m_1t} + c_2B_2e^{m_2t}, \end{array}\right. \tag{12.26}$$

where the c's are arbitrary constants and, in addition, $B_1/A_1 \neq B_2/A_2$. When $c_2 = 0$, we obtain the solution

$$\left\{\begin{array}{lcl} x & = & c_1A_1e^{m_1t} \\ y & = & c_1B_1e^{m_1t}. \end{array}\right. \tag{12.27}$$

When $c_1 = 0$, we obtain the solution

$$\left\{\begin{array}{lcl} x & = & c_2A_2e^{m_2t} \\ y & = & c_2B_2e^{m_2t}. \end{array}\right. \tag{12.28}$$

For any $c_1 > 0$, the solution (27) represents a path consisting of half of the

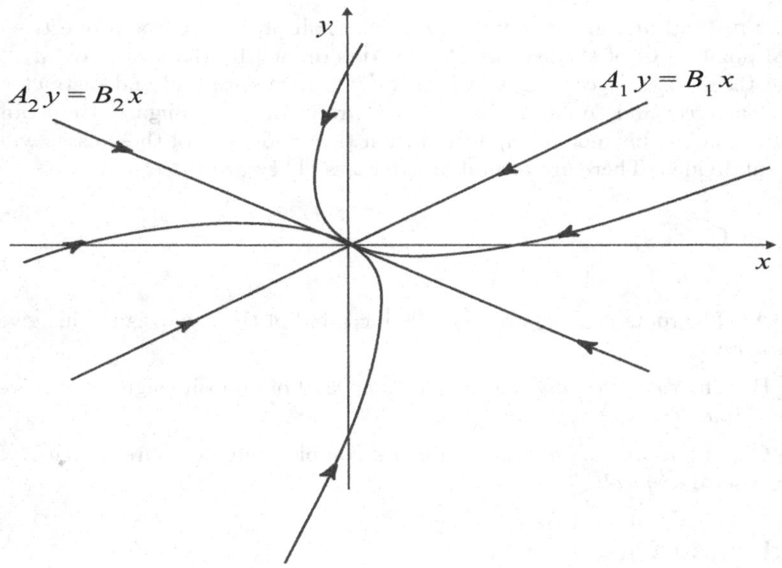

FIGURE 12.10
Analysis of Case A.

line $A_1y = B_1x$ with slope B_1/A_1. And for any $c_1 < 0$, the solution represents a path consisting of the other half of this line (the half on the other side of the origin). Since $m_1 < 0$, both of these half-line paths approach $(0,0)$ as $t \to \infty$. And, since $y/x = B_1/A_1$, both paths enter $(0,0)$ with slope B_1/A_1. See Figure 12.10. In just the same way, the solutions (28) represent two half-line paths lying on the line $A_2y = B_2x$ with slope B_2/A_2. These two paths also approach $(0,0)$ as $t \to \infty$, and enter it with slope B_2/A_2.

If both $c_1 \neq 0$ and $c_2 \neq 0$ then the general solution (26) represents curved (nonlinear) paths. Since $m_1 < 0$ and $m_2 < 0$, these paths also approach $(0,0)$ as $t \to \infty$. Also, since $m_1 - m_2 < 0$ and

$$\frac{y}{x} = \frac{c_1B_1e^{m_1t} + c_2B_2e^{m_2t}}{c_1A_1e^{m_1t} + c_2A_2e^{m_2t}} = \frac{(c_1B_1/c_2)e^{(m_1-m_2)t} + B_2}{(c_1A_1/c_2)e^{(m_1-m_2)t} + A_2},$$

it is clear that $y/x \to B_2/A_2$ as $t \to \infty$. So all of these paths enter $(0,0)$ with slope B_2/A_2. Figure 12.10 presents a qualitative picture of what is going on. Clearly our critical point is a node, and it is asymptotically stable.

If m_1 and m_2 are both positive, and if we choose the notation so that $m_1 > m_2 > 0$, then the situation is exactly the same except that all the paths now approach and enter $(0,0)$ as $t \to -\infty$. The picture of the paths given in Figure 12.10 is unchanged except that the arrows showing their directions are all reversed. We still have a node, but now it is *unstable*.

Analysis of Case B. If the roots m_1 and m_2 are real, distinct, and of opposite signs, then the critical point $(0,0)$ is a saddle point.

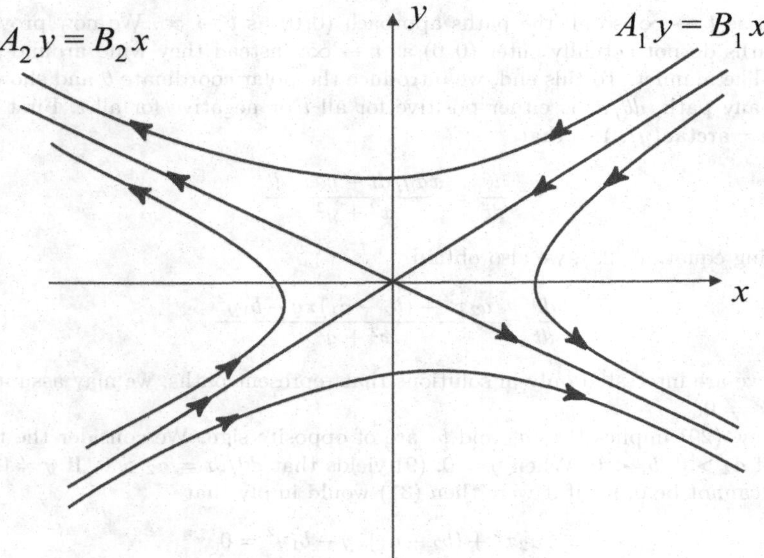

FIGURE 12.11
Analysis of Case B.

To see this, choose notation so that $m_1 < 0$ and $m_2 > 0$. The general solution of (23) can still be written in the form (26), and again we have particular solutions of the form (27) and (28). The two half-line paths represented by (27) still approach and enter $(0,0)$ as $t \to \infty$, but this time the two half-line paths represented by (28) approach and enter $(0,0)$ as $t \to -\infty$. If $c_1 \neq 0$ and $c_2 \neq 0$ then the general solution (26) still represents curved paths, but since $m_1 < 0 < m_2$, none of these paths approaches $(0,0)$ as $t \to \pm\infty$. Instead, as $t \to \infty$, each of these paths is asymptotic to one of the half-line paths represented by (28), and, as $t \to -\infty$, each is asymptotic to one of the half-line paths represented by (27). Figure 12.11 gives a qualitative picture of this behavior. Now the critical point is a saddle point, and it is obviously unstable.

Analysis of Case C. If the roots m_1 and m_2 are conjugate complex but not pure imaginary, then the critical point $(0,0)$ is a spiral.

To see this, we write m_1 and m_2 in the form $a \pm ib$, where a and b are nonzero real numbers. For later use, we record that the discriminant D of equation (25) is negative:

$$\begin{aligned} D &= (a_1 + b_2)^2 - 4(a_1 b_2 - a_2 b_1) \\ &= (a_1 - b_2)^2 + 4a_2 b_1 < 0 . \end{aligned} \qquad (12.29)$$

By the results of Section 12.2, the general solution of (23) is now

$$\begin{cases} x = e^{at}[c_1(A_1 \cos bt - A_2 \sin bt) + c_2(A_1 \sin bt + A_2 \cos bt)] \\ y = e^{at}[c_1(B_1 \cos bt - B_2 \sin bt) + c_2(B_1 \sin bt + B_2 \cos bt)] . \end{cases} \qquad (12.30)$$

Here the A's and the B's are definite constants and the c's are arbitrary constants.

First we assume that $a < 0$. Then formula (30) clearly implies that $x \to 0$ and

$y \to 0$ as $t \to \infty$, so all the paths approach $(0,0)$ as $t \to \infty$. We now prove that the paths do not actually enter $(0,0)$ as $t \to \infty$; instead they wind around it in a spiral-like manner. To this end, we introduce the polar coordinate θ and show that, along any path, $d\theta/dt$ is either positive for all t or negative for all t. First notice that $\theta = \arctan(y/x)$ so that

$$\frac{d\theta}{dt} = \frac{xdy/dt - ydx/dt}{x^2+y^2} .$$

By using equation (23), we also obtain

$$\frac{d\theta}{dt} = \frac{a_2x^2 + (b_2 - a_1)xy - b_1y^2}{x^2+y^2} . \tag{12.31}$$

Since we are interested only in solutions that represent paths, we may assume that $x^2 + y^2 \neq 0$.

Now (29) implies that a_2 and b_1 are of opposite sign. We consider the typical case of $a_2 > 0$, $b_1 < 0$. When $y = 0$, (9) yields that $d\theta/dt = a_2 > 0$. If $y \neq 0$, then $d\theta/dt$ cannot be 0; for if it were then (31) would imply that

$$a_2x^2 + (b_2 - a_1)xy - b_1y^2 = 0$$

or

$$a_2\left(\frac{x}{y}\right)^2 + (b_2 - a_1)\frac{x}{y} - b_1 = 0 \tag{12.32}$$

for some real number x/y. But this cannot be true because the discriminant of the quadratic equation (32) is D, which is negative by (29). Thus $d\theta/dt$ is always positive when $a_2 > 0$ and always negative when $a_2 < 0$.

By (30), x and y change sign infinitely often as $t \to \infty$. Thus all paths must spiral in to the origin (counterclockwise or clockwise according as $a_2 > 0$ or $a_2 < 0$). The critical point is therefore a spiral, and it is asymptotically stable.

Analysis of Case D. If the roots m_1 and m_2 are real and equal, then the critical point $(0,0)$ is a node.

To see this, we begin by supposing that $m_1 = m_2 = m < 0$. There are now two subcases that require separate discussions: **(i)** $a_1 = b_1 \neq 0$ and $a_2 = b_1 = 0$ and **(ii)** all other possibilities which lead to a double root of equation (25).

First consider subcase **(i)**. This situation is described in a footnote in Section 12.3. If a denotes the common value of a_1 and b_2, then equation (25) becomes $m^2 - 2am + a^2 = 0$ and thus $m = a$. The system (23) is thus

$$\begin{cases} \dfrac{dx}{dt} &= ax \\ \dfrac{dy}{dt} &= ay . \end{cases}$$

The general solution is then

$$\begin{cases} x &= c_1e^{mt} \\ y &= c_2e^{mt} , \end{cases} \tag{12.33}$$

where c_1 and c_2 are arbitrary constants. The paths defined by (33) are half-lines of all

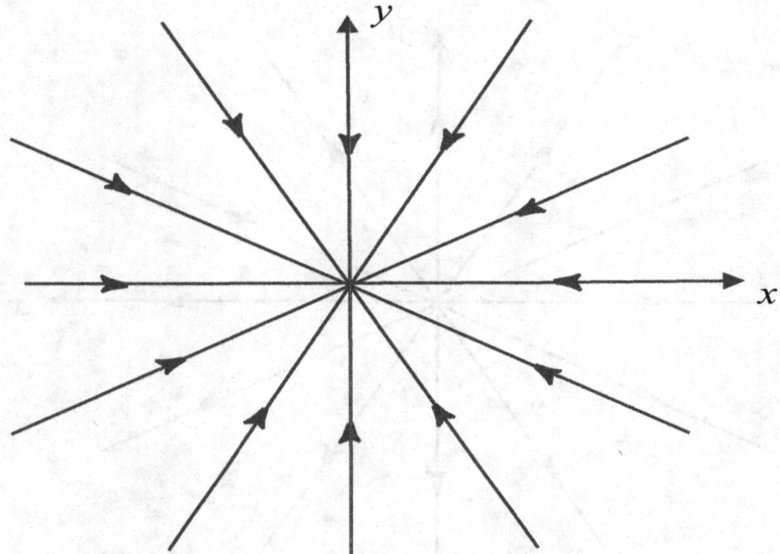

FIGURE 12.12
Analysis of Case D.

possible slopes (Figure 12.12), and since $m < 0$ we see that each path approaches and enters $(0,0)$ as $t \to \infty$. The critical point is therefore a node, and it is asymptotically stable. If $m > 0$, then we have just the same situation except that the paths enter $(0,0)$ as $t \to -\infty$, the arrows in Figure 12.12 are reversed (see Figure 12.13), and $(0,0)$ is unstable.

Now we consider subcase **(ii)**. The general solution of (23) can be written as

$$\left\{\begin{array}{lcl} x & = & c_1 A e^{mt} + c_2(A_1 + At)e^{mt} \\ y & = & c_1 B e^{mt} + c_2(B_1 + Bt)e^{mt}\,, \end{array}\right. \tag{12.34}$$

where the A's and B's are definite constants and the c's are arbitrary constants. When $c_2 = 0$ we obtain

$$\left\{\begin{array}{lcl} x & = & c_1 A e^{mt} \\ y & = & c_1 B e^{mt}\,. \end{array}\right. \tag{12.35}$$

We know that these solutions represent two half-line paths lying on the line $Ay = Bx$ with slope B/A. Since $m < 0$, both paths approach $(0,0)$ as $t \to \infty$ (Figure 12.14). Also, since $y/x = B/A$, both paths enter $(0,0)$ with slope B/A.

If $c_2 \neq 0$, then the solutions (34) represent curved paths, and since $m < 0$, it is clear from (34) that these paths approach $(0,0)$ as $t \to \infty$. Furthermore, it follows from the equation

$$\frac{y}{x} = \frac{c_1 B e^{mt} + c_2(B_1 + Bt)e^{mt}}{c_1 A e^{mt} + c_2(A_1 + At)e^{mt}} = \frac{c_1 B/c_2 + A_1 + Bt}{c_1 A/c_2 + A_1 + At}$$

that $y/x \to B/A$ as $t \to \infty$. Thus these curved paths all enter $(0,0)$ with slope B/A. We also note that $y/x \to B/A$ as $t \to -\infty$. Figure 12.14 gives a qualitative

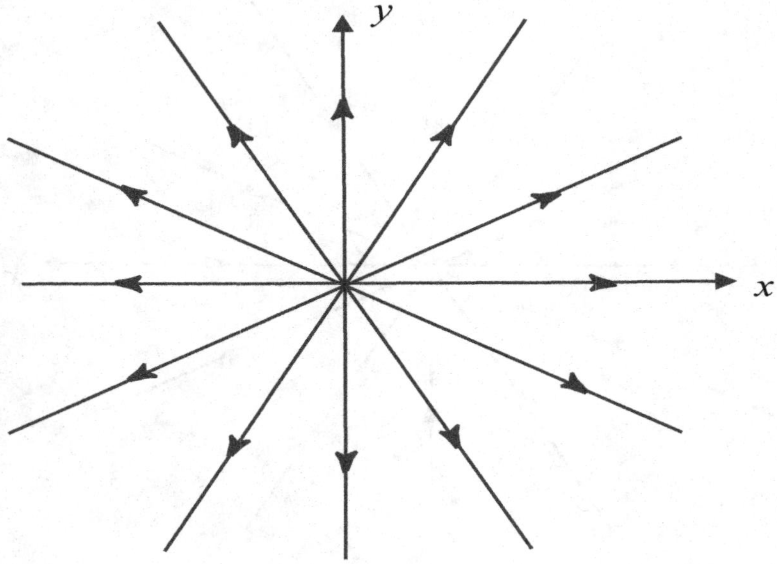

FIGURE 12.13
Case D with arrows reversed.

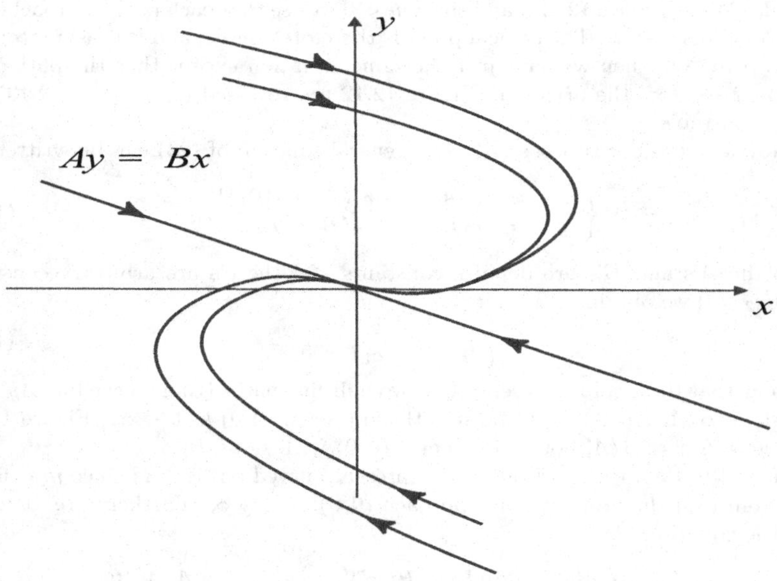

FIGURE 12.14
The paths in Case D.

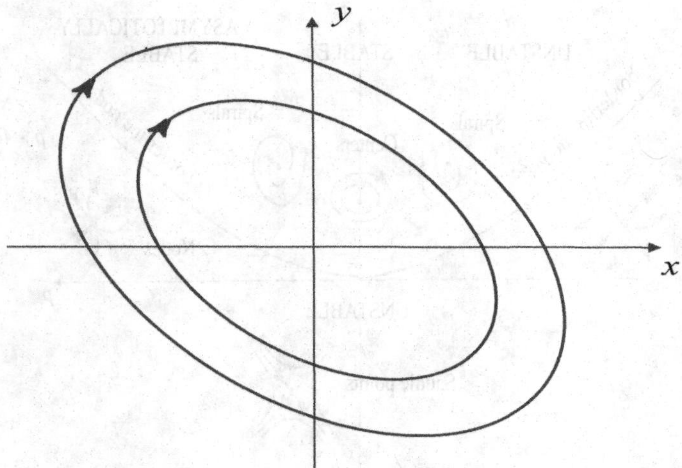

FIGURE 12.15
Analysis of Case E.

picture of the arrangement of these paths. It is clear that $(0,0)$ is a node that is asymptotically stable. If $m > 0$, then the situation is unchanged except that the directions of the paths are reversed and the critical point is unstable.

Analysis of Case E. If the roots m_1 and m_2 are pure imaginary, then the critical point $(0,0)$ is a center.

To see this, we first refer back to the discussion of **Case C**. Now m_1 and m_2 are of the form $a \pm ib$ with $a = 0$ and $b \neq 0$. The general solution of (23) is therefore given by (30) with the exponential factor no longer appearing. Thus $x(t)$ and $y(t)$ are periodic, and each path is a closed curve surrounding the origin. As Figure 12.15 suggests, these curves are actually ellipses; this assertion can actually be proved by solving the differential equation of the paths,

$$\frac{dy}{dx} = \frac{a_2x + b_2y}{a_1x + b_1y} . \tag{12.36}$$

Our critical point $(0,0)$ is evidently a center that is stable but not asymptotically stable.

We now summarize all our findings about stability in a single theorem:

> **Theorem 12.4.1** *The critical point $(0,0)$ of the linear system (23) is stable if and only if both roots of the auxiliary equation (25) have nonpositive real parts, and it is asymptotically unstable if and only if both roots have positive real parts.*

If we now write equation (25) in the form

$$(m - m_1)(m - m_2) = m^2 + pm + q = 0 ,$$

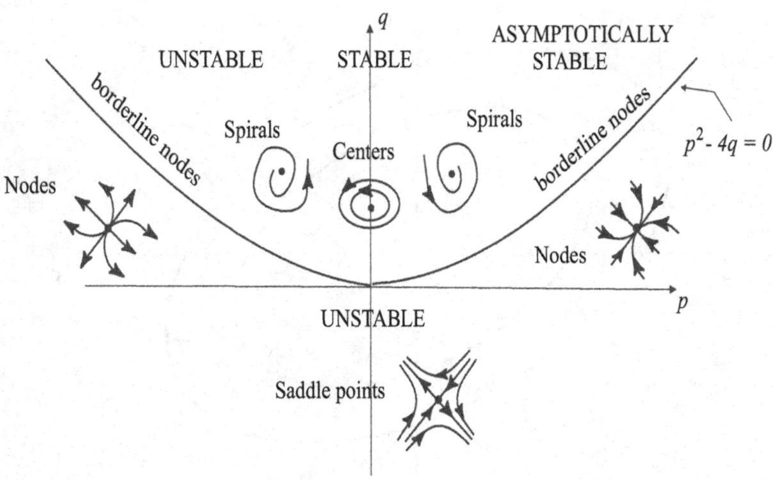

FIGURE 12.16
Critical point behavior in Case E.

so that $p = -(m_1 + m_2)$ and $q = m_1 m_2$, then our five cases can be described just as readily in terms of the coefficients p and q. In fact, if we interpret these cases in the p-q plane, then we arrive at a striking diagram (Figure 12.16) that displays at a glance the nature and stability properties of the critical point $(0, 0)$.

The first thing to notice is that the p-axis ($q = 0$) is excluded, since by condition (24) we know that $m_1 \cdot m_2 \neq 0$. In light of what we have learned about the five cases, all of the information contained in the diagram follows directly from the fact that

$$m_1 \, , \, m_2 = \frac{-p \pm \sqrt{p^2 - 4q}}{2} .$$

Thus, above the parabola $p^2 - 4q = 0$, we have $p^2 - 4q < 0$ so that m_1 and m_2 are conjugate complex numbers; these are pure imaginary if and only if $p = 0$. We have just described **Cases C** and **E** for the spirals and the centers.

Below the p-axis we have $q < 0$, which means that m_1 and m_2 are are real, distinct, and have opposite signs. This yields the saddle points of **Case B**. And, finally, the zone between these two regions (including the parabola but excluding the p-axis) is characterized by the relations $p^2 - 4q \geq 0$ and $q > 0$. Thus m_1 and m_2 are real and of the same sign. Here we have the nodes coming from **Cases C** and **D**.

Furthermore, it is clear that there is precisely one region of asymptotic stability: the first quadrant. We state the result formally as follows:

Theorem 12.4.2 *The critical point* $(0, 0)$ *of the linear system (23) is asymptotically stable if and only if the coefficients* $p = -(a_1 + b_2)$ *and* $q = a_1 b_2 - a_2 b_1$ *of the auxiliary equation (25) are both positive.*

Finally, it should be stressed that we have studied the paths of our linear system near a critical point by analyzing explicit solutions of the system. In the next two sections we enter more fully into the spirit of the subject and its natural technique by investigating similar problems for nonlinear systems which in general cannot be solved explicitly.

Exercises

1. Determine the nature and stability properties of the critical point $(0,0)$ for each of the following linear autonomous systems.

(a) $\begin{cases} \dfrac{dx}{dt} = 2x \\ \dfrac{dy}{dt} = 3y \end{cases}$

(b) $\begin{cases} \dfrac{dx}{dt} = -x - 2y \\ \dfrac{dy}{dt} = 4x - 5y \end{cases}$

(c) $\begin{cases} \dfrac{dx}{dt} = -3x + 4y \\ \dfrac{dy}{dt} = -2x + 3y \end{cases}$

(d) $\begin{cases} \dfrac{dx}{dt} = 5x + 2y \\ \dfrac{dy}{dt} = -17x - 5y \end{cases}$

(e) $\begin{cases} \dfrac{dx}{dt} = -4x - y \\ \dfrac{dy}{dt} = x - 2y \end{cases}$

(f) $\begin{cases} \dfrac{dx}{dt} = 4x - 3y \\ \dfrac{dy}{dt} = 8x - 6y \end{cases}$

(g) $\begin{cases} \dfrac{dx}{dt} = 4x - 2y \\ \dfrac{dy}{dt} = 5x + 2y \end{cases}$

2. If $a_1b_2 - a_2b_1 = 0$, then show that the system (23) has infinitely many critical points, none of which is isolated.

3. **(a)** If $a_1b_2 - a_2b_1 \neq 0$, then show that the system

$$\begin{cases} \dfrac{dx}{dt} = a_1x + b_1y + c_1 \\ \dfrac{dy}{dt} = a_2x + b_2y + c_2 \end{cases}$$

has a single isolated critical point (x_0, y_0).

(b) Show that the system in **(a)** can be written in the form of (23) by means of the change of variables $\overline{x} = x - x_0$, $\overline{y} = y - y_0$.

(c) Find the critical point of the system

$$\begin{cases} \dfrac{dx}{dt} = 2x - 2y + 10 \\ \dfrac{dy}{dt} = 11x - 8y + 49\,. \end{cases}$$

Write the system in the form (23) by changing variables, and determine the nature and stability properties of the critical point.

4. In Section 2.5, we studied the free vibrations of a mass attached to a spring by solving the equation

$$\frac{d^2x}{dt^2} + 2b\frac{dx}{dt} + a^2x = 0\,.$$

Here $b \geq 0$ and $a > 0$ are constants representing the viscosity of the medium and the stiffness of the spring, respectively. Consider the equivalent autonomous system

$$\left\{\begin{aligned} \frac{dx}{dt} &= y \\ \frac{dy}{dt} &= -a^2x - 2by\,. \end{aligned}\right. \qquad (*)$$

Observe that $(0,0)$ is the only critical point of this system.

(a) Find the auxiliary equation of $(*)$. What are p and q?

(b) In each of the following four cases, describe the nature and stability properties of the critical point, and give a brief physical interpretation of the corresponding motion of the mass.

(i) $b = 0$ **(iii)** $b = a$

(ii) $0 < b < a$ **(iv)** $b > a$

5. Solve equation (36) under the hypotheses of **Case E**, and show that the result is a one-parameter family of ellipses surrounding the origin. *Hint:* Recall that if $Ax^2 + Bxy + Cy^2 = D$ is the equation of a real curve, then the curve is an ellipse if and only if the discriminant $B^2 - 4AC$ is negative (see [THO, p. 546]).

12.5 Stability by Lyapunov's Direct Method

It is intuitively clear that, if the total energy in a physical system has a local minimum at a certain equilibrium point, then that point is stable. This idea was generalized by Alexander Mikhailovich Lyapunov (1857–1918) into a simple but powerful method for studying stability problems in a broader context. The current section is devoted to Lyapunov's method and its applications.

Consider an autonomous system

$$\left\{\begin{aligned} \frac{dx}{dt} &= F(x,y) \\ \frac{dy}{dt} &= G(x,y)\,. \end{aligned}\right. \qquad (12.37)$$

Assume that the system has an isolated critical point, which we take to be the origin $(0,0)$. Let $C = [x(t), y(t)]$ be a path of (37), and consider a function $E(x,y)$ that is continuous and has continuous first partial derivatives in a region containing this path. If a point (x,y) moves along the path in accordance with the equations $x = x(t)$, $y = y(t)$, then $E(x,y)$ can be regarded as a function of t along C. We denote this function by $E(t)$. The rate of change is

$$\begin{aligned}\frac{dE}{dt} &= \frac{\partial E}{\partial x}\frac{dx}{dt} + \frac{\partial E}{\partial y}\frac{dy}{dt} \\ &= \frac{\partial E}{\partial x}F + \frac{\partial E}{\partial y}G.\end{aligned} \tag{12.38}$$

This formula is the key to Lyapunov's technique. In order to exploit it, we need several definitions that specify the kinds of functions that we are interested in. We turn now to that preparatory work.

Suppose that $E(x,y)$ is a continuously differentiable function on a region containing the origin. If E vanishes at the origin, so that $E(0,0) = 0$, then it is said to be *of positive type* if $E(x,y) > 0$ for $(x,y) \neq 0$ and *of negative type* if $E(x,y) < 0$ for $(x,y) \neq 0$. Similarly, E is said to be *of positive semi-definite type* if $E(0,0) = 0$ and $E(x,y) \geq 0$ for $(x,y) \neq (0,0)$; and E is said to be *of negative semi-definite type* if $E(0,0) = 0$ and $E(x,y) \leq 0$ for $(x,y) \neq (0,0)$. If m, n are positive integers then of course the function $ax^{2m} + by^{2n}$, where a and b are positive constants, is of positive type. Since E is of negative type precisely when $-E$ is of positive type, we conclude that $ax^{2m} + by^{2n}$ is of negative type when $a, b < 0$. The functions $x^{2m}, y^{2n}, (x-y)^{2m}$ are not of positive type but are instead of positive semi-definite type.

If E is of positive type, then $z = E(x,y)$ is the equation of a surface (Figure 12.17) that resembles a paraboloid opening upward and that is tangent to the x-y plane at the origin.

A function $E(x,y)$ of positive type with the additional property that

$$\frac{\partial E}{\partial x}F + \frac{\partial E}{\partial y}G \tag{12.39}$$

is of negative semi-definite type is called a *Lyapunov function* for the system (37). By formula (38), the requirement that (39) be of negative semi-definite type means that $dE/dt \leq 0$—and therefore that E is nonincreasing—along paths of (37) near the origin. These functions generalize the concept of the total energy of a physical system. Their relevance for stability problems is made clear in the next theorem, which is Lyapunov's basic discovery.

Theorem 12.5.1 *If there exists a Lyapunov function $E(x,y)$ for the system (37), then the critical point $(0,0)$ is stable. Furthermore, if this function has the additional property that the function (39) is of negative type, then the critical point $(0,0)$ is asymptotically stable.*

This theorem is of such fundamental importance that we indicate its proof.

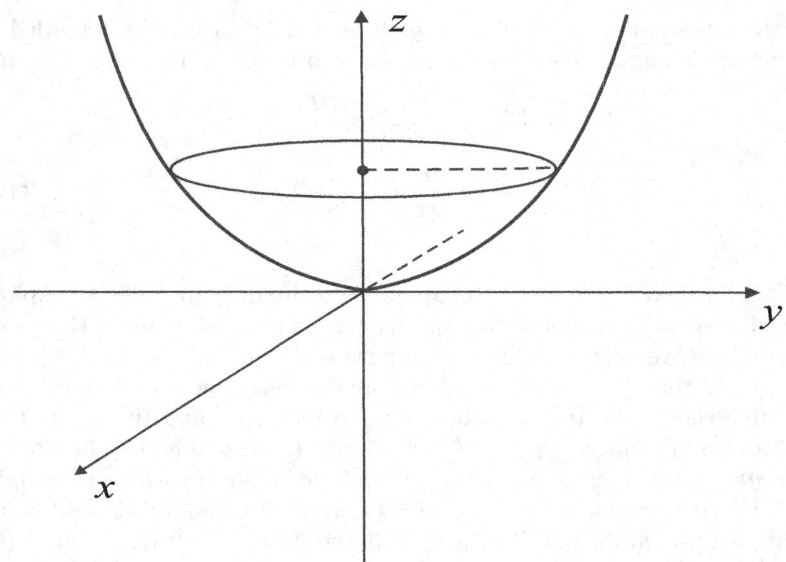

FIGURE 12.17
The function E of positive type.

Sketch of the Proof of Theorem 12.5.1: Let C_1 be a circle of radius $R > 0$ centered at the origin (Figure 12.18). Assume that C_1 is small enough that it lies entirely in the domain of definition of E. Since E is continuous and of positive type, it has a positive minimum m on C_1. Next, E is continuous at the origin and vanishes there, so we can find a positive number $0 < r < R$ such that $E(x, y) < m$ whenever (x, y) lies inside the circle C_2 of radius r and center the origin. Now let $C = [x(t), y(t)]$ be any path which is inside C_2 for $t = t_0$. Then $E(t_0) < m$, and since (39) is of negative semi-definite type, we may conclude that $dE/dt \leq 0$. This implies that $E(t) \leq E(t_0) < m$ for all $t > t_0$. We may conclude that the path C can never reach the circle C_1 for any $t > t_0$. This implies stability.

We omit the proof of the second part of the theorem. □

EXAMPLE 12.5.2 Consider the equation of motion of a mass m attached to a spring (and therefore obeying Hooke's Law):

$$m\frac{d^2x}{dt^2} + c\frac{dx}{dt} + kx = 0. \tag{12.40}$$

Here $c \geq 0$ is a constant representing the viscosity of the medium through which the mass moves (and thus giving rise to resistance), and $k > 0$ is the

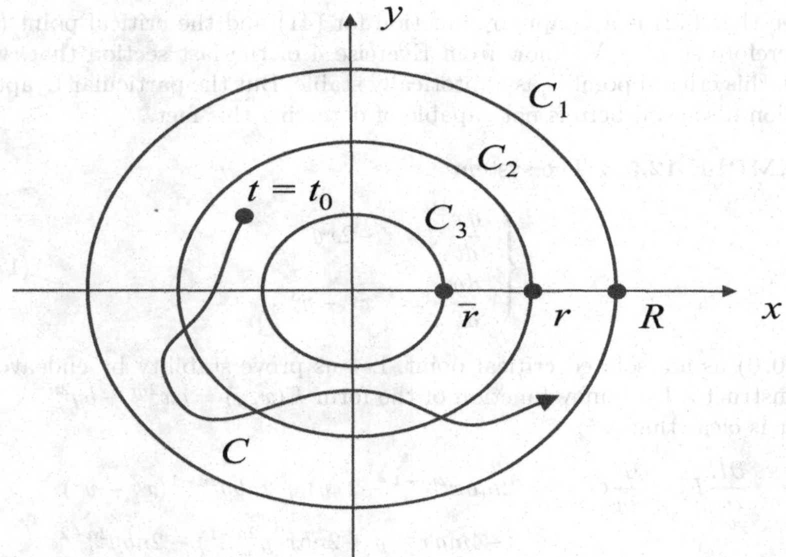

FIGURE 12.18
Proof of Lyapunov's theorem.

spring (Hooke's) constant. The autonomous system equivalent to (40) is

$$\left\{\begin{aligned} \frac{dx}{dt} &= y \\ \frac{dy}{dt} &= -\frac{k}{m}x - \frac{x}{m}y \,. \end{aligned}\right. \tag{12.41}$$

The only critical point of this system is $(0,0)$. The kinetic energy of the mass is $my^2/2$, and the potential energy (or energy stored in the spring) is

$$\int_0^x kx\,dx = \frac{1}{2}kx^2 \,.$$

Thus the total energy of the system is

$$E(x,y) = \frac{1}{2}my^2 + \frac{1}{2}kx^2 \,. \tag{12.42}$$

It is easy to see that (42) is of positive type. Since

$$\begin{aligned} \frac{\partial E}{\partial x}F + \frac{\partial E}{\partial y}G &= kxy + my\left(-\frac{k}{m}x - \frac{c}{m}y\right) \\ &= -cy^2 \leq 0 \,, \end{aligned}$$

we see that (42) is a Lyapunov function for (41) and the critical point $(0,0)$ is therefore stable. We know from Exercise 4 of the last section that, when $c > 0$, this critical point is asymptotically stable. But the particular Lyapunov function discussed here is not capable of detecting this fact. ▮

EXAMPLE 12.5.3 The system

$$\left\{ \begin{array}{rcl} \dfrac{dx}{dt} & = & -2xy \\ \dfrac{dy}{dt} & = & x^2 - y^3 \end{array} \right. \tag{12.43}$$

has $(0,0)$ as an isolated critical point. Let us prove stability by endeavoring to construct a Lyapunov function of the form $E(x,y) = ax^{2m} + by^{2n}$.

It is clear that

$$\begin{aligned} \frac{\partial E}{\partial x}F + \frac{\partial E}{\partial y}G &= 2max^{2m-1}(-2xy) + 2nby^{2n-1}(x^2 - y^3) \\ &= (-4max^{2m}y + 2nbx^2y^{2n-1}) - 2nby^{2n+2}\,. \end{aligned}$$

Our goal is to make the expression in parentheses vanish, and inspection shows that this can be accomplished by choosing $m = 1$, $n = 1$, $a = 1$, $b = 2$. With these choices we have $E(x,y) = x^2 + 2y^2$ (which is of positive type) and $(\partial E/\partial x)F + (\partial E/\partial y)G = -4y^4$ (which is of negative semi-definite type). By Lyapunov's theorem, the critical point $(0,0)$ of the system (43) is stable. ▮

As this example suggests, it may often be difficult to construct a Lyapunov function. The following result sometimes aids in the process.

Theorem 12.5.4 *The function $E(x,y) = ax^2 + bxy + cy^2$ is of positive type if and only if $a > 0$ and $b^2 - 4ac < 0$; it is of negative type if and only if $a < 0$ and $b^2 - 4ac < 0$.*

Proof If $y = 0$ then $E(x,0) = ax^2$. So $E(x,0) > 0$ for $x \neq 0$ if and only if $a > 0$. If $y \neq 0$ then

$$E(x,y) = y^2 \left[a\left(\frac{x}{y}\right)^2 + b\left(\frac{x}{y}\right) + c \right] .$$

When $a > 0$, then the bracketed expression (which is positive for large x/y) is positive for *all* x/y if and only if $b^2 - 4ac < 0$. This proves the first part of the theorem. The second part is analogous. □

Math Nugget

Alexander Mikhailovich Lyapunov (1857–1918) was a Russian mathematician and mechanical engineer. He performed the somewhat remarkable feat of producing a doctoral dissertation of lasting value. This classic work, still important today, was originally published in 1892 in Russian, but is now available in an English translation (*Stability of Motion*, Academic Press, New York, 1966). Lyapunov died tragically, by violence in Odessa; like many a middle-class intellectual of his time, he was a victim of the chaotic aftermath of the Russian revolution.

Exercises

1. Determine whether each of the following functions is of positive type, of negative type, or neither.

(a) $x^2 - xy - y^2$ **(c)** $-2x^2 + 3xy - y^2$
(b) $2x^2 - 3xy + 3y^2$ **(d)** $-x^2 - 4xy - 5y^2$

2. Show that a function of the form

$$ax^3 + bx^2y + cxy^2 + dy^3$$

cannot be either of positive type nor of negative type.

3. Show that $(0, 0)$ is an asymptotically stable critical point for each of the following systems.

$$\text{(a)} \begin{cases} \dfrac{dx}{dt} = -3x^3 - y \\ \dfrac{dy}{dt} = x^5 - 2y^3 \end{cases} \qquad \text{(b)} \begin{cases} \dfrac{dx}{dt} = -2x + xy^3 \\ \dfrac{dy}{dt} = x^2y^2 - y^3 \end{cases}$$

4. Show that the critical point $(0, 0)$ of the system (23) is unstable if there exists a function $E(x, y)$ with the following properties:

(a) $E(x, y)$ is continuously differentiable in some region containing the origin;
(b) $E(0, 0) = 0$;
(c) Every circle centered at $(0, 0)$ contains at least one point where $E(x, y)$ is positive;

(d) $(\partial E/\partial x)F + (\partial E/\partial y)G$ is of positive type.

5. Show that $(0,0)$ is an unstable critical point for the system

$$\left\{\begin{aligned} \frac{dx}{dt} &= 2xy + x^3 \\ \frac{dy}{dt} &= -x^2 + y^5 \end{aligned}\right.$$

6. Assume that $f(x)$ satisfies $f(0) = 0$ and $xf(x) > 0$ for $x \neq 0$.

(a) Show that

$$E(x,y) \equiv \frac{1}{2}y^2 + \int_0^x f(t)\,dt$$

is of positive type.

(b) Show that the equation

$$\frac{d^2x}{dt^2} + f(x) = 0$$

has $x = 0$, $y = dx/dt = 0$ as a stable critical point.

(c) Show that if $g(x) \geq 0$ in some neighborhood of the origin, then the equation

$$\frac{d^2x}{dt^2} + g(x)\frac{dx}{dt} + f(x) = 0$$

has $x = 0$, $y = dx/dt = 0$ as a stable critical point.

7. Use your symbol manipulation software, such as `Maple` or `Mathematica`, to write a routine that will seek a Lyapunov function for a given differential equation. Test it on the examples provided in the text.

12.6 Simple Critical Points of Nonlinear Systems

Consider an autonomous system

$$\left\{\begin{aligned} \frac{dx}{dt} &= F(x,y) \\ \frac{dy}{dt} &= G(x,y)\,. \end{aligned}\right. \tag{12.44}$$

Assume that the system has an isolated critical point at $(0,0)$. If $F(x,y)$, $G(x,y)$ can be expanded in power series in x and y, then (44) takes the form

$$\left\{\begin{aligned} \frac{dx}{dt} &= a_1x + b_1y + c_1x^2 + d_1xy + e_1y^2 + \cdots \\ \frac{dy}{dt} &= a_2x + b_2y + c_2x^2 + d_2xy + e_2y^2 + \cdots\,. \end{aligned}\right. \tag{12.45}$$

When $|x|$ and $|y|$ are small—that is, when the point (x, y) is close to the origin—the terms of second degree and higher are very small (as compared to the linear lead terms). It is therefore natural to discard these nonlinear terms and conjecture that the qualitative behavior of the paths of (45) near the critical point $(0, 0)$ is similar to that of the paths of the related linear system

$$\begin{cases} \dfrac{dx}{dt} &= a_1x + b_1y \\ \dfrac{dy}{dt} &= a_2x + b_2y\,. \end{cases} \tag{12.46}$$

We shall see that, quite frequently, this supposition is actually correct. The process of replacing the system (45) by the simpler linear system (46) is usually called *linearization.*

More generally, we shall consider systems of the form

$$\begin{cases} \dfrac{dx}{dt} &= a_1x + b_1y + f(x, y) \\ \dfrac{dy}{dt} &= a_2x + b_2y + g(x, y)\,. \end{cases} \tag{12.47}$$

We shall assume that

$$\det \begin{pmatrix} a_1 & b_1 \\ a_2 & b_2 \end{pmatrix} \neq 0\,; \tag{12.48}$$

thus the related linear system (46) has $(0, 0)$ as an isolated critical point. We shall also assume that f, g are continuously differentiable, and that

$$\lim_{(x,y)\to(0,0)} \frac{f(x, y)}{\sqrt{x^2 + y^2}} = 0 \quad \text{and} \quad \lim_{(x,y)\to(0,0)} \frac{g(x, y)}{\sqrt{x^2 + y^2}} = 0\,. \tag{12.49}$$

This last, rather technical, condition bears some discussion. First, it obviously implies that $f(0, 0) = 0$ and $g(0, 0) = 0$ so that $(0, 0)$ is a critical point of (47). In fact the critical point will be isolated. Further, condition (49) turns out to be sufficient to guarantee that the linearization (46) is a good approximation to the original system (45). We shall learn more about this last crucial point as the section develops. Under the conditions just described, we say that $(0, 0)$ is a *simple critical point* of the system (47).

EXAMPLE 12.6.1 Consider the system

$$\begin{cases} \dfrac{dx}{dt} &= -2x + 3y + xy \\ \dfrac{dy}{dt} &= -x + y - 2xy^2\,. \end{cases} \tag{12.50}$$

We first note that

$$\det \begin{pmatrix} a_1 & b_1 \\ a_2 & b_2 \end{pmatrix} = 1 \neq 0\,,$$

so condition (48) is satisfied. Furthermore, we may use polar coordinates to check easily that

$$\frac{|f(x,y)|}{\sqrt{x^2+y^2}} = \frac{|r^2 \sin\theta \cos\theta|}{r} \le r$$

and

$$\frac{|g(x,y)|}{\sqrt{x^2+y^2}} = \frac{|2r^3 \sin^2\theta \cos\theta|}{r} \le 2r^2 .$$

Thus $f(x,y)/r$ and $g(x,y)/r$ tend to 0 as $(x,y) \to (0,0)$ (or as $r \to 0$). We see that condition (49) is satisfied so that $(0,0)$ is a simple critical point of the system (50). ∎

The following theorem of Poincaré presents the main facts about the nature of simple critical points.

Theorem 12.6.2 *Let* $(0,0)$ *be a simple critical point of the nonlinear system (47). Consider the related linear system (46). If the critical point* $(0,0)$ *of (46) falls under any one of the three major cases described in Section 12.4, then the critical point* $(0,0)$ *of (47) is of the same type.*

Poincaré's theorem says that, under the standing hypotheses, the linearization method is successful. We can study the nonlinear system by instead studying the associated linearization, and obtain useful information about the original system.

To illustrate Poincaré's theorem, we re-examine the nonlinear system (50) of Example 12.6.1. The related linear system is

$$\left\{ \begin{array}{rcl} \dfrac{dx}{dt} & = & -2x + 3y \\ \dfrac{dy}{dt} & = & -x + y \,. \end{array} \right. \tag{12.51}$$

The auxiliary equation of (51) is $m^2 + m + 1 = 0$; it has roots

$$m_1, m_2 = \frac{-1 \pm \sqrt{3}i}{2} .$$

Since these roots are conjugate complex but *not* pure imaginary, we are in **Case C**. Thus the critical point $(0,0)$ of the linear system (51) is a spiral. By Theorem 12.6.2, the critical point of the nonlinear system (50) is also a spiral.

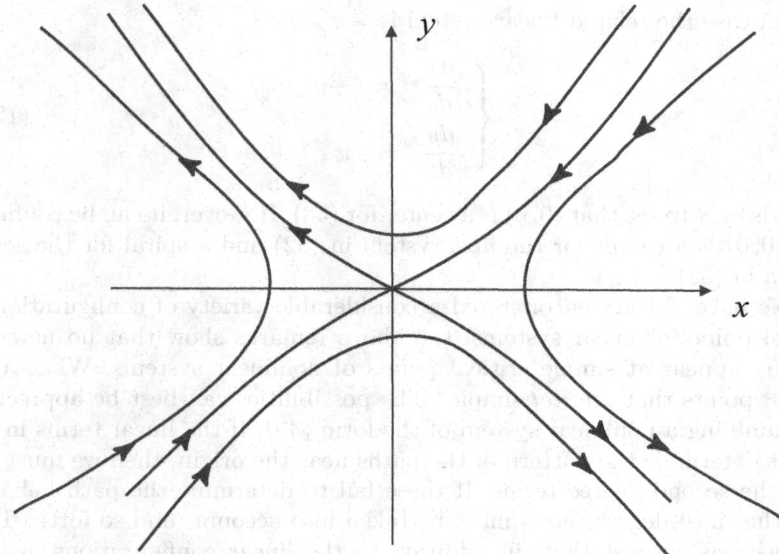

FIGURE 12.19
A nonlinear saddle point.

An important point that must be clearly understood is that, while the theorem guarantees that the *type* of the critical point for the nonlinear system will be the same as the type for the associated linear system, the actual appearance of the paths may be different. As an instance, Figure 12.11 shows a typical saddle point for a linear system; but Figure 12.19 represents how a nonlinear saddle point might appear. The point to notice is that the nonlinearity causes a certain amount of distortion, but the overall qualitative behavior is the same.

Of course we wonder about the two borderline cases, which are not treated in Theorem 12.6.2. What is true is this: If the related linear system (46) has a borderline node at the origin (**Case D**), then the nonlinear system (47) can have either a node or a spiral. And if (46) has a center at the origin (**Case E**), then (47) can have either a center or a spiral.

As an example, $(0, 0)$ is a critical point for each of the nonlinear systems

$$\left\{\begin{array}{lcl} \dfrac{dx}{dt} & = & -y - x^2 \\ \dfrac{dy}{dt} & = & x \end{array}\right. \quad \text{and} \quad \left\{\begin{array}{lcl} \dfrac{dx}{dt} & = & -y - x^3 \\ \dfrac{dy}{dt} & = & x\,. \end{array}\right. \tag{12.52}$$

In each case the related linear system is

$$\begin{cases} \dfrac{dx}{dt} = -y \\ \dfrac{dy}{dt} = x\,. \end{cases} \tag{12.53}$$

But it is easy to see that $(0, 0)$ is a center for (53). However, it can be confirmed that $(0, 0)$ is a center for the first system in (52) and a spiral for the second system in (52).

We have already encountered a considerable variety of configurations at critical points of linear systems; the above remarks show that no new phenomena appear at simple critical points of nonlinear systems. What about critical points that are *not* simple? The possibilities can best be appreciated by examining a nonlinear system of the form (45). If the linear terms in (45) do not determine the pattern of the paths near the origin, then we must consider the second degree terms. If these fail to determine the path behavior, then the third-degree terms must be taken into account, and so forth. These observations suggest that, in addition to the linear configurations, a great many others can arise. These can be of a vast variety and great complexity. Several are exhibited in Figure 12.20. It is perhaps surprising to realize that such involved path patterns as these can occur in connection with systems of rather simple appearance. For example, the three figures in the upper row of Figure 12.20 show the arrangements of the paths of

$$\begin{cases} \dfrac{dx}{dt} = 2xy \\ \dfrac{dy}{dt} = y^2 - x^2 \end{cases} \quad \begin{cases} \dfrac{dx}{dt} = x^3 - 2xy^2 \\ \dfrac{dy}{dt} = 2x^2y - y^3 \end{cases} \quad \begin{cases} \dfrac{dx}{dt} = x - 4y\sqrt{|xy|} \\ \dfrac{dy}{dt} = -y + 4x\sqrt{|xy|}\,. \end{cases}$$

We now discuss the question of stability for a simple critical point. The main result here is again due to Lyapunov: If (46) is asymptotically stable at the origin, then so is (47). We formally state the result:

Theorem 12.6.3 *Let* $(0, 0)$ *be a simple critical point of the nonlinear system (47). Consider the related linear system (46). If the critical point* $(0, 0)$ *of (46) is asymptotically stable, then the critical point* $(0, 0)$ *of (47) is also asymptotically stable.*

As an instance, consider the nonlinear system (50) of Example 12.6.1, with associated linear system (51). For (51) we have $p = 1 > 0$ and $q = 1 > 0$. Thus the critical point $(0, 0)$ is asymptotically stable—both for the linear system (51) and for the nonlinear system (50). We shall not prove this theorem, but instead illustrate it with an additional example.

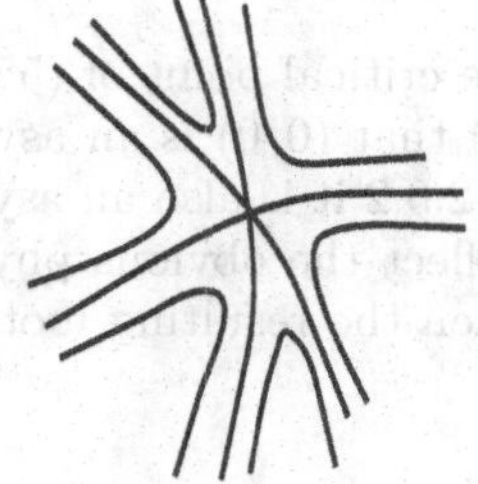

FIGURE 12.20
Different types of critical points.

EXAMPLE 12.6.4 We know from Section 5.1.2 that the equation of motion for the damped oscillations of a pendulum is given by

$$\frac{d^2x}{dt^2} + \frac{c}{m}\frac{dx}{dt} + \frac{g}{a}\sin x = 0 \,,$$

where c is a positive constant. The equivalent nonlinear system is

$$\left\{\begin{array}{lcl} \dfrac{dx}{dt} & = & y \\ \dfrac{dy}{dt} & = & -\dfrac{g}{a}\sin x - \dfrac{c}{m}y \,. \end{array}\right. \tag{12.54}$$

We now write (54) in the form

$$\left\{\begin{array}{lcl} \dfrac{dx}{dt} & = & y \\ \dfrac{dy}{dt} & = & -\dfrac{g}{a}x - \dfrac{c}{m}y + \dfrac{g}{a}(x - \sin x) \,. \end{array}\right. \tag{12.55}$$

It is easy to see that, if $x \neq 0$, then

$$\frac{|x - \sin x|}{\sqrt{x^2 + y^2}} \leq \frac{|x - \sin x|}{|x|} = \left|1 - \frac{\sin x}{x}\right| \to 0 \,.$$

Therefore

$$\frac{x - \sin x}{\sqrt{x^2 + y^2}} \to 0$$

as $(x, y) \to (0, 0)$. Since $(0, 0)$ is evidently an isolated critical point of the related linear system

$$\begin{cases} \dfrac{dx}{dt} = y \\ \dfrac{dy}{dt} = -\dfrac{g}{a}x - \dfrac{c}{m}y, \end{cases} \tag{12.56}$$

it follows that $(0, 0)$ is a simple critical point of (55). Inspection shows (since $p = c/m > 0$ and $q = g/a > 0$) that $(0, 0)$ is an asymptotically stable critical point of (56), so by Theorem 12.6.2 it is also an asymptotically stable critical point of (55). These results reflect the obvious physical fact that if the pendulum is slightly disturbed, then the resulting motion will die out over time. ∎

Exercises

1. Verify that $(0, 0)$ is an asymptotically stable critical point of

$$\begin{cases} \dfrac{dx}{dt} = -y - x^3 \\ \dfrac{dy}{dt} = x - y^3 \end{cases}$$

but is an unstable critical point of

$$\begin{cases} \dfrac{dx}{dt} = -y + x^3 \\ \dfrac{dy}{dt} = x + y^3 . \end{cases}$$

2. Sketch the family of curves whose polar equation is $r = a \sin 2\theta$ (see Figure 12.20). Express the differential equation of this family in the form $dy/dx = G(x, y)/F(x, y)$.
3. Verify that $(0, 0)$ is a simple critical point for each of the following systems, and determine its nature and stability properties.

(a) $$\begin{cases} \dfrac{dx}{dt} = x + y - 2xy \\ \dfrac{dy}{dt} = -2x + y + 3y^2 \end{cases}$$

(b) $$\begin{cases} \dfrac{dx}{dt} = -x - y - 3x^2y \\ \dfrac{dy}{dt} = -2x - 4y + y \sin x \end{cases}$$

4. The van der Pol equation

$$\frac{d^2x}{dt^2} + \mu(x^2 - 1)\frac{dx}{dt} + x = 0$$

is equivalent to the system

$$\begin{cases} \dfrac{dx}{dt} &= y \\ \dfrac{dy}{dt} &= -x - \mu(x^2 - 1)y \,. \end{cases}$$

Investigate the stability properties of the critical point $(0, 0)$ for the cases $\mu > 0$ and $\mu < 0$.

5. Show that if $(0, 0)$ is a simple critical point of (47), then it is necessarily isolated. *Hint:* Write conditions (6) in the form $f(x, y)/r = \epsilon_1 \to 0$ and $g(x, y)/r = \epsilon_2 \to 0$. In light of (48), use polar coordinates to deduce a contradiction from the assumption that the right sides of (47) both vanish at points arbitrarily close to the origin but different from it.

6. If $(0, 0)$ is a simple critical point of (47) and if $q = a_1b_2 - a_2b_1 < 0$, then Theorem 12.6.2 implies that $(0, 0)$ is a saddle point of (47) and is therefore unstable. Prove that if $p = -(a_1 + b_2) < 0$ and $q = a_1b_2 - a_2b_1 > 0$, then $(0, 0)$ is an unstable critical point of (47).

12.7 Nonlinear Mechanics: Conservative Systems

It is well known that energy is dissipated in the action of any real dynamical system, usually through some form of friction. In certain situations, however, this dissipation is so slow that it can be neglected over relatively short periods of time. In such cases we assume the law of conservation of energy, namely, that the sum of the kinetic energy and the potential energy is constant. A system of this sort is said to be *conservative.* Thus the rotating earth can be considered a conservative system over short intervals of time involving only a few centuries; but if we want to study its behavior through millions of years, then we must take into account the dissipation of energy by tidal friction.

The simplest conservative system consists of a mass m attached to a spring and moving in a straight line through a vacuum. If x denotes the displacement of m from its equilibrium position, and the restoring force exerted on m by the spring is $-kx$ for some $k > 0$, then we know that the equation of motion is

$$m\frac{d^2x}{dt^2} + kx = 0 \,.$$

A spring of this kind is called a *linear spring* because the restoring force is a linear function of x. If m moves through a resisting medium, and the resistance (or damping force) exerted on m is $-c(dx/dt)$ for some $c > 0$, then the equation of motion of this nonconservative system is

$$m\frac{d^2x}{dt^2} + c\frac{dx}{dt} + kx = 0\,.$$

Here we have *linear damping* because the damping force is a linear function of dx/dt. By analogy, if f and g are arbitrary functions with the property that $f(0) = 0$ and $g(0) = 0$, then the more general equation

$$m\frac{d^2x}{dt^2} + g\left(\frac{dx}{dt}\right) + f(x) = 0 \qquad (12.57)$$

can be interpreted as the equation of motion by a mass m under the action of a *restoring force* $-f(x)$ and a *damping force* $-g(dx/dt)$. In general these forces will be nonlinear, and equation (57) can be regarded as the basic equation of nonlinear mechanics. In the present section we shall briefly consider the special case of a nonlinear conservative system described by the equation

$$m\frac{d^2x}{dt^2} + f(x) = 0\,; \qquad (12.58)$$

here we are assuming that the damping force is 0 and there is consequently no dissipation of energy.

Equation (58) is equivalent to the autonomous system

$$\begin{cases} \dfrac{dx}{dt} & = & y \\[2ex] \dfrac{dy}{dt} & = & -\dfrac{f(x)}{m}\,. \end{cases} \qquad (12.59)$$

If we eliminate dt, then we obtain the differential equation of the paths of (59) in the phase plane:

$$\frac{dy}{dx} = -\frac{f(x)}{my}\,, \qquad (12.60)$$

and this can in turn be written in the form

$$my\,dy = -f(x)\,dx\,. \qquad (12.61)$$

If $x = x_0$ and $y = y_0$ when $t = t_0$ then, integrating (61) from t_0 to t yields

$$\frac{1}{2}my^2 - \frac{1}{2}my_0^2 = -\int_{x_0}^{x} f(t)\,dt$$

or

$$\frac{1}{2}my^2 + \int_0^x f(t)\,dt = \frac{1}{2}my_0^2 + \int_0^{x_0} f(t)\,dt\,. \qquad (12.62)$$

To interpret this last equation, we observe that $my^2/2 = [m/2](dx/dt)^2$ is the kinetic energy of the dynamical system and

$$V(x) = \int_0^x f(t)\,dt \tag{12.63}$$

is its potential energy. Equation (62) therefore expresses the law of conservation of energy,

$$\frac{1}{2}my^2 + V(x) = E\,, \tag{12.64}$$

where $E = [m/2]y_0^2 + V(x_0)$ is the constant total energy of the system. It is clear that (64) is the equation of the paths of (59), since we obtained it by solving (60). The particular path determined by a specifying a value for E is a curve of constant energy in the phase plane. The critical points of the system are the points $(x_c, 0)$, where the x_c are the roots of the equation $f(x) = 0$. As we observed in Section 12.2, these are the equilibrium points of the dynamical system described by (58). It is evident from (60) that the paths cross the x-axis at right angles and are horizontal when they cross the lines $x = x_c$. Equation (64) also shows that the paths are symmetric with respect to the x-axis.

If we write (64) in the form

$$y = \pm\sqrt{\frac{2}{m}[E - V(x)]}\,, \tag{12.65}$$

then the paths can be constructed by the following simple program.

(1) Establish an x-z plane with the z-axis on the same vertical line as the y-axis of the phase plane (Figure 12.21).

(2) Draw the graph of $z = V(x)$ and several horizontal lines $z = E$ in the x-z plane (one such line is shown in the figure), and observe the geometric meaning of the difference $E - V(x)$.

(3) For each x, multiply $E - V(x)$ as obtained in Step **(2)** by $2/m$ and use formula (65) to plot the corresponding values of y in the phase plane directly below. Note that, since $dx/dt = y$, it follows that the positive direction along any path is to the right above the x-axis and to the left below the x-axis.

EXAMPLE 12.7.1 We saw in Section 5.1.1 and Section 12.2 that the equation of motion of an undamped pendulum is

$$\frac{d^2x}{dt^2} + k\sin x = 0\,, \tag{12.66}$$

where $k > 0$ is a constant. Since this equation is of the form (58), it can be interpreted as describing the undamped rectilinear motion of a unit mass

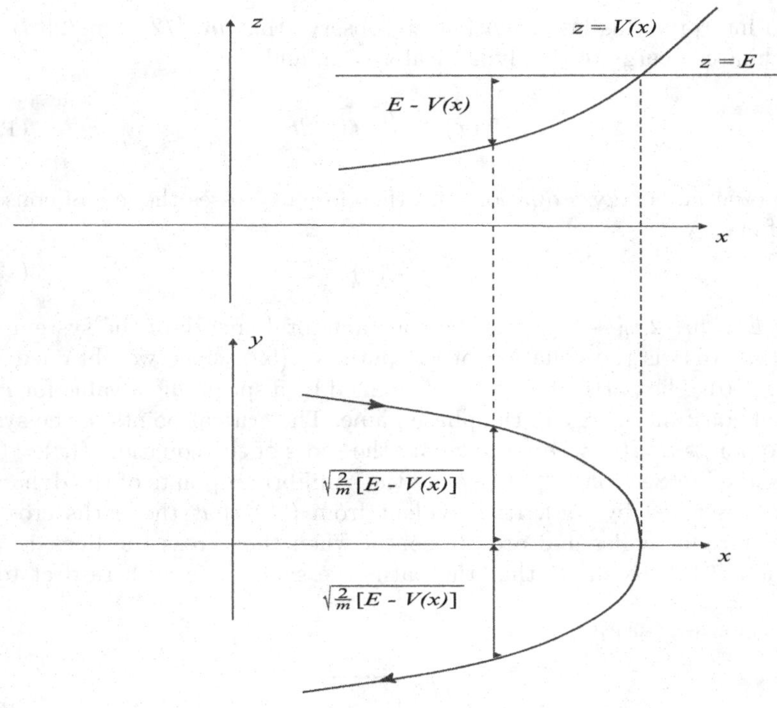

FIGURE 12.21
Construction of the paths.

under the influence of a nonlinear spring whose restoring force is $-k \sin x$. The autonomous system equivalent to (66) is

$$\begin{cases} \dfrac{dx}{dt} &= y \\ \dfrac{dy}{dt} &= -k \sin x \,. \end{cases}$$

Its critical points are $(0, 0)$, $(\pm\pi, 0)$, $(\pm 2\pi, 0)$, ... The differential equation of the paths is

$$\frac{dy}{dx} = -\frac{k \sin x}{y} \,.$$

By separating the variables and integrating, we see that the equation of the family of paths is

$$\frac{1}{2} y^2 + (k - k \cos x) = E \,.$$

This equation is evidently of the form (64), where $m = 1$ and

$$V(x) = \int_0^x f(t)\, dt = k - k \cos x$$

is the potential energy.

We now construct the paths by first drawing the graph of $z = V(x)$ and also several lines $z = E$ in the x-z plane (Figure 12.22, where $z = E = 2k$ is the only line shown). From this we read off the values $E - V(x)$ and sketch the paths in the phase plane directly below by using $y = \pm\sqrt{2[E - V(x)]}$. It is clear from this phase portrait that if the total energy E is between 0 and $2k$, then the corresponding paths are closed and equation (66) has periodic solutions. On the other hand, if $E > 2k$, then the path is not closed and the corresponding solution of (66) is not periodic. The value $E = 2k$ separates the two types of motion, and for this reason a path corresponding to $E = 2k$ is called a *separatrix*. The wavy paths outside the separatrices correspond to whirling motions of the pendulum, and the closed paths inside to oscillatory motions. It is evident that the critical points are alternately unstable saddle points and stable but not asymptotically stable centers. For the sake of contrast, it is interesting to consider the effect of transforming this conservative dynamical system into a nonconservative system by introducing a linear damping force. The equation of motion then takes the form

$$\frac{d^2x}{dt^2} + c\frac{dx}{dt} + k \sin x = 0\,, \quad c > 0\,,$$

and the configuration of the paths is suggested in Figure 12.23. We find that the centers in Figure 12.22 become asymptotically stable spirals, and also that every path—except the separatrices entering the saddle points as $t \to \infty$—ultimately winds into one of these spirals. ∎

Exercises

1. Most actual springs, coming from real life, are not linear. A nonlinear spring is called *hard* or *soft* according to whether the magnitude of the restoring force increases more rapidly or less rapidly than a linear function of the displacement. The equation

$$\frac{d^2x}{dt^2} + kx + \alpha x^3 = 0, \quad k > 0\,,$$

describes the motion of a hard spring if $\alpha > 0$ and a soft spring if $\alpha < 0$. Sketch the paths in each case.

2. Find the equation of the paths of

$$\frac{d^2x}{dt^2} - x + 2x^3 = 0\,.$$

Sketch these paths in the phase plane. Locate the critical points and determine the nature of each of these.

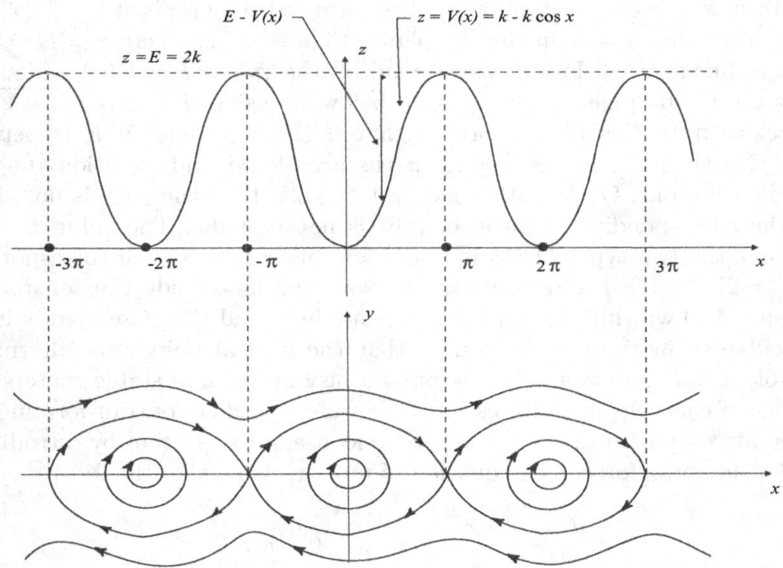

FIGURE 12.22
Construction of the paths.

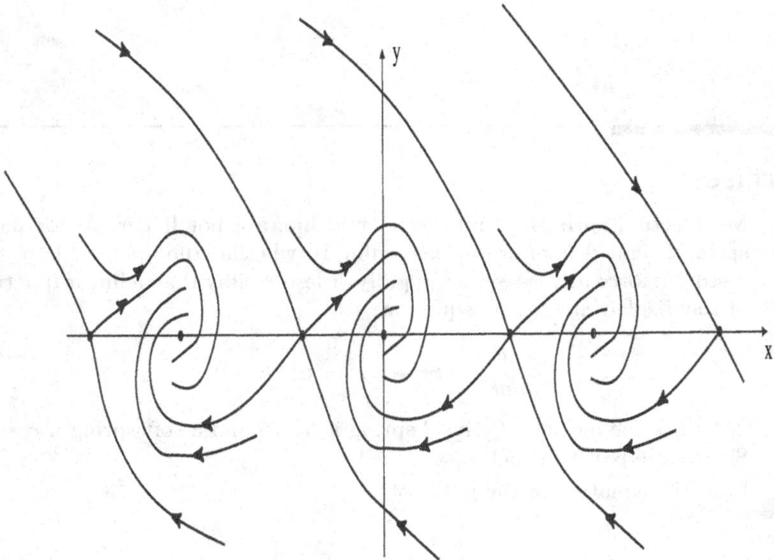

FIGURE 12.23
Path configuration.

3. Since, by equation (63), we have $dV/dx = f(x)$, the critical points of (59) are the points on the x-axis in the phase plane at which $V'(x) = 0$. In terms of the curve $z = V(x)$—as long as this curve is smooth and well-behaved—there are three possibilities: maxima, minima, and points of inflection. Sketch all three possibilities and determine the type of critical point associated with each. (A critical point of the third type is called a *cusp*.)

4. If $f(0) = 0$ and $xf(x) > 0$ for $x \neq 0$ then show that the paths of

$$\frac{d^2x}{dt^2} + f(x) = 0$$

are closed curves surrounding the origin in the phase plane. That is, show that the critical point $x = 0$, $y = dx/dt = 0$ is a stable but not asymptotically stable center. Describe this critical point with regard to its nature and stability if instead $f(0) = 0$ and $xf(x) < 0$ for $x \neq 0$.

12.8 Periodic Solutions: The Poincaré–Bendixson Theorem

Consider a nonlinear autonomous system

$$\left\{ \begin{array}{lcl} \dfrac{dx}{dt} & = & F(x,y) \\ \dfrac{dy}{dt} & = & G(x,y) \end{array} \right. \tag{12.67}$$

in which the functions F and G are continuously differentiable in the entire plane. We have thus far made a fairly detailed study of the paths of such a system in a neighborhood of certain types of critical points. But we really have no idea of the global behavior of the paths—that is, the behavior in the entire phase plane. In this section we apply some ideas of Poincaré and Bendixson to learn something of the global properties of paths of a system like (67).

The central problem of the global theory—at least in practice—is to determine when the system (67) has *closed* paths. As noted in Section 12.2, this question is important because of its connection with the issue of whether (67) has periodic solutions. Here a solution $x(t)$, $y(t)$ of (67) is said to be *periodic* if neither function is constant, if both are defined for all t, and if there exists a number $T > 0$ such that

$$x(t+T) = x(t) \quad \text{and} \quad y(t+T) = y(t) \text{ for all } t\,. \tag{12.68}$$

The least T with this property is called the *period* of the solution.

It is evident that any periodic solution of (67) defines a closed path that is traversed precisely once as t increases from t_0 to $t_0 + T$ for any given t_0.

Conversely, it is easy to see that if $C = [x(t), y(t)]$ is a closed path of (67), then $x(t)$, $y(t)$ is a periodic solution. Thus a search for periodic solutions of (67) reduces to a search for closed paths.

We know from Section 12.4 that a linear system has closed paths if and only if the roots of the auxiliary equation are pure imaginary, and in this case *every path is closed.* Thus, for a linear system, either every path is closed or else no path is closed. On the other hand, a nonlinear system may have a closed path that is isolated, in the sense that no other nearby path is closed. The following is a standard example of such a system:

$$\left\{\begin{aligned} \frac{dx}{dt} &= -y + x(1 - x^2 - y^2) \\ \frac{dy}{dt} &= x + y(1 - x^2 - y^2)\,. \end{aligned}\right. \tag{12.69}$$

To solve this system, we introduce polar coordinates (r, θ), where $x = r\cos\theta$ and $y = r\sin\theta$. If we differentiate the relations $r^2 = x^2 + y^2$ and $\theta = \arctan(y/x)$, then we obtain the useful formulas

$$x\frac{dx}{dt} + y\frac{dy}{dt} = r\frac{dr}{dt} \qquad \text{and} \qquad x\frac{dy}{dt} - y\frac{dx}{dt} = r^2\frac{d\theta}{dt}\,. \tag{12.70}$$

On multiplying the first equation of (69) by x and the second by y, and then adding, we find that

$$f\frac{dr}{dt} = r^2(1 - r^2)\,. \tag{12.71}$$

Similarly, if we multiply the second equation of (69) by x and the first by y, and then subtract, we obtain

$$r^2\frac{d\theta}{dt} = r^2\,. \tag{12.72}$$

The system (69) has a single critical point at $r = 0$. Since we are interested only in finding paths, we may assume that $r > 0$. In this case, (71) and (72) show that (69) becomes

$$\left\{\begin{aligned} \frac{dr}{dt} &= r(1 - r^2) \\ \frac{d\theta}{dt} &= 1\,. \end{aligned}\right. \tag{12.73}$$

These equations are decoupled, and we may solve them separately. The general solution of (73) is found to be

$$\left\{\begin{aligned} r &= \frac{1}{\sqrt{1 + ce^{-2t}}} \\ \theta &= t + t_0\,. \end{aligned}\right. \tag{12.74}$$

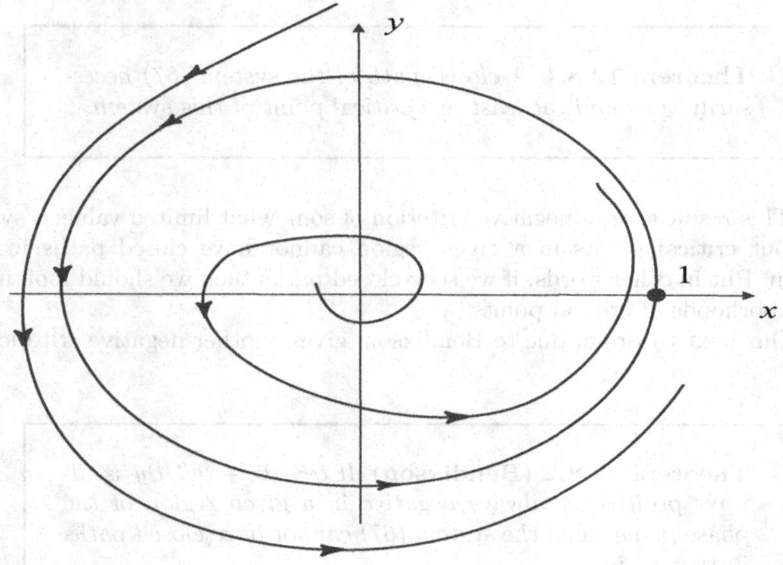

FIGURE 12.24
Geometric analysis of (74).

The corresponding general solution of (69) is

$$\begin{cases} x = \dfrac{\cos(t+t_0)}{\sqrt{1+ce^{-2t}}} \\ y = \dfrac{\sin(t+t_0)}{\sqrt{1+ce^{-2t}}}. \end{cases} \tag{12.75}$$

Let us analyze (74) geometrically (Figure 12.24). If $c = 0$, then we have the solutions $r = 1$, $\theta = t + t_0$, which trace out the closed circular path $x^2 + y^2 = 1$ in the counterclockwise direction. If $c < 0$, then it is clear that $r > 1$ and that $r \to 1$ as $t \to +\infty$. These observations show that there exists a single closed path ($r = 1$) which all other paths approach spirally from the outside or the inside as $t \to +\infty$.

In our discussions up to now, we have shown that the system (69) has a closed path by actually *finding* the path. In general, of course, we cannot hope to be able to do this. What we need are tests that make it possible for us to conclude that certain regions of the phase plane do or do not contain closed paths—without actually *having to find those paths.* Our first test is given in the following theorem of Poincaré.

Theorem 12.8.1 *A closed path of the system (67) necessarily surrounds at least one critical point of this system.*

This result gives a negative criterion of somewhat limited value: a system without critical points in a given region cannot have closed paths in that region. Put in other words, if we seek closed paths then we should look in the neighborhoods of critical points.

Our next theorem, due to Bendixson, gives another negative criterion.

Theorem 12.8.2 (Bendixson) *If $\partial F/\partial x + \partial G/\partial y$ is always positive or always negative in a given region of the phase plane, then the system (67) cannot have closed paths in that region.*

Proof Assume that the region contains a closed path $C = [x(t), y(t)]$ with interior R. Then Green's theorem [STE, p. 945] and our hypothesis yield

$$\int_C (F\,dy - G\,dx) = \iint\limits_R \left(\frac{\partial F}{\partial x} + \frac{\partial G}{\partial y}\right) dxdy \neq 0\,.$$

However, along C we have $dx = F\,dt$ and $dy = G\,dt$. Thus

$$\int_C (F\,dy - G\,dx) = \int_0^T (FG - GF)\,dt = 0\,.$$

This contradiction shows that our initial assumption is false, so the region under consideration cannot contain any closed path. □

Math Nugget

George Green (1793–1841) grew up in Sneinton, England, and was the son of the man who built the tallest, most powerful, and most modern windmill (for grinding grain) in all of Nottinghamshire. As a child, George Green had only 15 months of formal education; he left the Goodacre Academy at the age of 10 to help his father in the adjoining bakery.

After the publication of his first scientific paper in 1828, when George Green was 39 years old, Green became an undergraduate at Gonville and Caius College, Cambridge. After obtaining his degree, Green became a Fellow of the college and continued his scientific research in mathematics and physics. Unfortunately, his health failed and he returned to Sneinton, where he died in 1841. At the time of his death, George Green's contributions to science were little known nor recognized. Today we remember George Green for Green's theorem and the Green's function of partial differential equations.

At this time—after many mishaps, fires, and disasters—Green's Mill has been restored to full working order. One can visit the Mill and the Visitors' Center, learn of the history of the mill and of George Green, and purchase a variety of souvenirs. These include a sample of flour produced at the mill, a pencil emblazoned with Green's theorem, and a key fob.

What we really want are positive criteria for the existence of closed paths of (67). One of the great results is the classical *Poincaré–Bendixson theorem.*

Theorem 12.8.3 *Let R be a bounded region of the phase plane together with its boundary. Assume that R does not contain any critical points of the system (67). If $C = [x(t), y(t)]$ is a path of (67) that lies in R for some t_0 and remains in R for all $t \geq t_0$, then C is either itself a closed path or it spirals toward a closed path as $t \to \infty$. Thus, in either case, the system (67) has a closed path in R.*

To understand the statement of the Poincaré–Bendixson theorem, let us consider the situation in Figure 12.25. Here R consists of the two dotted curves together with the ring-shaped (or annular) region between them. Suppose that the vector field

$$\mathbf{V}(x,y) = F(x,y)\mathbf{i} + G(x,y)\mathbf{j}$$

points *into* R at every boundary point, as indicated in the figure. Then every path C that passes through a boundary point (at some $t = t_0$) must enter R and can never leave it. Under these circumstances, the theorem asserts that C must spiral toward a closed path C_0. Note this important point: we have chosen a ring-shaped (annular) region R to illustrate the theorem because (by Theorem 12.8.1) a closed path must surround a critical point (P in the figure) and R must therefore exclude all critical points.

EXAMPLE 12.8.4 The system (69) provides a simple application of these new ideas. It is clear that (69) has a critical point at $(0,0)$ and also that the region R between the circles $r = 1/2$ and $r = 2$ contains no critical points. In our earlier analysis we found that

$$\frac{dr}{dt} = r(1-r^2) \qquad \text{for } r > 0\,.$$

This shows that $dr/dt > 0$ on the inner circle and $dr/dt < 0$ on the outer circle, so the vector $\mathbf{V}$ points into R at all boundary points. Thus any path passing through a boundary point will enter R and remain in R as $t \to +\infty$. By the Poincaré–Bendixson theorem, we know that R contains a closed path C_0. We have already seen that the circle $r = 1$ is the closed path whose existence is guaranteed in this manner. ∎

The Poincaré–Bendixson theorem is quite compelling—at least from a theoretical point of view. But it is in general rather difficult to apply. A more practical criterion has been developed that assures the existence of closed paths, at least for equations of the form

$$\frac{d^2x}{dt^2} + f(x)\frac{dx}{dt} + g(x) = 0\,. \tag{12.76}$$

This is called *Liénard's equation.* When we speak of a closed path for this equation, we of course mean a closed path of the equivalent system

$$\left\{\begin{array}{rcl} \dfrac{dx}{dt} & = & y \\ \dfrac{dy}{dt} & = & -g(x) - f(x)y\,. \end{array}\right. \tag{12.77}$$

As we know, a closed path of (77) corresponds to a periodic solution of (76). The fundamental statement about the closed paths of (77) is the following theorem.

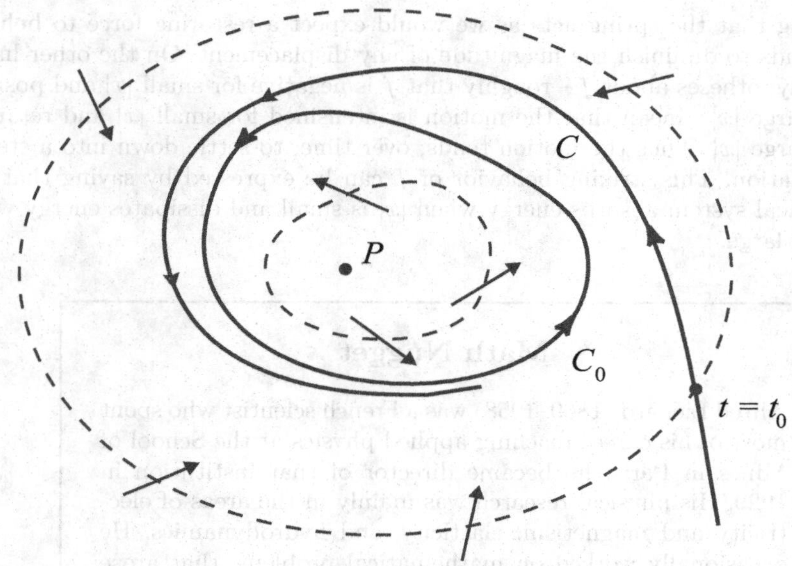

FIGURE 12.25
The Poincaré–Bendixson theorem.

Theorem 12.8.5 *Let the functions $f(x)$ and $g(x)$ satisfy the following conditions:*

(i) *Both are continuously differentiable;*

(ii) *g is an odd function with $g > 0$ for $x > 0$;*

(iii) *f is an even function;*

(iv) *the odd function $F(x) = \int_0^x f(t)\,dt$ has exactly one positive zero at $x = a$, is negative on the interval $(0, a)$, is positive and nondecreasing on (a, ∞), and $F(x) \to +\infty$ as $x \to +\infty$.*

Then equation (76) has a unique closed path surrounding the origin in the phase plane, and this path is approached spirally by every other path as $t \to +\infty$.

Let us explain this theorem, at least heuristically, in terms of the ideas of the previous section. Equation (76) is the equation of motion of a unit mass attached to a spring and subject to the dual influence of a restoring force $-g(x)$ and a damping force $-f(x)dx/dt$. The hypothesis about g amounts to saying that the spring acts as we would expect a restoring force to behave: it tends to diminish the magnitude of any displacement. On the other hand, the hypotheses about f—roughly that f is negative for small $|x|$ and positive for large $|x|$—mean that the motion is intensified for small $|x|$ and retarded for large $|x|$. Thus the motion tends, over time, to settle down into a steady oscillation. This striking behavior of f can be expressed by saying that the physical system absorbs energy when $|x|$ is small and dissipates energy when $|x|$ is large.

Math Nugget

Alfred Liénard (1869–1958) was a French scientist who spent most of his career teaching applied physics at the School of Mines in Paris; he became director of that institution in 1929. His physical research was mainly in the areas of electricity and magnetism, elasticity, and hydrodynamics. He occasionally worked on mathematical problems that arose from his more concrete scientific investigations, and in 1933 he was elected president of the French Mathematical Society. Liénard was an unassuming bachelor whose life was devoted entirely to his work and to his students.

EXAMPLE 12.8.6 We illustrate Liénard's theorem with an application to the van der Pol equation

$$\frac{d^2x}{dt^2} + \mu(x^2 - 1)\frac{dx}{dt} + x = 0 . \tag{12.78}$$

Here, for physical reasons, μ is assumed to be a positive constant. In this example, $f(x) = \mu(x^2 - 1)$ and $g(x) = x$. Thus conditions **(i)** and **(ii)** are clearly satisfied by inspection. We may calculate that

$$F(x) = \mu\left(\frac{1}{3}x^3 - x\right) = \frac{1}{3}\mu x(x^2 - 3) .$$

Thus F has a single positive zero at $x = \sqrt{3}$. It is negative on $(0, \sqrt{3})$ and is positive on $(\sqrt{3}, \infty)$. Also $F(x) \to +\infty$ as $x \to +\infty$. Finally, $F'(x) = \mu(x^2-1) > 0$ for $x > 1$, so F is increasing for $x > \sqrt{3}$. In sum, all the conditions of Liénard's theorem are met. We conclude that equation (78) has a unique closed path (periodic solution) that is approached spirally (asymptotically) by every other path (nontrivial solution). ∎

Math Nugget

Ivar Otto Bendixson (1861–1935) was a Swedish mathematician who published one important memoir in 1901; it served to supplement some of Poincaré's earlier work. Bendixson served as professor (and later as president) at the University of Stockholm. He was an energetic and long-time member of the Stockholm City Council.

Exercises

1. Verify that the nonlinear autonomous system

$$\begin{cases} \dfrac{dx}{dt} &= 3x - y - xe^{x^2+y^2} \\ \dfrac{dy}{dt} &= x + 3y - ye^{x^2+y^2} \end{cases}$$

has a periodic solution.

2. For each of the following differential equations, use a theorem of this section to determine whether or not the given differential equation has a periodic solution.

(a) $\dfrac{d^2x}{dt^2} + (5x^4 - 9x^2)\dfrac{dx}{dt} + x^5 = 0$

(b) $\dfrac{d^2x}{dt^2} - (x^2 + 1)\dfrac{dx}{dt} + x^5 = 0$

(c) $\dfrac{d^2x}{dt^2} - \left(\dfrac{dx}{dt}\right)^2 - (1 + x^2) = 0$

(d) $\dfrac{d^2x}{dt^2} + \dfrac{dx}{dt} + \left(\dfrac{dx}{dt}\right)^5 - 3x^3 = 0$

(e) $\dfrac{d^2x}{dt^2} + x^6\dfrac{dx}{dt} - x^2\dfrac{dx}{dt} + x = 0$

3. Show that any differential equation of the form

$$a\frac{d^2x}{dt^2} + b(x^2-1)\frac{dx}{dt} + cx = 0 \qquad (a, b, c > 0)$$

can be transformed into the van der Pol equation by a change of the independent variable.

4. Consider the nonlinear autonomous system

$$\begin{cases} \dfrac{dx}{dt} &= 4x + 4y - x(x^2 + y^2) \\ \dfrac{dy}{dt} &= -4x + 4y - y(x^2 + y^2) . \end{cases}$$

(a) Transform the system into polar coordinate form.

(b) Apply the Poincaré–Bendixson theorem to show that there is a closed path between the circles $r = 1$ and $r = 3$.

(c) Find the general nonconstant solution $x = x(t)$, $y = y(t)$ of the original system, and use this solution to find a periodic solution corresponding to the closed path whose existence was established in **(b)**.

(d) Sketch the closed path and at least two other paths in the phase plane.

Historical Note

Poincaré

Jules Henri Poincaré (1854–1912) was universally recognized at the beginning of the twentieth century as the greatest mathematician of his generation. Already a prodigy when he was quite young, he was watched with love and admiration by the entire country of France as he developed into one of the pre-eminent mathematicians of all time.

He began his academic career at Caen in 1879, but just two years later he was appointed to a professorship at the Sorbonne. He remained there for the rest of his life, lecturing on a different subject each year. His lectures were recorded, edited, and published by his students. In them, he treated with great originality and technical mastery all the known fields of pure and applied mathematics. Altogether, Poincaré produced more than 30 technical books on mathematical physics and celestial mechanics, 6 books of a more popular nature, and almost 500 research articles on mathematics.

Poincaré had a prodigious memory, and enjoyed doing mathematics in his head while he paced back and forth in his study. He would write down his ideas only after he had them completely worked out in his head. He was elected to the French Academy of Sciences at the very young age of 32. The academician who proposed Poincaré for membership said that "his work is above ordinary praise, and reminds us inevitably of what Jacobi wrote of Abel—that he had settled questions which, before him, were unimagined." It is arguable that Poincaré invented both topology and dynamical systems—both very active fields of research today.

Poincaré gave an invited lecture at the 1904 World's Fair in St. Louis. That event was held to celebrate the 100th anniversary of the Louisiana Purchase. Its location was Forest Park, which at that time lay far on the western outskirts of the city. With great foresight, the city fathers planned that Washington University (home of the author of this text) would move, after the Fair, 12 miles west from its then location on the Mississippi River to a new home next to Forest Park. So the administration buildings for the World's Fair were built with extra care and to an especially high standard, because they were to be given over after the Fair to the University. In fact the mathematics department is currently housed in Cupples Hall, which was one of those buildings. Poincaré's lecture was given in Holmes Lounge, just a few steps from Cupples. In this lecture—delivered just one year before Einstein's blockbuster publication of his own ideas—Poincaré explained many of the basic ideas of special relativity. He also published his ideas in a now obscure periodical called *The Monist.* In the end, Poincaré found his ideas to be much more limited than Einstein's, and he came to admire Einstein's work tremendously. He helped to nominate Einstein for his first real academic position.

By some measures, mathematical knowledge today doubles every ten years. It is difficult for any person to keep up with more than a few fields. Poincaré is one of the few men in history of whom it can be said that he had creative command of the whole of mathematics—and much of mathematical physics—as it existed in his day (the same is often said of his contemporary Hilbert). He was a remarkable genius, and his ideas will be remembered and studied for some time.

Anatomy of an Application

Mechanical Analysis of a Block on a Spring

We present a simple model from the theory of mechanics to illustrate the idea of phase plane analysis. Imagine a rectangular block of mass m that is attached to a horizontal spring. The block lies flat on a conveyor belt that moves at constant speed v in order to carry the block away. See Figure 12.26. Let $x(t)$ be the length of the spring (which varies with t, because it stretches and contracts). Let x_e be the equilibrium length of the spring.

Now let us do a bit of physical analysis. Let k be the spring constant (Hooke's constant—representing the stiffness of the spring). This means that the force exerted by the spring is proportional to the amount it is stretched, with constant of proportionality k. Let $\mathcal{F}(\dot{x})$ denote the frictional force exerted on the block by the conveyor belt. So the amount of friction depends on the

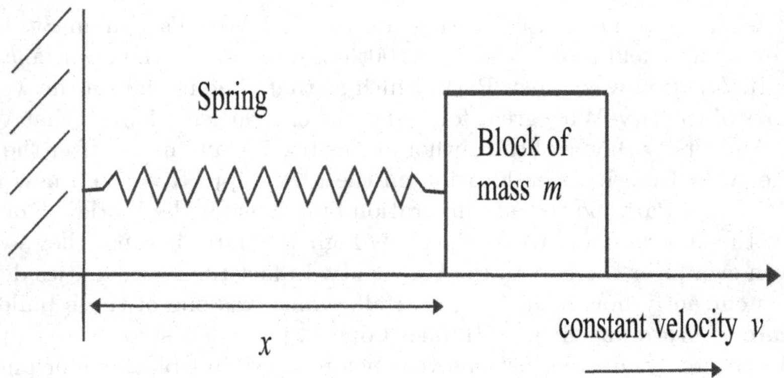

FIGURE 12.26
A block on a conveyer belt.

velocity $\dot{x}$. Now Hooke's spring law and Newton's second law tell us that

$$m \cdot \ddot{x} = F(\dot{x}) - k(x - x_e)\,. \tag{12.79}$$

If the force exerted by the spring is *less* than the frictional force, then the block will not move. Thus it makes sense to model the frictional force by

$$F(\dot{x}) = \begin{cases} F_0 & \text{for} \quad \dot{x} < v \\ -F_0 & \text{for} \quad \dot{x} > v\,. \end{cases}$$

Here F_0 is a constant force. Notice that when $\dot{x} = v$, the block moves at the same speed as the conveyor belt. This occurs when $k \cdot |x - x_e| \leq F_0$. In other words, when $x_e - F_0/k \leq x \leq x_e + F_0/k$, we have the solution $\dot{x} = v$.

It is convenient now to introduce dimensionless coordinates. We set

$$\overline{x} = \frac{x}{x_e} \qquad \text{and} \qquad \overline{t} = \frac{t}{\sqrt{m/k}}\,.$$

Our differential equation (79) then becomes

$$\ddot{\overline{x}} = \overline{F} - \overline{x} + 1\,, \tag{12.80}$$

where

$$\overline{F}(\dot{x}) = \begin{cases} \overline{F}_0 & \text{for} \quad \dot{\overline{x}} < \overline{v} \\ -\overline{F}_0 & \text{for} \quad \dot{\overline{x}} > \overline{v}\,. \end{cases}$$

As we see, there are two dimensionless parameters

$$\overline{F}_0 = \frac{F}{kx_e} \qquad \text{and} \qquad \overline{v} = \frac{v}{x_e} \cdot \sqrt{\frac{m}{k}}\,.$$

Now, in order to bring phase plane analysis into play, we write the differential equation (80) as the system

$$\left\{ \begin{array}{lcl} \dot{x} & = & y \\ \dot{y} & = & F(y) - x + 1 . \end{array} \right.$$

Of course we leave off the bars for reading convenience. This system has just one equilibrium point, at $x = 1 + F_0$, $y = 0$. Because $y = 0 < \overline{v}$, the system is linear in a neighborhood of this point, with

$$\begin{pmatrix} \dot{x} \\ \dot{y} \end{pmatrix} = \begin{pmatrix} 0 & 1 \\ -1 & 0 \end{pmatrix} \begin{pmatrix} x \\ y \end{pmatrix} + \begin{pmatrix} 0 \\ F_0 + 1 \end{pmatrix} .$$

The Jacobian matrix

$$\begin{pmatrix} 0 & 1 \\ -1 & 0 \end{pmatrix}$$

of this system has eigenvalues $\pm i$, hence the equilibrium point is a linear center. In point of fact, we can multiply together the two equations in our system to obtain

$$\dot{x}(x - 1 - F) + y\dot{y} = 0 .$$

This is easily integrated to yield

$$\left\{ \begin{array}{lll} (x - (1 + F_0))^2 + y^2 = c & \text{for} & y < v \\ (x - (1 - F_0))^2 + y^2 = c & \text{for} & y > v . \end{array} \right.$$

We see, then, that the solutions for $y \neq v$ are concentric circles. The equilibrium point remains a center, even when we take into account the nonlinear terms.

Now let us examine the phase portrait in Figure 12.27 and try to learn more about the motion of our mechanical block. The force function F is discontinuous at $y = v$; hence, by the differential equations, the slope of the integral paths will be discontinuous there. If $x < 1 - F_0$ or $x > 1 - F_0$ then the trajectory crosses the line $y = v$ transversally. But when $1 - F_0 \leq x \leq 1 + F_0$ then the line $y = v$ itself is a solution. What conclusion may we draw?

An integral curve that meets the line $y = v$ with x in the range $1 - F_0 \leq x \leq 1 + F_0$ will follow this trajectory until $x = 1 + F_0$, at which point it will move off the limit cycle through D. As a specific example, consider the trajectory that begins at point A. This is an initially stationary block with the spring sufficiently stretched that it can immediately overcome the force of the conveyor belt. The solution will follow the circular trajectory, as the figure shows, until it reaches point B. At that moment, the direction of the frictional force changes; so the solution follows a *different* circular trajectory up to the point C. At that moment, the block is stationary relative to the conveyor belt, and it will be carried along until the spring is stretched enough that the force of the spring can overcome the force of the friction. This happens at point D. After that, the solution remains on the period solution (i.e., the

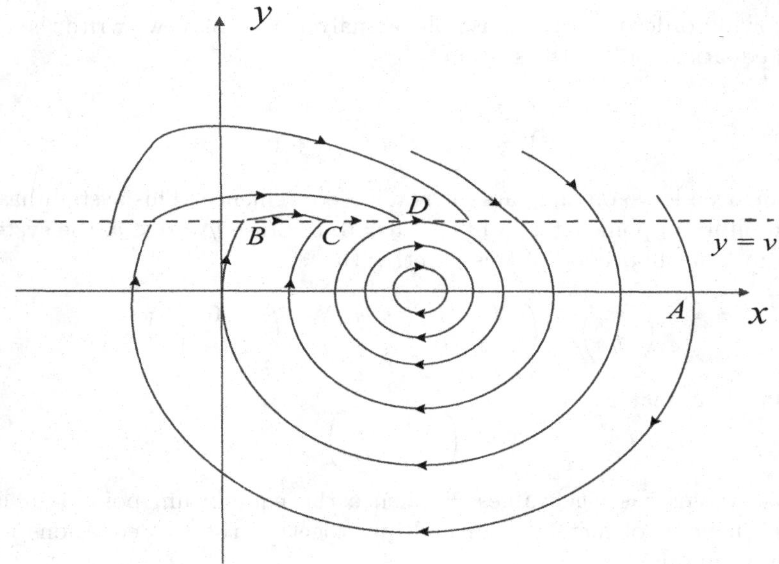

FIGURE 12.27
The phase portrait.

circle) through D. On this periodic, circular solution, the speed of the block is always less than the speed of the conveyor belt; thus the friction remains constant. The block stays in simple harmonic motion.

Problems for Review and Discovery

A. Drill Exercises

1. Sketch and describe the phase portrait of each of the following systems.

(a) $$\begin{cases} \dfrac{dx}{dt} & = & -1 \\ \dfrac{dy}{dt} & = & 1 \end{cases}$$

(b) $$\begin{cases} \dfrac{dx}{dt} & = & 0 \\ \dfrac{dy}{dt} & = & y \end{cases}$$

(c) $$\begin{cases} \dfrac{dx}{dt} & = & y \\ \dfrac{dy}{dt} & = & x \end{cases}$$

(d) $$\begin{cases} \dfrac{dx}{dt} & = & -y \\ \dfrac{dy}{dt} & = & -x \end{cases}$$

2. Find all solutions of the nonautonomous system

$$\begin{cases} \dfrac{dx}{dt} &= y-1 \\ \dfrac{dy}{dt} &= x \end{cases}$$

and sketch (in the x-y plane) some of the curves defined by these solutions.

3. For each of the following nonlinear systems, do the following:

(i) Find the critical points;
(ii) Find the differential equation of the paths;
(iii) Solve this equation to find the paths;
(iv) Sketch a few of the paths and show the direction of increasing t.

(a) $\begin{cases} \dfrac{dx}{dt} &= 2yx^2 \\ \dfrac{dy}{dt} &= x(y^2-1) \end{cases}$ **(c)** $\begin{cases} \dfrac{dx}{dt} &= e^x \sin y \\ \dfrac{dy}{dt} &= e^x \end{cases}$

(b) $\begin{cases} \dfrac{dx}{dt} &= x(y^2+1) \\ \dfrac{dy}{dt} &= -y(y^2+1) \end{cases}$ **(d)** $\begin{cases} \dfrac{dx}{dt} &= x^2y^2 \\ \dfrac{dy}{dt} &= -y \end{cases}$

4. Each of the following linear systems has the origin as an isolated critical point. For each system,

(i) Find the general solution;
(ii) Find the differential equation of the paths;
(iii) Solve the equation found in part **(ii)** and sketch a few of the paths, showing the direction of increasing t;
(iv) Discuss the stability of the critical points.

(a) $\begin{cases} \dfrac{dx}{dt} &= -y \\ \dfrac{dy}{dt} &= x \end{cases}$ **(b)** $\begin{cases} \dfrac{dx}{dt} &= -3x \\ \dfrac{dy}{dt} &= -y \end{cases}$ **(c)** $\begin{cases} \dfrac{dx}{dt} &= 2y \\ \dfrac{dy}{dt} &= -4x \end{cases}$

5. Determine the nature and stability properties of the critical point $(0,0)$ for each of the following linear autonomous systems.

(a) $\begin{cases} \dfrac{dx}{dt} &= x \\ \dfrac{dy}{dt} &= 2y \end{cases}$ **(e)** $\begin{cases} \dfrac{dx}{dt} &= -x-y \\ \dfrac{dy}{dt} &= 3x-y \end{cases}$

(b) $\begin{cases} \dfrac{dx}{dt} &= -x+y \\ \dfrac{dy}{dt} &= x-2y \end{cases}$ **(f)** $\begin{cases} \dfrac{dx}{dt} &= x-2y \\ \dfrac{dy}{dt} &= 3x-y \end{cases}$

(c) $\begin{cases} \dfrac{dx}{dt} &= -2x + 2y \\ \dfrac{dy}{dt} &= -x + y \end{cases}$ (g) $\begin{cases} \dfrac{dx}{dt} &= 3x - y \\ \dfrac{dy}{dt} &= x + y \end{cases}$

(d) $\begin{cases} \dfrac{dx}{dt} &= 5x + y \\ \dfrac{dy}{dt} &= -7x - 2y \end{cases}$

6. Determine whether each of the following functions is of positive type, of negative type, or neither.

(a) $-x^2 + xy + y^2$ **(c)** $-x^2 + 2xy - 2y^2$
(b) $x^2 - 2xy + 4y^2$ **(d)** $-x^2 - 2xy + 3y^2$

7. Show that the indicated critical point is asymptotically stable for each of the following systems.

(a) $\begin{cases} \dfrac{dx}{dt} &= 1 + 2y \\ \dfrac{dy}{dt} &= 1 - 3x^2 \end{cases}$ $(\sqrt{3}/3, -1/2)$

(b) $\begin{cases} \dfrac{dx}{dt} &= y(2 - x - y) \\ \dfrac{dy}{dt} &= -y - x - 2yx \end{cases}$ $(0, 0)$

(c) $\begin{cases} \dfrac{dx}{dt} &= -x \\ \dfrac{dy}{dt} &= -2y \end{cases}$ $(0, 0)$

8. Show that $(3, -2)$ is an unstable critical point for the system

$$\begin{cases} \dfrac{dx}{dt} &= -(x - y)(1 - y - x) \\ \dfrac{dy}{dt} &= x(2 + y) \, . \end{cases}$$

9. Find the equation of the paths of

$$\frac{d^2x}{dt^2} + x - x^2 = 0 \, .$$

Sketch these paths in the phase plane. Locate the critical points and determine the nature of each of these.

10. For each of the following differential equations, use a theorem from this chapter to determine whether or not the given differential equation has a periodic solution.

(a) $\dfrac{d^2x}{dt^2} + x^3 \dfrac{dx}{dt} + x^4 = 0$

(b) $\dfrac{d^2x}{dt^2} + (x^2+1)\dfrac{dx}{dt} + 3x^5 = 0$

(c) $\dfrac{d^2x}{dt^2} - x\left(\dfrac{dx}{dt}\right)^2 - (2+x^4) = 0$

(d) $\dfrac{d^2x}{dt^2} + x^2\dfrac{dx}{dt} + \left(\dfrac{dx}{dt}\right)^3 - 3x = 0$

B. Challenge Problems

1. For Duffing's system of equations

$$\begin{aligned}\frac{dx}{dt} &= y \\ \frac{dy}{dt} &= -x + \frac{x^3}{6},\end{aligned}$$

find an equation of the form $\Theta(x, y) = 0$ that is satisfied by the paths. Plot several level curves of the function Θ; these will of course be paths of the Duffing system. Indicate the direction of motion on each path.

2. The motion of a certain undamped pendulum is described by the system

$$\begin{aligned}\frac{dx}{dt} &= y \\ \frac{dy}{dt} &= 2\sin x.\end{aligned}$$

If the pendulum is set in motion with initial angular displacement $\alpha = 0.1$ and no initial velocity, then the initial conditions are $x(0) = 0.1$ and $y(0) = 0$.

Plot x versus t. Study this graph to obtain an estimate of the amplitude R and the period T of the motion of the pendulum. Now repeat all these steps if the angular displacement is instead $\alpha = 0.25$. What about $\alpha = 0.05$?

Discuss how the amplitude and period of the pendulum depend on the initial angular displacement. Use graphics to express these relationships. What can you say about the asymptotic behavior as α tends to 0? What happens when α is rather large? Is there a critical value for α at which the behavior of the pendulum changes?

3. Consider the system

$$\begin{aligned}\frac{dx}{dt} &= -x + y \\ \frac{dy}{dt} &= -y.\end{aligned}$$

Show that the eigenvalues for this system are the repeated pair $-1, -1$. Thus the critical point $(0, 0)$ is an asymptotically stable node.

Now compare this result with the system

$$\begin{aligned}\frac{dx}{dt} &= -x + y \\ \frac{dy}{dt} &= -\eta x - y.\end{aligned}$$

Verify that if $\eta > 0$ then the asymptotically stable node is now an asymptotically stable spiral point. However, if $\eta < 0$ then the stable node remains a stable node.

C. Problems for Discussion and Exploration

1. Show that if a trajectory $x = \alpha(t), y = \beta(t)$ for an autonomous system is defined for all time (positive and negative) and is closed then it is periodic.

2. Show that, for the system

$$\begin{aligned} \frac{dx}{dt} &= F(x,y) \\ \frac{dy}{dt} &= G(x,y)\,, \end{aligned}$$

with F, G smooth, there is at most one trajectory passing through any given point (x_0, y_0). (*Hint:* Picard's existence and uniqueness theorem is relevant.)

3. Show that if a trajectory begins at a noncritical point of the system

$$\begin{aligned} \frac{dx}{dt} &= F(x,y) \\ \frac{dy}{dt} &= G(x,y)\,, \end{aligned}$$

with F, G smooth, then it cannot reach a critical point in a finite length of time (in other words, critical points are "infinitely far away").

13

Qualitative Properties and Theoretical Aspects

- The general solution
- Existence and uniqueness theorems
- The Wronskian
- Linear independence
- The Sturm theorems

13.1 A Bit of Theory

Until now, in our study of second-order and higher-order differential equations, we have spoken of "independent solutions" without saying exactly what we mean. We now take the time to discuss this matter carefully. The reader may find it useful to refer to the appendix on linear algebra in order to put these ideas in context.

A collection of functions $\phi_1, \ldots, \phi_k$ defined on an interval $[a, b]$ is called *linearly independent* if there *do not* exist constants $a_1, \ldots, a_k$ (not all zero) such that

$$a_1\phi_1 + \cdots + a_k\phi_k \equiv 0 .$$

If the functions are not linearly independent then they are linearly *dependent.*

It is intuitively clear, for example, that $\phi_1(x) = \sin x$ and $\phi_2(x) = \cos x$ are linearly independent on $[0, 2\pi]$. A glance at their graphs (Figure 13.1) shows that they have maxima and minima and zeros at different places; how could it be that

$$a_1 \sin x + a_2 \cos x \equiv 0\,?$$

But let us give a rigorous argument. Suppose that there *were* constants a_1, a_2, not both zero, such that

$$a_1 \sin x + a_2 \cos x \equiv 0 .$$

Differentiating, we find that

$$a_1 \cos x - a_2 \sin x \equiv 0 .$$

Solving this system of linear equations simultaneously reveals that $a_1 = 0, a_2 = 0$, and that is a contradiction.

DOI: 10.1201/9781003214526-13

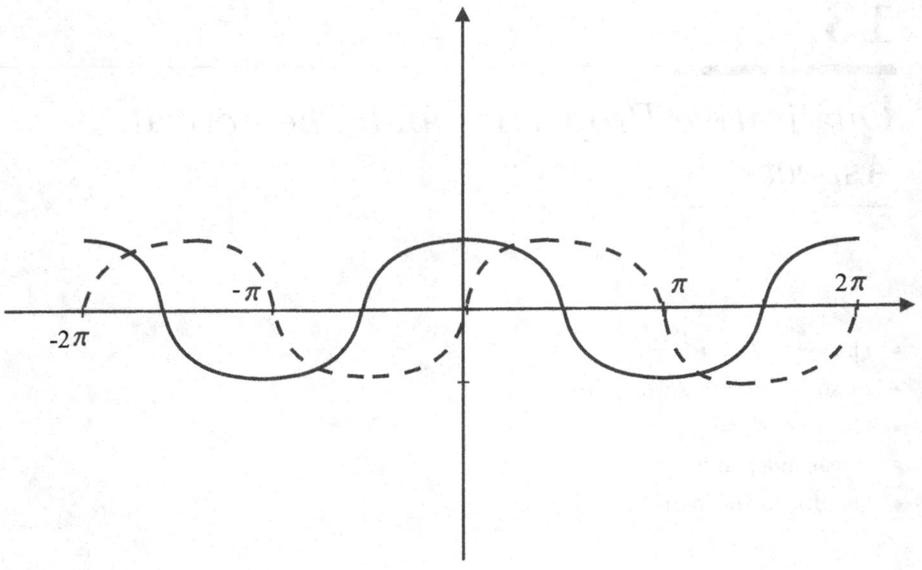

FIGURE 13.1
The sine and cosine functions.

There is a neat test for linear independence of two solutions of a second-order linear differential equation

$$y'' + p(x)y + q(x)y = 0\,. \tag{13.1}$$

It is called the *Wronskian*. Suppose that y_1, y_2 are solutions of this differential equation. We form the determinant

$$W(x) = \det \begin{pmatrix} y_1 & y_2 \\ y_1' & y_2' \end{pmatrix} = y_1 y_2' - y_2 y_1'\,.$$

The basic fact about the Wronskian is this:

Proposition 13.1.1 *Let y_1, y_2 be solutions of the second-order linear differential equation*

$$y'' + p(x)y' + q(x)y = 0\,.$$

Then the Wronskian $W(x)$ of y_1 and y_2 is either nowhere vanishing or vanishes identically. In the former case, the two solutions are linearly independent. In the latter case, they are linearly dependent.

Proof We know that

$$y_2'' + p(x)y_2' + q(x)y_2 = 0$$

and

$$y_1'' + p(x)y_1' + q(x)y_1 = 0\,.$$

We multiply the first of these equations by y_1 and the second by y_2 and subtract. The result is

$$[y_1y_2'' - y_2y_1''] + p(x)W(x) = 0\,.$$

But the expression in brackets is $W'(x)$ because $W = y_1y_2' - y_2y_1'$ hence

$$W' = [y_1'y_2' + y_1y_2''] - [y_2'y_1' + y_2y_1''] = y_1y_2'' - y_2y_1''\,.$$

Thus our equation is

$$W'(x) + p(x)W(x) = 0\,.$$

This is a first-order linear equation that we may solve easily. The solution is

$$W(x) = C \cdot e^{-\int p(x)\,dx}\,.$$

Now the exponential is never 0. If $C \neq 0$ then the first option in our conclusion holds. If $C = 0$ then the second option holds.

Of course if $W \equiv 0$, then this is a guarantee that the linear system

$$\begin{aligned} a_1y_1 + a_2y_2 &= 0 \\ a_1y_1' + a_2y_2' &= 0 \end{aligned}$$

has a nontrivial solution. So, as we saw in the case of sine and cosine, the functions are linearly dependent. If instead W is never zero, then the system has only the trivial solution. So y_1, y_2 are linearly independent. □

The next result gives a solid mathematical foundation to an idea that we have been using consistently and effectively throughout the text until now.

Theorem 13.1.2 *Consider the differential equation*

$$y'' + p(x)y' + q(x)y = r(x)\,. \tag{13.2}$$

with p, q, r continuous functions on an interval $[a, b]$. Fix a point $x_0 \in [a, b]$. Let y_0, y_1 be real numbers. Then (2) has a unique solution that satisfies $y(x_0) = y_0$ and $y'(x_0) = y_1$.

Idea of the Proof We shall give a more complete proof of this result later on. For now, we reason heuristically. We know that a first-order linear equation

$$y' + p(x)y = r(x) \tag{13.3}$$

has general solution

$$y = e^{-\int p(x)\,dx} \int e^{\int p(x)\,dx} r(x)\,dx + C \cdot e^{-\int p(x)\,dx}\,.$$

Thus it is clear that there is only one solution to (3) satisfying $y(x_0) = y_0$—just because there is only one value of C that will make this condition true.

Now equation (2) can be reduced to a system of first-order equations with the substitution $w(x) = y'(x)$. Thus (2) becomes

$$\begin{aligned} w' + pw + y &= r \\ y' &= w\,. \end{aligned}$$

The initial conditions specified in the theorem now become $y(x_0) = y_0$ and $w(x_0) = y_1$.

The argument that produced the solution of (3) will now produce the general solution (y, w) of this pair of equations, and it will be clear that there is only one such solution satisfying these two initial conditions. □

Math Nugget

Hoëné Wronski (1778–1853) was an impecunious Pole of erratic personality who spent most of his life in France. The Wronskian determinant, which plays a significant role both in this section and throughout the book, was Wronski's only contribution to theoretical mathematics. He was, in addition, the only Polish mathematician of the nineteenth century whose name is remembered today. This fact is remarkable, for the twentieth century has produced a significant number of important Polish mathematicians (including Antoni Zygmund, the "mathematical grandfather" of the author).

Now one of the main points of this section is the following theorem:

Theorem 13.1.3 *Consider the differential equation*

$$y''(x) + p(x) \cdot y'(x) + q(x) \cdot y(x) = 0\,. \tag{13.4}$$

Suppose that y_1 *and* y_2 *are linearly independent solutions of* (4). *Then*

$$Ay_1 + By_2$$

is the general solution of (4), *in the sense that any solution* y *of* (4) *has the form*

$$y = Ay_1 + By_2\,.$$

Proof Let y be any solution of (4). Fix a point x_0 in the domain of y. We shall show that constants A and B can be found so that

$$y(x_0) = Ay_1(x_0) + By_2(x_0)$$

and

$$y'(x_0) = Ay_1'(x_0) + By_2'(x_0)\,.$$

Since both y and $Ay_1 + By_2$ are solutions of (4) satisfying the same initial conditions, it then follows from Theorem 13.1.2 that $y = Ay_1 + By_2$.

To find A and B, we think of the system of equations

$$\begin{aligned} y_1(x_0)A + y_2(x_0)B &= y(x_0) \\ y_1'(x_0)A + y_2'(x_0)B &= y'(x_0) \end{aligned}$$

as two linear equations in the two unknowns A and B. Of course this system will have a nontrivial solution if and only if the determinant of its coefficients is nonzero. Thus we need

$$\det \begin{pmatrix} y_1(x_0) & y_2(x_0) \\ y_1'(x_0) & y_2'(x_0) \end{pmatrix} \neq 0$$

But this is just the nonvanishing of the Wronskian, and that is guaranteed by the linear independence of y_1 and y_2. □

EXAMPLE 13.1.4 Show that $y = A\sin x + B\cos x$ is the general solution of the differential equation

$$y'' + y = 0$$

and find the particular solution that satisfies $y(0) = 2$ and $y'(0) = 3$.

Solution: Of course we know that $y_1(x) = \sin x$ and $y_2(x) = \cos x$ are solutions, and we may verify this by direct substitution into the equation.

The Wronskian of y_1 and y_2 is

$$W(x) = \det \begin{pmatrix} \sin x & \cos x \\ \cos x & -\sin x \end{pmatrix} \equiv -1 \neq 0\,.$$

Hence y_1 and y_2 are linearly independent. It follows from Theorem 13.1.3 above that $y = A\sin x + B\cos x$ is the general solution of the given differential equation.

We leave it as an exercise to find the particular solution that satisfies the given initial conditions. □

Exercises

1. Show that $y_1(x) = e^x$ and $y_2(x) = e^{-x}$ are linearly independent solutions of $y'' - y = 0$.

2. Show that $y(x) = c_1x + c_2x^2$ is the general solution of

$$x^2y'' - 2xy' + 2y = 0$$

on any interval not containing 0. Find the particular solution that satisfies $y(1) = 3$ and $y'(1) = 5$.

3. Show that $y(x) = c_1e^x + c_2e^{2x}$ is the general solution of the differential equation

$$y'' - 3y' + 2y = 0\,.$$

Find the particular solution for which $y(0) = -1$ and $y'(0) = 1$.

4. Show that $y(x) = c_1e^{2x} + c_2xe^{2x}$ is the general solution of the differential equation

$$y'' - 4y' + 4y = 0\,.$$

5. By either inspection or by experiment, find two linearly independent solutions of the equation $x^2y'' - 2y = 0$ on the interval $[1, 2]$. Find the particular solution satisfying the initial conditions $y(1) = 1$ and $y'(1) = 8$.

6. In each of the following problems, verify that the functions y_1 and y_2 are linearly independent solutions of the given differential equation on the interval $[0, 2]$. Then find the particular solution satisfying the stated initial conditions.

(a) $y'' + y' - 2y = 0$ $\quad y_1(x) = e^x$
$\quad y_2(x) = e^{-2x} \quad y(0) = 8,\ y'(0) = 2$

(b) $y'' + y' - 2y = 0$ $\quad y_1(x) = e^x$
$\quad y_2(x) = e^{-2x} \quad y(1) = 0,\ y'(1) = 0$

(c) $y'' + 5y' + 6y = 0$ $\quad y_1(x) = e^{-2x}$
$\quad y_2(x) = e^{-3x} \quad y(0) = 1,\ y'(0) = 1$

(d) $y'' + y' = 0$ $\quad y_1(x) \equiv 1$
$\quad y_2(x) = e^{-x} \quad y(2) = 0,\ y'(2) = e^{-2}$

7. **(a)** Use one (or both) methods described in the section on reduction of order (Section 1.9) to find all solutions of $y'' + (y')^2 = 0$.

(b) Verify that $y_1(x) \equiv 1$ and $y_2(x) = \ln x$ are linearly independent solutions of the equation in part **(a)** on any interval in the positive real half-line. Is $y = c_1 + c_2 \ln x$ the general solution? Why or why not?

8. Use the Wronskian to prove that two solutions of the homogeneous equation

$$y'' + P(x)y' + Q(x)y = 0$$

on an interval $[a, b]$ are linearly *dependent* if

(a) they have a common zero x_0 in the interval;

or

(b) they have maxima or minima at the same point x_0 in the interval.

9. Consider the two functions $f(x) = x^3$ and $g(x) = x^2|x|$ on the interval $[-1, 1]$.

(a) Show that their Wronskian $W(f, g)$ vanishes identically.

(b) Show that f and g are not linearly independent.

(c) Why do the results of (a) and (b) not contradict Proposition 13.1.1?

10. (a) Show that, by applying the substitution $y = uv$ to the homogeneous equation

$$y'' + P(x)y' + Q(x)y = 0,$$

we obtain a homogeneous, second-order linear equation for v with no v' term present. Find u and the equation for v in terms of the original coefficients P and Q.

(b) Use the method of part (a) to find the general solution of $y'' + 2xy' + (1 + x^2)y = 0$.

13.2 Picard's Existence and Uniqueness Theorem

13.2.1 The Form of a Differential Equation

A fairly general first-order differential equation will have the form

$$\frac{dy}{dx} = F(x, y). \tag{13.5}$$

Here F is a continuously differentiable function on some domain $(a, b) \times (c, d)$. We think of y as the dependent variable (that is, the function that we seek) and x as the independent variable. For technical reasons, we assume that the function F is bounded,

$$|F(x, y)| \leq M \tag{13.6}$$

and, in addition, that F satisfies a *Lipschitz condition*:

$$|F(x, s) - F(x, t)| \leq C \cdot |s - t|. \tag{13.7}$$

[In many treatments it is standard to assume that F is bounded and $\partial F/\partial y$ is bounded. It is easy to see, using the mean value theorem, that these two conditions imply (6), (7).]

According to Picard's theorem, which we shall formally enunciate below, these two hypotheses will guarantee that the initial value problem

$$\frac{dy}{dx} = F(x, y), \quad y(x_0) = y_0$$

has a unique solution. This result is of fundamental importance. It can be used to establish existence and uniqueness theorems for differential equations of all orders, and in many different contexts. The theorem is used to establish the existence of geodesics in differential geometry, of transversal flows, and of many basic constructions in all areas of geometric analysis.

Let us now formally enunciate Picard's theorem.

Theorem 13.2.1 *Let F be a continuously differentiable function on some domain $(a,b) \times (c,d)$. We assume that the function F is bounded,*

$$|F(x,y)| \leq M$$

and, in addition, that F satisfies a Lipschitz condition:

$$|F(x,s) - F(x,t)| \leq C \cdot |s - t| \, .$$

Let $x_0 \in (a,b)$ and $y_0 \in (c,d)$. Then there is an $h > 0$ such that $(x_0 - h, x_0 + h) \subseteq (a,b)$ and a continuously differentiable function y on $(x_0 - h, x_0 + h)$ that solves the initial value problem

$$\frac{dy}{dx} = F(x,y) \; , \; y(x_0) = y_0 \, . \tag{13.8}$$

The solution is unique in the sense that, if $\widetilde{y}$ is another continuously differentiable function on some interval $(x_0 - \widetilde{h}, x_0 + \widetilde{h})$ that solves the initial value problem (8), then $y \equiv \widetilde{y}$ on $(x_0 - h, x_0 + h) \cap (x_0 - \widetilde{h}, x_0 + \widetilde{h})$.

EXAMPLE 13.2.2 Consider the equation

$$\frac{dy}{dx} = x^2 \sin y - y \ln x \, . \tag{13.9}$$

Then this equation fits the paradigm of equation (5) with $F(x,y) = x^2 \sin y - y \ln x$ provided that $1 \leq x \leq 2$ and $0 \leq y \leq 3$ (for instance). ∎

Picard's idea is to set up an iterative scheme for producing the solution of the initial value problem. The most remarkable fact about Picard's technique is that it always works: As long as F satisfies the Lipschitz condition, then the problem will possess one and only one solution. If F is only continuous (no Lipschitz condition), then it is still possible to show that a solution exists; but it will no longer (in general) be unique.

13.2.2 Picard's Iteration Technique

While we shall not actually give a complete proof that Picard's technique works, we shall set it up and indicate the sequence of functions it produces that converges (uniformly) to the solution of our problem.

Picard's approach is inspired by the fact that the differential equation and initial condition taken together are equivalent to the single integral equation

$$y(x) = y_0 + \int_{x_0}^{x} F[t, y(t)]\, dt\,. \tag{13.10}$$

We invite the reader to differentiate both sides of this equation, using the fundamental theorem of calculus, to derive the original differential equation (5). Of course the initial condition $y(x_0) = y_0$ is built into (10). This integral equation gives rise to the iteration scheme that we now describe.

We assume that $x_0 \in (a, b)$ and that $y_0 \in (c, d)$. We set

$$y_1(x) = y_0 + \int_{x_0}^{x} F(t, y_0)\, dt\,.$$

For x near to x_0, this definition makes sense. Now we define

$$y_2(x) = y_0 + \int_{x_0}^{x} F(t, y_1(t))\, dt$$

and, for any integer $j \geq 1$,

$$y_{j+1}(x) = y_0 + \int_{x_0}^{x} F(t, y_j(t))\, dt\,. \tag{13.11}$$

It turns out that the sequence of functions $\{y_1, y_2, \ldots\}$ will converge uniformly on an interval of the form $(x_0 - h, x_0 + h) \subseteq (a, b)$.

13.2.3 Some Illustrative Examples

Picard's iteration method is best apprehended by way of some examples that show how the iterates arise and how they converge to a solution. We now proceed to develop such illustrations.

EXAMPLE 13.2.3 Consider the initial value problem

$$y' = 2y, \qquad y(0) = 1\,.$$

Of course this could easily be solved by the method of first-order linear equations, or by separation of variables. Our purpose here is to illustrate how the Picard method works.

First notice that the stated initial value problem is equivalent to the integral equation

$$y(x) = 1 + \int_0^x 2y(t)\, dt\,.$$

Following the paradigm (11), we thus find that

$$y_{j+1}(x) = 1 + \int_0^x 2y_j(t)\, dt\,.$$

Using $y_0(x) \equiv 1$ (because 1 is the initial value), we then find that

$$\begin{aligned} y_1(x) &= 1 + \int_0^x 2\,dt = 1 + 2x\,, \\ y_2(x) &= 1 + \int_0^x 2(1+2t)\,dt = 1 + 2x + 2x^2\,, \\ y_3(x) &= 1 + \int_0^x 2(1+2t+2t^2)\,dt = 1 + 2x + 2x^2 + \frac{4x^3}{3}\,. \end{aligned}$$

In general, we find that

$$y_j(x) = 1 + 2x + 2x^2 + \frac{4x^3}{3} + \cdots + \frac{(2x)^j}{j!} = \sum_{\ell=0}^{j} \frac{(2x)^\ell}{\ell!}\,.$$

It is plain that these are the partial sums for the power series expansion of $y = e^{2x}$. We conclude that the solution of our initial value problem is given by $y = e^{2x}$. ∎

EXAMPLE 13.2.4 Let us use Picard's method to solve the initial value problem

$$y' = 2x - y, \qquad y(0) = 1\,.$$

The equivalent integral equation is

$$y(x) = 1 + \int_0^x [2t - y(t)]\,dt$$

and (11) tells us that

$$y_{j+1}(x) = 1 + \int_0^x [2t - y_j(t)]\,dt\,.$$

Taking $y_0(x) \equiv 1$, we then find that

$$\begin{aligned} y_1(x) &= 1 + \int_0^x (2t-1)\,dt = 1 - x + x^2\,, \\ y_2(x) &= 1 + \int_0^x \left(2t - [1 - t + t^2]\right)\,dt \\ &= 1 - x + \frac{3x^2}{2} - \frac{x^3}{3}\,, \\ y_3(x) &= 1 + \int_0^x \left(2t - [1 - t + 3t^2/2 - t^3/3]\right)\,dt \\ &= 1 - x + \frac{3x^2}{2} - \frac{x^3}{2} + \frac{x^4}{4\cdot 3}\,, \\ y_4(x) &= 1 + \int_0^x \left(2t - [1 - t + 3t^2/2 - t^3/2 + t^4/4\cdot 3]\right)\,dt \\ &= 1 - x + \frac{3x^2}{2} - \frac{x^3}{2} + \frac{x^4}{4\cdot 2} - \frac{x^5}{5\cdot 4\cdot 3}\,. \end{aligned}$$

In general, we find that

$$\begin{aligned} y_j(x) &= 1 - x + \frac{3x^2}{2!} - \frac{3x^3}{3!} + \frac{3x^4}{4!} - + \cdots + (-1)^j \frac{3x^j}{j!} + (-1)^{j+1} \frac{2x^{j+1}}{(j+1)!} \\ &= [2x-2] + 3 \cdot \left[\sum_{\ell=0}^{j} (-1)^\ell \frac{x^\ell}{\ell!} \right] + (-1)^{j+1} \frac{2x^{j+1}}{(j+1)!} \\ &= [2x-2] + 3 \cdot \left[\sum_{\ell=0}^{j} \frac{(-x)^\ell}{\ell!} \right] + (-1)^{j+1} \frac{2x^{j+1}}{(j+1)!} . \end{aligned}$$

Thus we see that the iterates $y_j(x)$ converge to the solution $y(x) = [2x-2] + 3e^{-x}$ for the initial value problem. ∎

13.2.4 Estimation of the Picard Iterates

To get an idea of why the assertion at the end of Section 13.2.2—that the functions y_j converge uniformly—is true, let us do some elementary estimations. Choose $h > 0$ so small that $h \cdot C < 1$, where C is the constant from the Lipschitz condition in (7). We shall assume in the following calculations that $|x - x_0| < h$.

Now we proceed with the iteration. Let $y_0(t)$ be identically equal to the initial value y_0. Then

$$\begin{aligned} |y_0(t) - y_1(t)| = |y_0 - y_1(t)| &= \left| \int_{x_0}^{x} F(t, y_0)\, dt \right| \\ &\le \int_{x_0}^{x} |F(t, y_0)|\, dt \\ &\le M \cdot |x - x_0| \\ &\le M \cdot h . \end{aligned}$$

We have of course used the boundedness condition (6).

Next we have

$$\begin{aligned} |y_1(x) - y_2(x)| &= \left| \int_{x_0}^{x} F(t, y_0(t))\, dt - \int_{x_0}^{x} F(t, y_1(t))\, dt \right| \\ &\le \int_{x_0}^{x} |F(t, y_0(t)) - F(t, y_1(t))|\, dt \\ &\le \int_{x_0}^{x} C \cdot |y_0(t) - y_1(t)|\, dt \\ &\le C \cdot (M \cdot h) \cdot h \\ &= M \cdot C \cdot h^2 . \end{aligned}$$

One can continue this procedure to find that

$$|y_2(x) - y_3(x)| \le M \cdot C^2 \cdot h^3 = M \cdot h \cdot (Ch)^2 = M \cdot C^2 \cdot h^3$$

and, more generally,

$$|y_j(x) - y_{j+1}(x)| \le M \cdot C^j \cdot h^{j+1} < M \cdot h \cdot (Ch)^j = M \cdot C^j \cdot h^{j+1} .$$

Now, if $0 < K < L$ are integers, then

$$\begin{aligned} |y_K(x) - y_L(x)| &\le |y_K(x) - y_{K+1}(x)| + |y_{K+1}(x) - y_{K+2}(x)| \\ &\quad + \cdots + |y_{L-1}(x) - y_L(x)| \\ &\le M \cdot h \cdot \left([Ch]^K + [Ch]^{K+1} + \cdots + [Ch]^{L-1}\right) . \end{aligned}$$

Since $|Ch| < 1$ by design, the geometric series $\sum_j [Ch]^j$ converges. As a result, the expression on the right of our last display is as small as we please, for K and L large, just by the Cauchy criterion for convergent series. It follows that the sequence $\{y_j\}$ of approximate solutions converges uniformly to a function $y = y(x)$. In particular, y is continuous.

Furthermore, we know that

$$y_{j+1}(x) = y_0 + \int_{x_0}^{x} F(t, y_j(t))\, dt .$$

Letting $j \to \infty$, and invoking the uniform convergence of the y_j, we may pass to the limit and find that

$$y(x) = y_0 + \int_{x_0}^{x} F(t, y(t))\, dt .$$

This says that y satisfies the integral equation that is equivalent to our original initial value problem. This equation also shows that y is continuously differentiable. Thus y is the function that we seek.

It can be shown that this y is in fact the *unique* solution to our initial value problem. We shall not provide the details of the proof of this assertion.

We repeat that in case F is not Lipschitz—say that F is only continuous—then it is still possible to show that a solution y exists. But it may no longer be unique. An example follows.

EXAMPLE 13.2.5 Consider the initial value problem

$$y' = \sqrt{|y|}, \quad y(0) = 0 .$$

Then both $y(x) \equiv 0$ and

$$y(x) = \begin{cases} 0 & \text{if} \quad x < 0 \\ x^2/4 & \text{if} \quad x \ge 0 . \end{cases}$$

are solutions. So the solution of this initial value problem is not unique. ∎

13.3 Oscillations and the Sturm Separation Theorem

In the best of all possible situations, we would like to be able to *explicitly write down* the general solution to the equation

$$y'' + p \cdot y' + q \cdot y = 0 .$$

However, we often cannot do so. In general, the solutions to most differential equations—even important equations of physics and engineering—cannot be found, at least not as a formula that we can write down. What we do instead in such circumstances is that we endeavor to learn qualitative properties of the solutions.

This is a new idea for us, and we introduce it gradually. As a finger exercise, we begin by examining the equation

$$y'' + y = 0 . \tag{13.12}$$

Of course the general solution $y = A \sin x + B \cos x$ is very well known to us. *But let us pretend that we do not know how to find this general solution, and let us see what we can learn about the solutions of (12) just by examining the differential equation.*

Imagine a solution y_1 of (12) that satisfies the initial conditions $y_1(0) = 0$ and $y_1'(0) = 1$. Let us think about the graph of this function. We know that it begins at the origin $(0, 0)$ with slope 1. Thus, as we move to the right from the origin, we see that the curve is rising, and $y_1 > 0$ (Figure 13.2). The differential equation

$$y_1'' = -y_1$$

then tells us that $y_1'' = (y_1')'$ is negative, so the slope is decreasing—and at an increasing rate (Figure 13.3). Thus the slope is eventually 0 at some point $x = m$ (Figure 13.4). But then the slope will continue to decrease, so it is negative. Thus the function is falling, y_1 is decreasing, and therefore (by the differential equation again) the slope is decreasing *at a decreasing rate* (Figure 13.5). The curve will eventually cross the x-axis at a point that we shall call π (yes, *that* π). See Figure 13.6.

Since y_1'' depends only on y_1, we see then that y_1' too depends only on y_1. It follows then that y_1 will fall off, after x passes m, at the same rate that it rose before it reached m. Thus the graph is symmetric about the line $x = m$. We conclude that $m = \pi/2$ and that the slope of the curve at $x = \pi$ must be -1. See Figure 13.7.

A similar argument can be used to show that the next portion of the curve is an inverted copy of the first piece, and so on indefinitely. In short, we find that y_1 is 2π-periodic. In fact it is $\sin x$, as Theorem 13.1.2 tells us (because of the initial conditions).

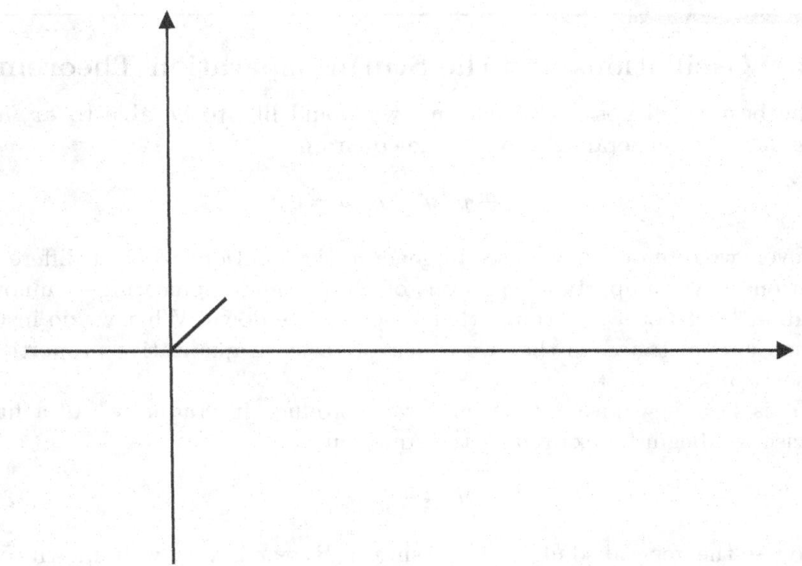

FIGURE 13.2
Discerning the sine function.

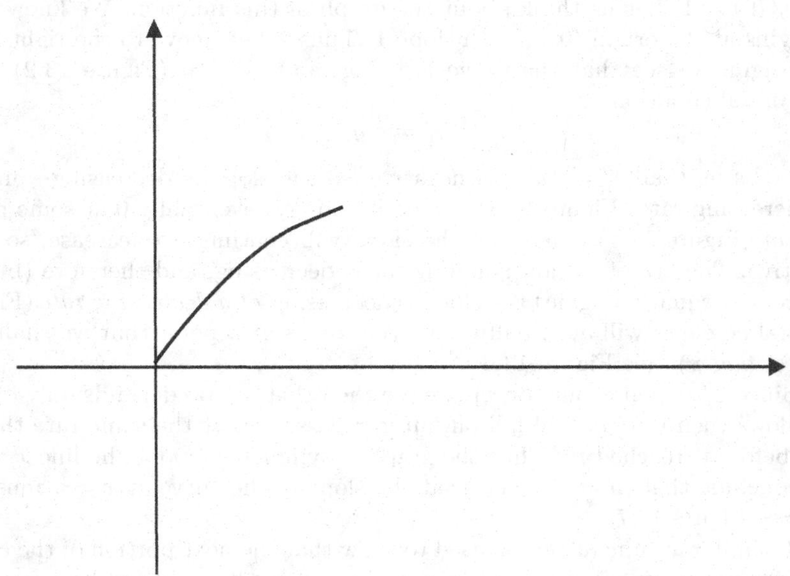

FIGURE 13.3
More on the sine function.

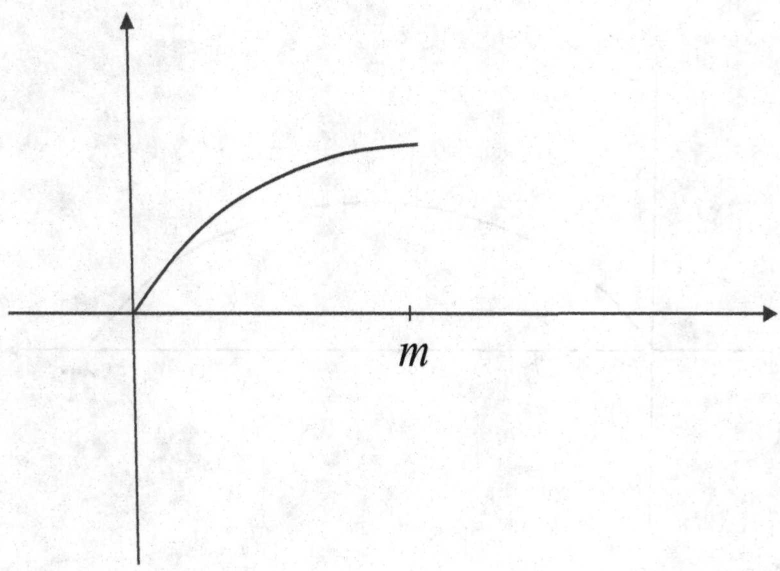

FIGURE 13.4
The sine function on $[0, \pi/2]$.

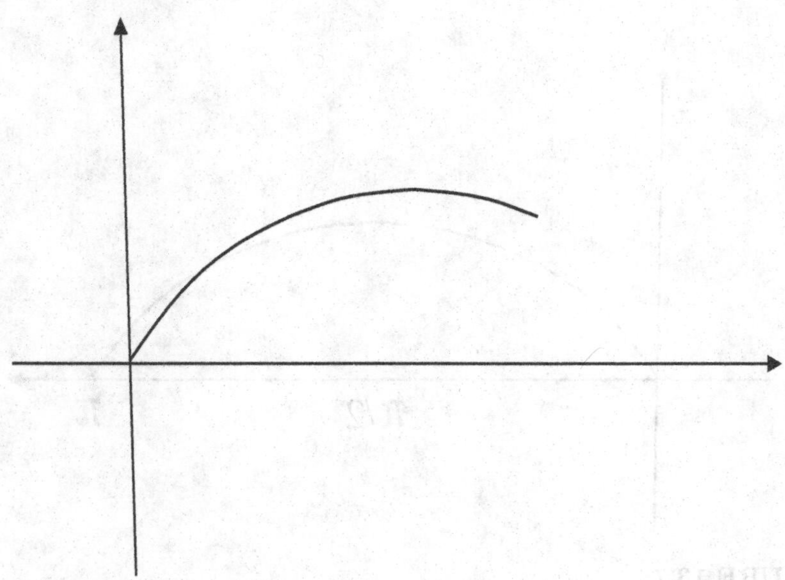

FIGURE 13.5
The sine function starts to decrease.

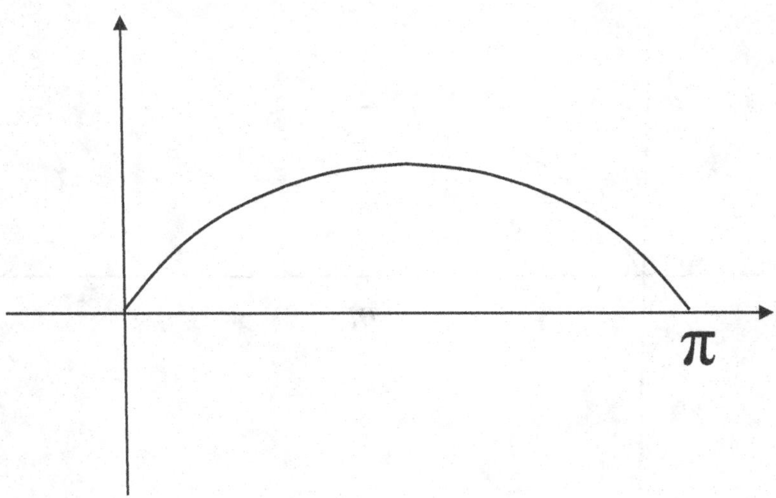

FIGURE 13.6
The sine function vanishes at π.

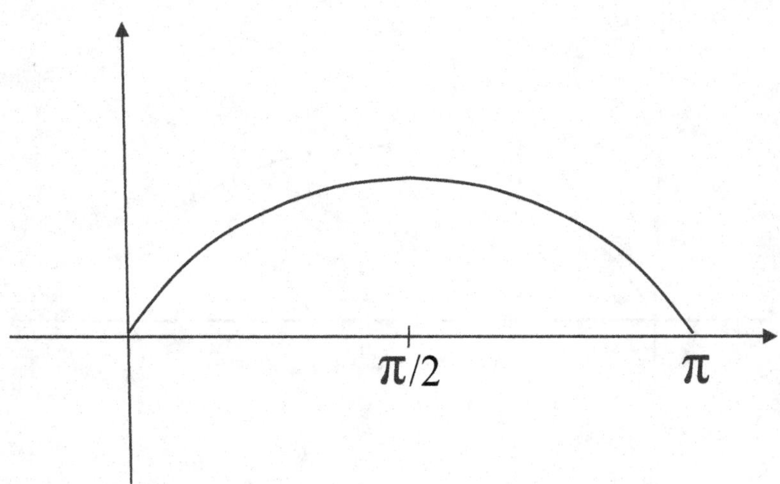

FIGURE 13.7
Analysis of one quarter of the sine curve.

We can do a similar analysis for the solution y_2 that satisfies $y_2(0) = 1$, $y_2'(0) = 0$. We find, again by analyzing the differential equation, that the curve falls off with increasingly negative slope. It eventually crosses the x-axis at some point n. Then it continues to decrease until it reaches a minimum at some new point p. What can we say about n and p, and can we relate them to the critical values that we found for the solution y_1?

We first notice that, differentiating the differential equation, we find that

$$y''' + y' = 0$$

or

$$(y')'' + (y') = 0\,.$$

Thus, if y is a solution of the equation, then so is y'. Observe that $y_1'(0) = 1$ and (by the differential equation) $(y_1')'(0) = y_1''(0) = -y_1(0) = 0$. Thus y_1' satisfies the same initial conditions as y_2. It follows from Theorem 13.1.2 that $y_1' = y_2$. Similar reasoning shows that $y_2' = -y_1$. (Of course you should be thinking all along here that y_1 is just the sine function and y_2 is just the cosine function. But we are trying to discover the properties of y_1 and y_2 without that knowledge.)

Next observe that

$$y_1 \cdot y_2' + y_2 \cdot y_1' = 0$$

hence

$$[(y_1)^2 + (y_2)^2]' = 0\,.$$

It follows that the function $(y_1)^2 + (y_2)^2$ is constant. Since $(y_1(0))^2 + (y_2(0))^2 = 1$, we find that

$$y_1^2 + y_2^2 \equiv 1\,. \tag{13.13}$$

Now y_1 reaches its maximum at a point where $y_1' = 0$. Hence $y_2 = 0$. As a result, by (13), $y_1 = 1$ at that point. Of course that maximum occurs at $x = \pi/2$. Hence $y_1(\pi/2) = 1$, $y_2(\pi/2) = 0$.

Certainly the Wronskian of y_1 and y_2 is

$$W(x) = \det \begin{pmatrix} y_1 & y_2 \\ y_2 & -y_1 \end{pmatrix} = -y_1^2 - y_2^2 = -1\,.$$

Hence y_1 and y_2 are linearly independent and $y = Ay_1 + By_2$ is the general solution of the differential equation.

We may also use (see Exercise 1) elementary differential equation techniques to prove that

(1) $y_1(a+b) = y_1(a)y_2(b) + y_2(a)y_1(b)$
(2) $y_2(a+b) = y_2(a)y_2(b) - y_1(a)y_1(b)$
(3) $y_1(2a) = 2y_1(a)y_2(a)$
(4) $y_2(2a) = y_2^2(a) - y_1^2(a)$
(5) $y_1(a+2\pi) = y_1(a)$
(6) $y_2(a+2\pi) = y_2(a)$

In particular, one derives easily from **(6)** and the results we obtained earlier about zeros of y_1 and y_2 that the zeros of y_1 are at $0, \pm\pi, \pm 2\pi, \ldots$ and the zeros of y_2 are at $\pm\pi/2$, $\pm 3\pi/2$, $\pm 5\pi/2, \ldots$.

We now enunciate three results that generalize this analysis of sine and cosine (i.e., the solutions of $y'' + y = 0$). These celebrated results of Jacques Sturm (1803–1855) are part of the bedrock of the subject of ordinary differential equations.

Theorem 13.3.1 *Let y_1, y_2 be two linearly independent solutions of*

$$y'' + p \cdot y' + q \cdot y = 0\,.$$

Then the zeros of y_1 and y_2 are distinct and occur alternately—in the sense that y_1 vanishes precisely once between any two successive zeros of y_2 and vice versa.

Proof We know that the Wronskian of y_1, y_2 never vanishes. Since it is real-valued, it must have constant sign.

First we observe that y_1, y_2 cannot have a common zero. If they did, the Wronskian would vanish at that point; and that is impossible. Now let a, b be successive zeros of y_2. We shall prove that y_1 vanishes between those two points. Notice that the Wronskian at a and b reduces to $y_1(x) \cdot y_2'(x)$. Thus both y_1 and y_2' are nonvanishing at a and at b. Also $y_2'(a)$ and $y_2'(b)$ must have opposite signs, because if y_2 is increasing at a then it must be decreasing at b, and vice versa (Figure 13.8).

Since the Wronskian (which is $y_1(x) \cdot y_2'(x)$) has constant sign, we may now conclude that $y_1(a)$ and $y_1(b)$ have opposite signs. By continuity, we conclude that y_1 vanishes at some point between a and b. Observe that y_1 cannot vanish twice between a and b; if it did, then we repeat the above argument with the roles of y_1 and y_2 reversed to conclude that y_2 has another zero between a and b. That is, of course, impossible. □

We saw in our analysis of the equation $y'' + y = 0$ that it is useful to have an equation in which the y' term is missing. In fact this can always be arranged.

Suppose that we are given an ordinary differential equation of the form

$$y'' + p \cdot y' + q \cdot y = 0\,. \tag{13.14}$$

We shall implement the change of notation $y = u \cdot v$. Thus $y' = u \cdot v' + u' \cdot v$ and $y'' = u \cdot v'' + 2 \cdot u' \cdot v' + u'' \cdot v$. Substituting these expressions into (14) yields

$$v \cdot u'' + (2v' + pv) \cdot u' + (v'' + pv' + qv) \cdot u = 0\,. \tag{13.15}$$

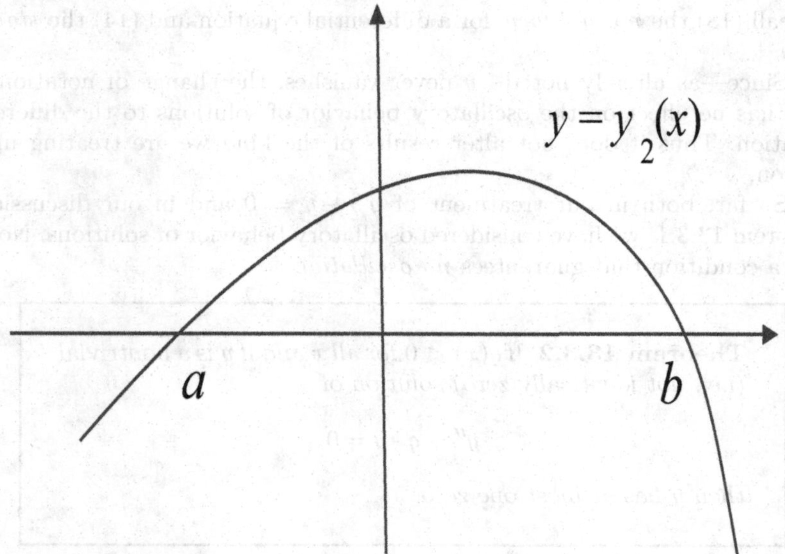

FIGURE 13.8
The Sturm theorem.

We set the coefficient of u' equal to zero, thus

$$2v' + pv = 0 .$$

This is of course a first-order linear equation, which we may easily solve. The result is

$$v = e^{-(1/2)\int p(x)\,dx} . \tag{13.16}$$

We take this as our choice of v. There is then no problem to choose u, because v (being an exponential) never vanishes, so we may take $u = y/v$. And of course our differential equation reduces to

$$v \cdot u'' + (v'' + pv' + qv) \cdot u = 0 . \tag{13.17}$$

From (16), we may calculate that

$$v' = -\frac{1}{2} \cdot p \cdot e^{-(1/2)\int p(x)\,dx}$$

and

$$v'' = -\frac{1}{2} \cdot p' \cdot e^{-(1/2)\int p(x)\,dx} + \frac{1}{4} \cdot p^2 \cdot e^{-(1/2)\int p(x)\,dx} .$$

Substituting these quantities into (17) (and dividing out by the resulting common factor of $e^{-(1/2)\int p(x)\,dx}$) gives

$$u'' + \left(q - \frac{1}{4}p^2 - \frac{1}{2}p'\right) u = 0 . \tag{13.18}$$

We call (18) the *normal form* for a differential equation and (14) the *standard form.*

Since—as already noted—v never vanishes, the change of notation $y = u \cdot v$ has no effect on the oscillatory behavior of solutions to the differential equation. Thus it does not alter results of the kind we are treating in this section.

So far, both in our treatment of $y'' + y = 0$ and in our discussion of Theorem 13.3.1, we have considered oscillatory behavior of solutions. Now we give a condition that guarantees *no oscillation.*

Theorem 13.3.2 *If $q(x) < 0$ for all x and if y is a nontrivial (i.e., not identically zero) solution of*

$$y'' + q \cdot y = 0$$

then y has at most one zero.

Proof Let x_0 be a zero of the function y, so $y(x_0) = 0$. By Theorem 13.1.2, it cannot be that $y'(x_0) = 0$, for that would imply that $y \equiv 0$. So say that $y'(x_0) > 0$ (the case $y'(x_0) < 0$ is handled similarly). Then, to the right of x_0, y is positive. But then

$$y'' = -q \cdot y > 0 \,.$$

This means that y' is increasing immediately to the right of x_0. But we can continue to apply this reasoning, and extend arbitrarily far to the right the interval on which this assertion is true. In particular, y can never vanish to the right of x_0.

Similar reasoning shows that y is negative immediately to the left of x_0, and falls off as we move left. So y does not vanish to the left of x_0.

In conclusion: Either y does not vanish at all or else, if it vanishes at one point, it can vanish nowhere else. □

Since our primary interest in this section is in oscillation of solutions, the last theorem leads us to concentrate our attention on equations $y'' + q \cdot y$ with $q \geq 0$. However $q \geq 0$ is not sufficient to guarantee oscillation. To get an idea of what is going on, consider a solution y of $y'' + q \cdot y = 0$ with $q > 0$. For specificity, consider a portion of the graph of y where $y > 0$. Then $y'' = -q \cdot y$ is of course negative, hence the graph is concave down and the slope y' is decreasing as x moves from left to right. If this slope ever becomes negative, then y will continue to decrease at a steady rate until the graph crosses the x-axis—resulting in a zero (Figure 13.9). Of course this is exactly what happens when q is a constant, say $q \equiv 1$, as in the last section.

The alternative possibility—which actually *can* happen—is that y' decreases but never reaches zero (Figure 13.10). Then y' remains positive and the curve continues to rise, as the figure suggests. The aggregate of these

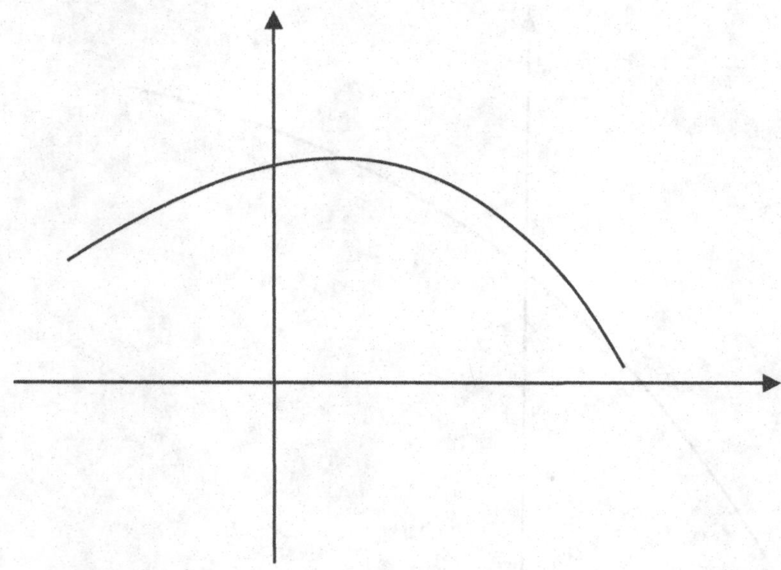

FIGURE 13.9
The case of nonoscillation.

remarks suggests that y will have zeros as x increases just as long as q does not decrease too rapidly. That is the content of the next theorem.

Theorem 13.3.3 *Let y be any nontrivial (i.e., not identically 0) solution of $y'' + q \cdot y = 0$. Assume that $q(x) > 0$ for all $x > 0$. If*

$$\int_1^\infty q(x)\, dx = +\infty \tag{13.19}$$

then y has infinitely many zeros on the positive x-axis.

Proof Suppose instead that y vanishes only finitely many times on $(0, \infty)$. Thus there is a point $\alpha > 1$ such that $y(x) \neq 0$ for all $x \geq \alpha$. Since y is of one sign on $[\alpha, \infty)$, we may as well suppose (after multiplying by -1 if necessary) that $y > 0$ on $[\alpha, \infty)$. We shall obtain a contradiction by showing that in fact $y' < 0$ at some point to the right of α. By the negativity of y'', this would imply that y has a zero to the right of α—and that is false.

Put

$$v(x) = -\frac{y'(x)}{y(x)}$$

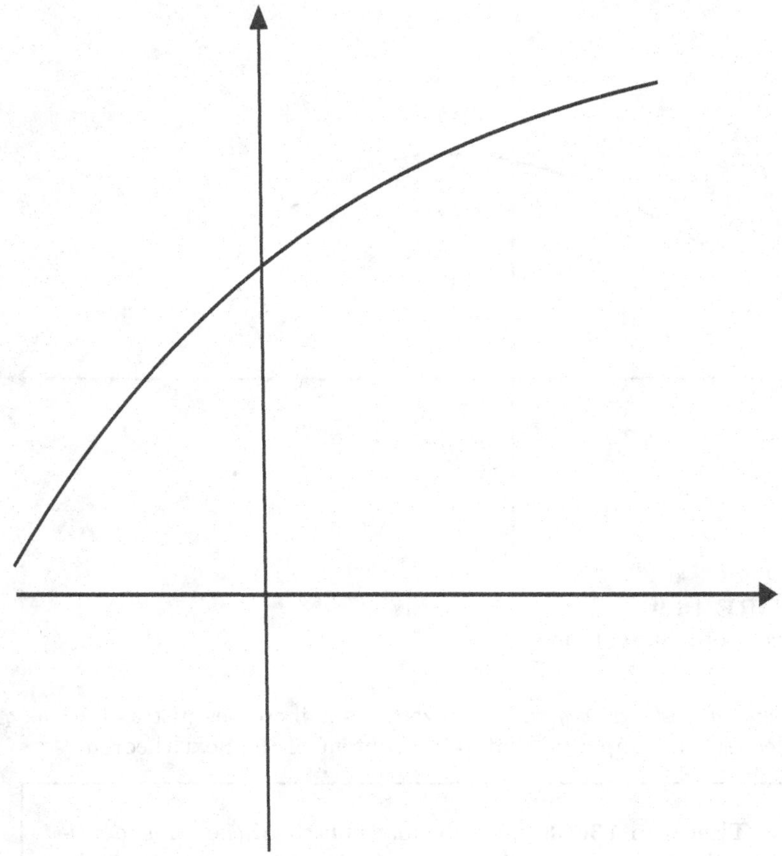

FIGURE 13.10
More on nonoscillation.

for $x \geq \alpha$. Then a simple calculation shows that

$$v'(x) = -\frac{y''}{y} + \frac{(y')^2}{y^2} = q + v^2 .$$

We integrate this last equation from α to x, for $x > \alpha$. The result is

$$v(x) - v(\alpha) = \int_\alpha^x q(x)\,dx + \int_\alpha^x v(x)^2\,dx .$$

Now our hypothesis (19) shows that the first integral on the right tends to $+\infty$ as $x \to +\infty$. The second integral is of course positive. The result is that $v(x)$ is positive if x is large enough. Thus (by the definition of v) y and y' have opposite signs if x is sufficiently large and positive. In conclusion, y' is negative. That is the desired contradiction. □

EXAMPLE 13.3.4 Consider the equation

$$y'' + y = 0 .$$

Then $q \equiv 1$ so the hypothesis of the Theorem 13.3.3 is certainly satisfied. The solutions of the equation are $y_1(x) = \cos x$ and $y_2(x) = \sin x$. Both of these have infinitely many zeroes on the right halfline.

Math Nugget

Jacques Charles François Sturm (1803–1855) was a Swiss mathematician who spent most of his life in Paris. For a time he was tutor to the wealthy and distinguished Broglie family. After holding several other positions, he at last succeeded Poisson in the Chair of Mechanics at the Sorbonne. His main work concentrated in the area that is now known as Sturm–Liouville theory. This part of the subject of differential equations has assumed an ever-more prominent role in both pure mathematics and mathematical physics. It is widely used today.

Exercises

1. Use the techniques presented in this section to prove formulas **(1)**–**(6)**.
2. Show that the zeros of the functions $a \sin x + b \cos x$ and $c \sin x + d \cos x$ are distinct and occur alternately provided that $ad - bc \neq 0$.
3. Find the normal form of Bessel's equation

$$x^2 y'' + xy' + (x^2 - p^2)y = 0 .$$

   Use it to show that every nontrivial solution has infinitely many positive zeros.
4. The hypothesis of Theorem 13.3.2 is false for Euler's equation

$$y'' + (k/x^2)y = 0$$

   if $k > 0$. However, the conclusion is sometimes true and sometimes false, depending on the size of the positive constant k. Show that every nontrivial solution has an infinite number of positive zeros if $k > 1/4$ and only finitely many positive zeros if $k \leq 1/4$.

5. Use your symbol manipulation software, such as **Maple** or **Mathematica**, to confirm the conclusions of the Sturm separation theorem for the differential equations

(a) $y'' + [\ln x]y' + [\sin x]y = 0$
(b) $y'' - e^x y' + x^3 y = 0$
(c) $y'' + [\cos x]y' - [\sin x]y = 0$
(d) $y'' - \dfrac{x}{x^4+1}y' - \dfrac{x}{x^2+1}y = 0$

13.4 The Sturm Comparison Theorem

In this section, we continue to develop the ideas of Jacques Sturm. Our first result rules out the possibility of infinitely many oscillations on a closed interval.

Theorem 13.4.1 *Let y be a nontrivial (i.e., not identically zero) solution of the equation $y'' + q \cdot y = 0$ on the closed interval $[a, b]$. Assume that q is a positive function. Then y has at most a finite number of zeros in the interval $[a, b]$.*

Remark: This result should be compared and contrasted with Theorem 13.3.3 in the last section. That theorem had hypotheses about q on the entire half-line $(0, \infty)$. In the present theorem we are studying behavior on a closed, bounded interval. If you think of our differential equation as describing, for example, a heat flow, then you might expect different physical characteristics in these two distinct situations.

Proof Seeking a contradiction, we assume that there are infinitely many zeros of the function y in the interval $[a, b]$. Since the interval is closed and bounded, there will be a sequence x_j of the zeros which accumulate at some point $x_0 \in [a, b]$. Now the solution y is continuous and differentiable, so we know that

$$y(x_0) = \lim_{j\to\infty} y(x_j) = \lim_{j\to\infty} 0 = 0$$

and

$$y'(x_0) = \lim_{j\to\infty} \frac{y(x_j) - y(x_0)}{x_j - x_0} = \lim_{j\to\infty} \frac{0-0}{x_j - x_0} = 0 \,.$$

We conclude that $y(x_0) = 0$ and $y'(x_0) = 0$. By Theorem 13.1.2, the function y is forced to be the trivial solution $y \equiv 0$. That contradicts our hypotheses. □

Let us return now to the Sturm separation theorem. We know that the zeros of two linearly independent solutions of

$$y'' + q \cdot y = 0$$

alternate. But now consider these two differential equations:

$$y'' + y = 0 \qquad \text{and} \qquad y'' + 4y = 0 .$$

Independent solutions of the first equation are $\sin x$ and $\cos x$; independent solutions of the second equation are $\sin 2x$ and $\cos 2x$. Of course the latter oscillate more rapidly (twice as rapidly) as the former. These observations suggest the following theorem:

> **Theorem 13.4.2** *Suppose that q and $\widetilde{q}$ are positive functions with $q > \widetilde{q}$. Let y be a nontrivial solution of the differential equation*
>
> $$y'' + q \cdot y = 0$$
>
> *and let $\widetilde{y}$ be a nontrivial solution of the differential equation*
>
> $$\widetilde{y}'' + \widetilde{q} \cdot \widetilde{y} = 0 .$$
>
> *Then y vanishes at least once between any two successive zeros of $\widetilde{y}$.*

Proof Let a and b be successive zeros of $\widetilde{y}$, with no zeros in between them. Seeking a contradiction, we assume that y does not vanish on the open interval (a, b). We may assume that both y and $\widetilde{y}$ are positive on (a, b) (otherwise replace the function by its negative). Note in particular that this means that $\widetilde{y}'(a) \geq 0$ and $\widetilde{y}'(b) \leq 0$.

Then a Wronskian-like expression calculated with y and $\widetilde{y}$ is

$$W(x) = y(x) \cdot \widetilde{y}'(x) - \widetilde{y}(x) \cdot y'(x)$$

and

$$\begin{aligned} W' &= y\widetilde{y}'' - \widetilde{y}y'' \\ &= y(-\widetilde{q}\widetilde{y}) - \widetilde{y}(-qy) \\ &= (q - \widetilde{q})y\widetilde{y} > 0 \end{aligned}$$

since $q > \widetilde{q}$ by hypothesis. It follows by integration of this last equality that

$$W(b) - W(a) > 0 \quad \text{or} \quad W(b) > W(a) . \tag{13.20}$$

Observe that, because of the vanishing of $\widetilde{y}$, the Wronskian simplifies to $y \cdot \widetilde{y}'$ at a and b. Thus $W(a) \geq 0$ and $W(b) \leq 0$. These two facts contradict (20). □

A consequence of this last theorem is that if $q > k^2$, k a constant, then any solution of $y'' + qy = 0$ must vanish between any two zeros of $y = \sin k(x - \alpha)$ (which is a solution of $y'' + k^2 y = 0$). Thus any solution of $y'' + qy = 0$ vanishes on *any* interval of length π/k.

As an example of this last equation, consider *Bessel's equation*

$$x^2 y'' + xy' + (x^2 - p^2)y = 0 .$$

This equation is important in mathematical physics—especially in problems involving vibrations, as of a drum (see Section 5.1).

In normal form, Bessel's equation is

$$u'' + \left(1 + \frac{1-4p^2}{4x^2}\right) u = 0 .$$

Here $y = u/\sqrt{x}$. We now have the following basic result.

Theorem 13.4.3 *Let y_p be a nontrivial solution of Bessel's equation on the positive x-axis. If $0 \le p < 1/2$ then every interval of length π contains at least one zero of y_p. If $p = 1/2$, then the distance between successive zeros of y_p is exactly π. If $p > 1/2$ then every interval of length π contains at most one zero of y_p.*

Proof If $0 \le p < 1/2$ then

$$1 + \frac{1-4p^2}{4x^2} > 1$$

so we may apply the previous theorem to Bessel's equation and to $y'' + y = 0$. The first assertion follows.

If $p > 1/2$ then

$$1 + \frac{1-4p^2}{4x^2} < 1$$

and we may apply the previous theorem to $y'' + y = 0$ and to Bessel's equation. The last assertion follows.

If $p = 1/2$, then the normal form of Bessel's equation reduces to

$$u'' + u = 0 ,$$

and the solutions of this equation are $A \sin x + B \cos x$. This completes our analysis of Bessel's equation. □

Exercises

1. Let x_1, x_2 be successive positive zeros of a nontrivial solution y_p of Bessel's equation

$$x^2y'' + xy' + (x^2 - p^2)y = 0 .$$

   (a) If $0 \le p < 1/2$, then show that $x_2 - x_1$ is less than π and approaches π as $x_1 \to +\infty$.

   (b) If $p > 1/2$, then show that $x_2 - x_1$ is greater than π and approaches π as $x_1 \to +\infty$.

2. If y is a nontrivial solution of $y'' + q(x)y = 0$, then show that y has an infinite number of positive zeros if $q(x) > k/x^2$ for all $x > 0$ and for some $k > 1/4$. It also has just a finite number of positive zeros if $q(x) < 1/(4x^2)$.

3. Every nontrivial solution of $y''+(\sin^2 x+1)y = 0$ has an infinite number of positive zeros. Endeavor to formulate a general result that would include this as a special case.

4. Use your symbol manipulation software, such as `Maple` or `Mathematica`, to confirm the conclusions of the Sturm comparison theorem for each of the following pairs of equations. In each instance take $x > 0$.

(a) $y'' + (1 + x)y = 0$ and $y'' + (1 + x)^2y = 0$
(b) $y'' + e^x y = 0$ and $y'' + \ln xy = 0$
(c) $y'' + [\sin x + 2]^2 y = 0$ and $y'' + [5 - \cos x]y = 0$
(d) $y'' + (x^2 + 1)y = 0$ and $y'' + (x^3 + 4)y = 0$

Anatomy of an Application

The Green's Function

Given a boundary value problem for a second-order linear equation on an interval $[a, b]$—say

$$y''(x) + \alpha(x)y'(x) + \beta(x)y(x) = g(x), \quad y(a) = 0\,, \, y(b) = 0\,, \qquad (13.21)$$

it is useful to have a formula that expresses the solution y in terms of the input function g. First, such a formula would eliminate the need to solve the equation repeatedly. One could just plug a new data function directly into the formula. Second, such a formula would allow one to study qualitative properties of the solution, and how those properties depend on the data.

What would such a formula look like? Since the differential equation involves *derivatives*, one might expect that a solution formula would involve *integrals*. And that is indeed the case. In the present Anatomy we produce an important solution formula involving the *Green's function*.

First, it is convenient to write equation (21) in different notation. Namely, we want to consider a boundary value problem having this form:

$$[p(x)y'(x)]' + q(x)y(x) = f(x), \quad y(a) = 0, \, y(b) = 0\,. \qquad (13.22)$$

In fact problem (22) is neither more nor less general than problem (21). For obviously problem (22) is of the form (21). For the converse, given an equation of the form (21), set $p(x) = \exp[\int \alpha(x)\,dx]$, $q(x) = p(x) \cdot \beta(x)$, and $f(x) = p(x) \cdot g(x)$.

Now we may attack (22) by first finding independent solutions u_1, u_2 of the homogeneous equation

$$[p(x)y'(x)]' + q(x)y(x) = 0\,. \tag{13.23}$$

Then we may apply the method of variation of parameters to see that the general solution of (22) is

$$y(x) = Au_1(x) + Bu_2(x) + \int_{s=a}^{x} \frac{f(s)}{W(s)} \cdot [u_1(s)u_2(x) - u_1(x)u_2(s)]\, ds\,. \tag{13.24}$$

Here A and B are arbitrary constants and W is the Wronskian of u_1 and u_2. The two boundary conditions entail

$$\begin{aligned} Au_1(a) + Bu_2(a) &= 0 \\ Au_1(b) + Bu_2(b) &= \int_a^b \frac{f(s)}{W(s)}[u_1(b)u_2(s) - u_1(s)u_2(b)]\, ds\,. \end{aligned}$$

Provided that $u_1(a)u_2(b) \neq u_1(b)u_2(a)$, we can solve these simultaneous equations and put the results back into (24) to obtain

$$\begin{aligned} y(x) &= \int_a^b \frac{f(s)}{W(s)} \frac{[u_1(b)u_2(s) - u_1(s)u_2(b)] \cdot [u_1(a)u_2(x) - u_1(x)u_2(a)]}{u_1(a)u_2(b) - u_1(b)u_2(a)}\, ds \\ &\quad + \int_a^x \frac{f(s)}{W(s)}[u_1(s)u_2(x) - u_1(x)u_2(s)]\, ds\,. \end{aligned}$$

Things are a bit clearer if we define $w_1(x) = u_1(a)u_2(x) - u_1(x)u_2(a)$ and $w_2(x) = u_1(b)u_2(x) - u_1(x)u_2(b)$. Both u_1 and u_2 are solutions of the homogeneous equation (23) and $w_1(a) = w_2(b) = 0$. Of course u_1, u_2, w_1, w_2 do *not* depend on the data function f. We may write our solution $y(x)$ more compactly as

$$y(x) = \int_a^b \frac{f(s)}{W(s)} \frac{w_2(s) \cdot w_1(x)}{\sqrt{-w_1(b) \cdot w_2(a)}}\, ds + \int_a^x \frac{f(s)}{W(s)}[u_1(s)u_2(x) - u_1(x)u_2(s)]\, ds\,. \tag{13.25}$$

Thus we have expressed the solution y of the original boundary value problem in the form

$$y(x) = \int_a^b f(s)G(x,s)\, ds\,,$$

where the "kernel" $G(x,s)$ does not depend on f. Let us analyze this formula.

Observe that the role of x and s in the kernel in formula (25) for y is nearly symmetric. Also the roles of x and s are opposite on the intervals $[a, x]$ and $[x, b]$. This leads us to guess that the Green's function $G(x, s)$ should have the form

$$G(x,s) = \begin{cases} w_1(s)w_2(x) & \text{if} \quad a \leq s < x \\ w_1(x)w_2(s) & \text{if} \quad x < s \leq b\,. \end{cases}$$

It will be convenient to write $G^-(x,s) = w_1(s)w_2(x)$ and $G^+(x,s) = w_1(x)w_2(s)$. Thus G is just G^- and G^+ patched together. Notice that they agree at $s = x$. Also notice that, if we now define

$$y(x) = \int_a^b f(s)G(x,s)\,ds$$

(hoping that this will be the solution of our boundary value problem), then we are encouraged because $y(a) = 0$ (because $G(a,s) = 0$). Likewise $y(b) = 0$.

Calculating the partial derivative of G with respect to x, we find that

$$\begin{aligned}
\lim_{s\to x^-} \frac{\partial}{\partial x}G(x,s) - \lim_{s\to x^+} \frac{\partial}{\partial x}G(x,s) &= w_1(x)w_2'(x) - w_1'(x)w_2(x) \\
&= W(x) \\
&= \frac{C}{p(x)}\,.
\end{aligned}$$

This is a standard formula for the Wronskian.

Now let us write

$$y(x) = \int_a^x G^-(x,s)f(s)\,ds + \int_x^b G^+(x,s)f(s)\,ds\,.$$

Then, by the fundamental theorem of calculus,

$$\begin{aligned}
y'(x) &= \int_a^x \frac{\partial}{\partial x}G^-(x,s)f(s)\,ds + G^-(x,x)f(x) \\
&\quad + \int_x^b \frac{\partial}{\partial x}G^+(x,s)f(s)\,ds - G^+(x,x)f(x)\,.
\end{aligned}$$

This simplifies, by the continuity of the Green's function, to

$$\begin{aligned}
y'(x) &= \int_a^x \frac{\partial}{\partial x}G^-(x,s)f(s)\,ds + \int_x^b \frac{\partial}{\partial x}G^+(x,s)f(s)\,ds \\
&= \int_a^x w_1(s)\frac{\partial}{\partial x}w_2(x)f(s)\,ds + \int_x^b \frac{\partial}{\partial x}w_1(x)w_2(s)f(s)\,ds\,.
\end{aligned}$$

Therefore

$$\begin{aligned}
(p(x)y'(x))' &= \int_a^x w_1(s)\frac{\partial}{\partial x}\left(p(x)\frac{\partial}{\partial x}w_2(x)\right)f(s)\,ds \\
&\quad + p(x)w_1(x)\frac{\partial}{\partial x}w_2(x)f(x) \\
&\quad + \int_x^b \frac{\partial}{\partial x}\left(p(x)\frac{\partial}{\partial x}w_1(x)\right)w_2(s)f(s)\,ds
\end{aligned}$$

$$\begin{aligned}&-p(x)\frac{\partial}{\partial x}w_1(x)w_2(x)f(x)\\ &= \int_a^x w_1(s)\frac{\partial}{\partial x}\left(p(x)\frac{\partial}{\partial x}w_2(x)\right)f(s)\,ds\\ &\quad+\int_x^b \frac{\partial}{\partial x}\left(p(x)\frac{\partial}{\partial x}w_1(x)\right)w_2(s)f(s)\,ds + Cf(x)\,.\end{aligned}$$

Now we substitute this last equality into our differential equation (22) and obtain

$$\begin{aligned}&\int_a^x w_1(s)\frac{\partial}{\partial x}\left(p(x)\frac{\partial}{\partial x}w_2(x)\right)f(s)\,ds\\ &\quad+\int_x^b \frac{\partial}{\partial x}\left(p(x)\frac{\partial}{\partial x}w_1(x)\right)w_2(s)f(s)\,ds + Cf(x)\\ &\quad+q(x)\left\{\int_a^x w_1(s)w_2(x)f(s)\,ds + \int_x^b w_1(x)w_2(s)f(s)\,ds\right\}\\ &= f(x)\,.\end{aligned}$$

Since w_1 and w_2 are solutions of the homogeneous equations, all the integrals in this last equation vanish. If we choose $C = 1$, then we have a valid identity and our Green's integral solves the given boundary value problem.

EXAMPLE Let us consider the familiar boundary value problem

$$y'' - y = f(x), \quad y(a) = 0,\, y(b) = 0\,. \tag{13.26}$$

Solutions of the homogeneous equation are $u_1(x) = e^x$ and $u_2(x) = e^{-x}$. Linear combinations of these that satisfy the boundary conditions are $w_1(x) = c_1 \sinh x$ and $w_2(x) = c_2 \sinh(1-x)$. Following the paradigm we have set up for constructing the Green's function, we now see that

$$G(x,s) = \begin{cases} c_1c_2\sinh(1-x)\sinh s & \text{if} \quad 0 \le s < x \\ c_1c_2\sinh(1-s)\sinh x & \text{if} \quad x < s \le 1\,. \end{cases}$$

Next we calculate the constant C:

$$\begin{aligned}&\lim_{s\to x^-}\frac{\partial}{\partial x}G(x,s) - \lim_{s\to x^+}\frac{\partial}{\partial x}G(x,s)\\ &= -c_1c_2\cosh(1-x)\sinh x - c_1c_2\sinh(1-x)\cosh x = -c_1c_2\sinh 1\,.\end{aligned}$$

Since $p(x) \equiv 1$ for this particular differential equation, we see that $c_1c_2 = -1/\sinh 1$. Thus the Green's function turns out to be

$$G(x,s) = \begin{cases} -\dfrac{\sinh(1-x)\sinh(s)}{\sinh 1} & \text{if} \quad 0 \le s < x \\ -\dfrac{\sinh(1-s)\sinh(x)}{\sinh 1} & \text{if} \quad x < s \le 1\,. \end{cases}$$

In summary, the solution of the nonhomogeneous boundary value problem (26) is

$$\begin{aligned} y(x) &= \int_0^1 f(s)G(x,s)\,ds \\ &= -\int_0^x f(s)\frac{\sinh(1-x)\sinh(s)}{\sinh 1}\,ds \\ &\quad -\int_x^1 f(s)\frac{\sinh(1-s)\sinh(x)}{\sinh 1}\,ds. \end{aligned}$$

To take a specific example, if we want to solve the nonhomogeneous boundary value problem

$$y'' - y = 1, \quad y(a) = 0,\, y(b) = 0\,,$$

we could certainly proceed by using the methods of Chapter 4. But, instead (if, for example, we wanted to put the procedure on a computer), we could simply evaluate the expression

$$y(x) = -\int_0^x 1\cdot\frac{\sinh(1-x)\sinh(s)}{\sinh 1}\,ds - \int_x^1 1\cdot\frac{\sinh(1-s)\sinh(x)}{\sinh 1}\,ds\,.$$

We invite the reader to do so. The result will be

$$y(x) = \frac{1-e^{-1}}{e-e^{-1}}e^{x} + \frac{e-1}{e-e^{-1}}e^{-x} - 1\,. \quad \blacksquare$$

Problems for Review and Discovery

A. Drill Exercises

1. Show that $y_1(x) = e^x$ and $y_2(x) = e^{3x}$ are linearly independent solutions of the differential equation $y'' - 4y' + 3y = 0$. Write the general solution. Write the particular solution that satisfies $y(0) = 2, y'(0) = -2$.
2. Find two linearly independent solutions of the differential equation $y'' + xy' - y = 0$ (find the first solution by guessing—it will be a polynomial of low degree). Verify, using the Wronskian, that your two solutions are indeed linearly independent. Write the general solution of the differential equation. Write the particular solution that satisfies $y(1) = 1, y'(1) = -2$.
3. Verify that $y_1(x) = \sqrt{x}$ and $y_2(x) = 1/x$ are linearly independent solutions of the equation

   $$2x^2y'' + 3xy' - y = 0\,.$$

   Find the particular solution that satisfies $y(2) = 0, y'(2) = -1$.

4. Consider the differential equation

$$(Pu')' + Qu = 0\,. \qquad (*)$$

Define $\theta = \arctan(u/Pu')$. Show that, at each point x where a solution of $(*)$ has a maximum or minimum, it holds that $d\theta/dx = Q(x)$.

5. Relate the Sturm comparison theorem as stated in the text to this ostensibly stronger form of the theorem:

Theorem: Let $p(x) \geq p_1(x) > 0$ and $q_1(x) \geq q(x)$ on the domain of the pair of differential equations

$$\begin{aligned} \frac{d}{dx}\left(p(x)\frac{du}{dx}\right) + q(x)u &= 0\,, \\ \frac{d}{dx}\left(p_1(x)\frac{du_1}{dx}\right) + q_1(x)u_1 &= 0\,. \end{aligned}$$

Then, between any two zeros of a nontrivial solution $u(x)$ of the first differential equation there lies at least one zero of every solution of the second differential equation (except in the trivial case when $u(x) \equiv u_1(x)$, $p(x) \equiv p_1(x)$, and $q(x) \equiv q_1(x)$).

(*Hint:* Introduce the new dependent variables $t = \int_a^x dx/p(s)$ and $t_1 = \int_a^x ds/p_1(s)$.)

6. For any solution u of $u'' + q(x)u = 0$ with $q(x) < 0$, show that the product $u(x)u'(x)$ is an increasing function. Conclude that a nontrivial solution can have at most one zero.

B. Challenge Problems

1. The differential equation $y'' + (1 - x^2)y = 0$ does not exactly fit the hypotheses of the Sturm theorems. Can you make a change of variables to force it to fit? What conclusions can you draw?

2. The differential equation $y'' + (x^2 - 1)y = 0$ does not exactly fit the hypotheses of the Sturm theorems. Can you make a change of variables to force it to fit? What conclusions can you draw?

3. Give an example of two functions y_1 and y_2 on the interval $[-1, 1]$ such that

   - The restrictions of y_1 and y_2 to the interval $[-1, 0]$ are linearly dependent.
   - The restrictions of y_1 and y_2 to the interval $[0, 1]$ are linearly independent.

   What can you say about the linear independence or dependence of y_1 and y_2 on the entire interval $[-1, 1]$?

4. What does the Sturm comparison theorem tell you about the solutions of the differential equation $y'' + [1 + e^x]y = 0$?

C. Problems for Discussion and Exploration

1. Formulate a notion of what it means for n functions $y_1, y_2, \ldots, y_n$ to be linearly independent. Check that the functions $\cos x, \sin x, e^x$ are linearly independent. Check that the functions $x, x^2 - x, 3x^2 + x$ are *not* linearly independent.

2. Prove that a collection of vectors $\mathbf{v}_1, \ldots, \mathbf{v}_k$ in $\mathbb{R}^N$ is linearly dependent if $k > N$. Prove that the collection is linearly independent if and only if each vector $\mathbf{w}$ in $\mathbb{R}^N$ can be written as a *linear combination*

$$\mathbf{w} = a_1\mathbf{v}_1 + a_2\mathbf{v}_2 + \cdots + a_k\mathbf{v}_k$$

for real constants $a_1, \ldots, a_k$ not all zero.

3. Let $\mathbf{P}_k$ be the collection of polynomials $p(x)$ of degree not exceeding k. Then any such p may be written as a *linear combination* of the "unit polynomials" $1, x, x^2, \ldots, x^k$:

$$p(x) = a_0 \cdot 1 + a_1 x + a_2 x^2 + \cdots + a_k x^k$$

for real constants $a_0, \ldots, a_k$ not all zero. Find another collection of "unit polynomials" with the same property. (*Hint:* This collection should also have $k + 1$ elements.)

4. Check that the functions e^x, e^{-2x}, e^{3x} are linearly independent solutions of the differential equation

$$y''' - 2y'' - 5y' + 6y = 0\,.$$

Write the general solution of this ordinary differential equation. Use the method of undetermined coefficients to find the general solution to

$$y''' - 2y'' - 5y' + 6y = x^2 - 2x\,.$$

Now find the particular solution that satisfies $y(0) = 1, y'(0) = 0, y''(0) = -1$.

5. The differential equation $y'' - [6/x^2]y = 0$ has at least one polynomial solution. Apply one of the Sturm theorems to say something about the roots of that solution.

APPENDIX: Review of Linear Algebra

One of the problems with learning linear algebra as a student is that the subject is taught out of context. It is not easy to show meaningful applications of linear algebra *inside* a linear algebra course. In fact linear algebra is the natural language in which to express many of the important ideas of differential equations. So it is appropriate here to bring the subject back to life for us so that we can use it. In the present section, we give a very quick treatment of some of the most basic ideas. Our discussion will not be complete, and we shall provide few proofs.

A1. Vector Spaces

A *vector space* is a set V together with two operations, *addition* and *scalar multiplication.* The space V is of course assumed to be closed under these operations.

Addition is hypothesized to be both commutative and associative. We of course write the addition of two elements $\mathbf{v}, \mathbf{w} \in V$ as $\mathbf{v} + \mathbf{w}$ and the closure property means that $\mathbf{v} + \mathbf{w} \in V$. We suppose that there is an additive identity $\mathbf{0} \in V$, so that $\mathbf{v} + \mathbf{0} = \mathbf{0} + \mathbf{v} = \mathbf{v}$ for each element $\mathbf{v} \in V$. Last, if $\mathbf{v} \in V$, then there is an additive inverse $-\mathbf{v} \in V$ so that $\mathbf{v} + (-\mathbf{v}) = \mathbf{0}$.

If $c \in \mathbb{R}$ and $\mathbf{v} \in V$ then we write $c\mathbf{v}$ or sometimes $c \cdot \mathbf{v}$ to denote the scalar multiplication of $\mathbf{v}$ by the scalar c. The closure property means that $c\mathbf{v} \in V$. We hypothesize that

$$c(\mathbf{v} + \mathbf{w}) = c\mathbf{v} + c\mathbf{w}$$

and, for scalars $c, c' \in \mathbb{R}$,

$$c(c'\mathbf{v}) = (cc')\mathbf{v}$$

and

$$(c + c')\mathbf{v} = c\mathbf{v} + c'\mathbf{v} .$$

We assume that the scalar 1 acts according to the law

$$1\mathbf{v} = \mathbf{v} .$$

Example A1.1 Let $V = \{(x, y, z) : x \in \mathbb{R}, y \in \mathbb{R}, z \in \mathbb{R}\}$. Let addition be defined by

$$(x, y, z) + (x', y', z') = (x + x', y + y', z + z')$$

and scalar multiplication by

$$c(x, y, z) = (cx, cy, cz) .$$

Of course the additive identity is $\mathbf{0} = (0, 0, 0)$. Then it is easy to verify that V is a vector space. We usually denote this vector space by $\mathbb{R}^3$ and call it *Euclidean 3-space.* ∎

Example A1.2 Let W be the collection of all continuous functions with domain the interval $[0, 1]$. Let the addition operation be ordinary addition of functions. Let scalar multiplication be the usual notion of the multiplication of a function by a scalar or number. Then W is a vector space. We call this *the space of continuous functions on the interval* $[0, 1]$. ∎

A2. The Concept of Linear Independence

Let V be a vector space and $\mathbf{v}_1, \mathbf{v}_2, \ldots, \mathbf{v}_k$ elements of V. We say that this collection of vectors is *linearly dependent* if there are constants $c_1, c_2, \ldots, c_k$, *not all zero* such that

$$c_1\mathbf{v}_1 + c_2\mathbf{v}_2, \ldots, c_k\mathbf{v}_k = \mathbf{0} .$$

Of course we mandate that not all the c_js be zero, otherwise the idea would be trivial.

Example A2.1 Consider the vector space from Example A1.1. Let

$$\mathbf{v}_1 = (0, 1, 1), \ \mathbf{v}_2 = (5, 3, 4), \ \mathbf{v}_3 = (10, 0, 2) .$$

Then it is easy to check that

$$-3\mathbf{v}_1 + \mathbf{v}_2 - \frac{1}{2}\mathbf{v}_3 = \mathbf{0} .$$

Thus $\mathbf{v}_1, \mathbf{v}_2$, and $\mathbf{v}_3$ are linearly dependent. ∎

A collection of vectors $\mathbf{v}_1, \mathbf{v}_2, \ldots, \mathbf{v}_k$ is said to be *linearly independent* if it is *not* linearly dependent. Put in different words, if

$$c_1\mathbf{v}_1 + c_2\mathbf{v}_2 + \cdots + c_k\mathbf{v}_k = 0$$

implies that $c_1 = c_2 = \cdots = c_k = 0$ then the vectors $\mathbf{v}_1, \mathbf{v}_2, \ldots, \mathbf{v}_k$ are linearly independent.

Example A2.2 Consider the vector space from Example A1.1. Let

$$\mathbf{v}_1 = (1, 2, 1),\ \mathbf{v}_2 = (-3, 0, 4),\ \mathbf{v}_3 = (6, -2, 2)\,.$$

If there are constants c_1, c_2, c_3 such that

$$c_1\mathbf{v}_1 + c_2\mathbf{v}_2 + c_3\mathbf{v}_3 = \mathbf{0}$$

then we have

$$(c_1 - 3c_2 + 6c_3, 2c_1 - 2c_3, c_1 + 4c_2 + 2c_3) = (0, 0, 0)$$

hence

$$\begin{array}{rcrcrcl} c_1 & - & 3c_2 & + & 6c_3 & = & 0 \\ 2c_1 & & & - & 2c_3 & = & 0 \\ c_1 & + & 4c_2 & + & 2c_3 & = & 0\,. \end{array} \qquad (*)$$

But the determinant of the coefficients of the system $(*)$ is $74 \neq 0$. Hence the only solution to $(*)$ is the trivial one $c_1 = c_2 = c_3 = 0$. We conclude that $\mathbf{v}_1, \mathbf{v}_2, \mathbf{v}_3$ are linearly independent. ∎

The last example illustrates a general principle that is of great utility:

Consider the vector space $V = \mathbb{R}^n$, equipped with the familiar notion of vector addition and scalar multiplication as in Example A1.1. Let

$$\begin{array}{rcl} \mathbf{v}_1 & = & (v_1^1, v_2^1, \ldots, v_n^1) \\ \mathbf{v}_2 & = & (v_1^2, v_2^2, \ldots, v_n^2) \\ & \cdots & \\ \mathbf{v}_n & = & (v_1^n, v_2^n, \ldots, v_n^n) \end{array}$$

be elements of V. Here v_j^i denotes the jth entry of the ith vector. Then $\mathbf{v}_1, \mathbf{v}_2, \ldots, \mathbf{v}_n$ are linearly independent if and only if

$$\det \begin{pmatrix} v_1^1 & v_2^1 & \cdots & v_n^1 \\ v_1^2 & v_2^2 & \cdots & v_n^2 \\ & \cdots & & \\ v_1^n & v_2^n & \cdots & v_n^n \end{pmatrix} \neq 0\,.$$

Example A2.3 Let V be the vector space of all continuous functions on the interval $[0,1]$, as discussed in Example A1.2. Consider the vectors $\mathbf{v}_1 = \cos x, \mathbf{v}_2 = \sin x, \mathbf{v}_3 = x^2$. It is intuitively obvious that there does not exist a set of nonzero constants such that

$$c_1\mathbf{v}_1 + c_2\mathbf{v}_2 + c_3\mathbf{v}_3 = 0\,,$$

in other words, that $\mathbf{v}_1, \mathbf{v}_2, \mathbf{v}_3$ are linearly independent. But let us verify this assertion rigorously.

If indeed it is the case that

$$c_1 \cos x + c_2 \sin x + c_3 x^2 = 0 \tag{13.27}$$

then we may differentiate the equation to obtain

$$-c_1 \sin x + c_2 \cos x + 2c_3 x = 0 \tag{13.28}$$

and yet again to yield

$$-c_1 \cos x - c_2 \sin x + 2c_3 = 0\,. \tag{13.29}$$

Adding (1) and (3) yields

$$c_3 x^2 + 2c_3 = 0\,.$$

Since this is an identity in x, we must conclude that $c_3 = 0$.

Now we substitute this value into (1) and (2) to find that

$$\begin{aligned} c_1 \cos x + c_2 \sin x &= 0 \\ -c_1 \sin x + c_2 \cos x &= 0\,. \end{aligned}$$

Multiplying the first equation by c_2 and the second equation by c_1 and subtracting yields that

$$(c_1^2 + c_2^2) \sin x = 0\,.$$

Again, since this is an identity in x, we must conclude that $c_1^2 + c_2^2 = 0$ hence $c_1 = 0$ and $c_2 = 0$. Thus $\mathbf{v}_1, \mathbf{v}_2, \mathbf{v}_3$ are linearly independent. ∎

A3. Bases

Let V be a vector space. Let $\mathbf{v}_1, \mathbf{v}_2, \ldots, \mathbf{v}_k \in V$. If

(1) the vectors $\mathbf{v}_1, \mathbf{v}_2, \ldots, \mathbf{v}_k$ are linearly independent

(2) for any $\mathbf{w} \in V$ there are constants $a_1, a_2, \ldots a_k$ such that

$$\mathbf{w} = a_1\mathbf{v}_1 + a_2\mathbf{v}_2 + \cdots + a_k\mathbf{v}_k \qquad (\star)$$

then we say that $\mathbf{v}_1, \mathbf{v}_2, \ldots, \mathbf{v}_k$ form a *basis* for the vector space V.

We often describe condition **(2)** by saying that the vectors $\mathbf{v}_1, \mathbf{v}_2, \ldots, \mathbf{v}_k$ *span* the vector space V. We call the expression $(\star)$ "writing $\mathbf{w}$ as a linear combination of $\mathbf{v}_1, \mathbf{v}_2, \ldots, \mathbf{v}_k$." A basis for a vector space V is a minimal spanning set.

Example A3.1 Consider the vector space $V = \mathbb{R}^3$ as in Example A1.1. The vectors $\mathbf{v}_1 = (1, 0, 1), \mathbf{v}_2 = (0, 1, 1)$ do *not* form a basis of V. These vectors are certainly linearly independent, but they do not span all of V; for example, the vector $(1, 1, 1)$ cannot be written as a linear combination of $\mathbf{v}_1$ and $\mathbf{v}_2$.

On the other hand, the vectors $\mathbf{v}_1 = (1, 0, 1), \mathbf{v}_2 = (0, 1, 1), \mathbf{v}_3 = (1, 1, 0)$ *do* form a basis for V. We can see this because, if $\mathbf{w} = (w_1, w_2, w_3)$ is any element of V, then the equation

$$c_1\mathbf{v}_1 + c_2\mathbf{v}_2 + c_3\mathbf{v}_3 = \mathbf{w}$$

leads to the system

$$\begin{array}{ccccccc} c_1 & & + & & c_3 & = & w_1 \\ & & c_2 & + & c_3 & = & w_2 \\ c_1 & + & c_2 & & & = & w_3 \,. \end{array}$$

Since the determinant of the matrix of coefficients is not 0, the system can always be solved. So $\mathbf{v}_1, \mathbf{v}_2, \mathbf{v}_3$ form a basis for V. ∎

If V is a vector space, and if $\mathbf{v}_1, \mathbf{v}_2, \ldots, \mathbf{v}_k$ forms a basis for V, then any other basis for V will also have k elements. This special number k is called the *dimension* of V. We noted in the last example that $(1, 0, 1)$, $(0, 1, 1)$ and $(1, 1, 0)$ are a basis for $V = \mathbb{R}^3$. But $(1, 0, 0), (0, 1, 0)$, and $(0, 0, 1)$ also form a basis for V. In fact there are infinitely many different bases for this vector space. No two-vector set will be a basis, because it cannot span. No four-vector set will be a basis, because it cannot be linearly independent.

A4. Inner Product Spaces

Let V be a vector space. An *inner product* on V is a mapping

$$\langle \bullet, \bullet \rangle : V \times V \to \mathbb{R}$$

with these properties:

(a) $\langle \mathbf{v}, \mathbf{v} \rangle \geq 0$;

(b) $\langle \mathbf{v}, \mathbf{v} \rangle = 0$ if and only if $\mathbf{v} = 0$;

(c) $\langle \alpha\mathbf{v} + \beta\mathbf{w}, \mathbf{u} \rangle = \alpha\langle \mathbf{v}, \mathbf{u} \rangle + \beta\langle \mathbf{v}, \mathbf{u} \rangle$ for any vectors $\mathbf{u}, \mathbf{v}, \mathbf{w} \in V$ and scalars α, β.

Example A4.1 Let $V = \mathbb{R}^3$ be the vector space from Example A1.1. Define an inner product on V by

$$\langle (x, y, z), (x', y', z') \rangle = xx' + yy' + zz' .$$

Then it is easy to verify that properties **(a)**–**(c)** are satisfied. For instance, if $\mathbf{v} = (x, y, z)$, then

$$\langle \mathbf{v}, \mathbf{v} \rangle = 0 \quad \Leftrightarrow \quad x^2 + y^2 + z^2 = 0 \quad \Leftrightarrow \quad (x, y, z) = 0 .$$

That verifies property **(b)**. Properties **(a)** and **(c)** are equally straightforward. ∎

Example A4.2 Let V be the vector space of continuous functions on the interval $[0, 1]$, just as in Example A1.2. Define an inner product on V by

$$\langle f, g \rangle = \int_0^1 f(x) \cdot g(x) \, dx .$$

Then it is straightforward to verify properties **(a)**–**(c)**. For instance, if $f \in V$, then

$$\langle f, f \rangle = \int_0^1 f(x) \cdot f(x) \, dx = \int_0^1 [f(x)]^2 \, dx \geq 0 ,$$

hence property **(a)** is true. The other two properties are just as easy. ∎

Two nonzero vectors $\mathbf{v}$ and $\mathbf{w}$ are said to be *orthogonal* if $\langle \mathbf{v}, \mathbf{w} \rangle = 0$. For example, $(1, 0, -1)$ and $(1, 0, 1)$ in $\mathbb{R}^3$ are orthogonal in the inner product described in Example A4.1. If $\mathbf{v}_1, \mathbf{v}_2, \ldots, \mathbf{v}_k$ are pairwise orthogonal then they are, perforce, linearly independent. To see this, suppose that

$$c_1\mathbf{v}_1 + c_2\mathbf{v}_2 + \cdots c_k\mathbf{v}_k = \mathbf{0} .$$

Take the inner product of both sides with $\mathbf{v}_j$. The result is

$$c_j\langle \mathbf{v}_j, \mathbf{v}_j \rangle = \mathbf{0}$$

or

$$c_j = 0 .$$

Since this property holds for any j, we see that the vectors $\mathbf{v}_j$ are linearly independent.

A *norm* on a vector space V is a function

$$\mathbf{v} \longmapsto \|\mathbf{v}\|$$

satisfying these properties:

(i) $\|\mathbf{v}\| \geq 0$;

(ii) $\|\mathbf{v}\| = 0$ if and only if $\mathbf{v} = 0$;

(iii) $\|c\mathbf{v}\| = |c| \cdot \|\mathbf{v}\|$ for any scalar c;

(iv) $\|\mathbf{v} + \mathbf{w}\| \leq \|\mathbf{v}\| + \|\mathbf{w}\|$ for any vectors $\mathbf{v}, \mathbf{w} \in V$.

A particularly useful norm may be defined from the inner product: we set

$$\|\mathbf{v}\| = \sqrt{\langle \mathbf{v}, \mathbf{v} \rangle}\,.$$

With this particular norm, we have the very useful *Cauchy–Schwarz–Bunyakovsky inequality*

$$|\langle \mathbf{v}, \mathbf{w} \rangle| \leq \|\mathbf{v}\| \cdot \|\mathbf{w}\|\,.$$

A5. Linear Transformations and Matrices

Let V, W be vector spaces. A function $T : V \to W$, with domain V and range W, is said to be *linear* (or a *linear transformation*) if

(a) $T(\mathbf{v} + \mathbf{w}) = T(\mathbf{v}) + T(\mathbf{w})$;

(b) $T(c\mathbf{v}) = cT(\mathbf{v})$ for any scalar c.

It follows immediately that a linear transformation T will satisfy

$$T(c\mathbf{v} + c'\mathbf{w}) = cT(\mathbf{v}) + c'T(\mathbf{w})\,.$$

Example A5.1 Let $V = \mathbb{R}^3$ and $W = \mathbb{R}^4$. Define

$$T(x, y, z) = (x - y, z + 2y, y - x, z + x)\,.$$

Then it is easy to confirm that T is a linear transformation from V to W. For example,

$$\begin{aligned} T(c(x, y, z)) &= T(cx, cy, cz) \\ &= (cx - cy, cz + 2cy, cy - cx, cz + cx) \\ &= c(x - y, z + 2y, y - x, z + x) \\ &= cT(x, y, z)\,. \end{aligned}$$

The additive property is also straightforward to check. ∎

Example A5.2 Let V be the vector space of continuously differentiable functions (i.e., continuous functions having a continuous derivative) defined on the interval $[0, 1]$ and let W be the vector space of continuous functions defined on $[0, 1]$. Define $T : V \to W$ by

$$T(f) = f' - 2f,$$

where the prime $'$ denotes a derivative. Then T is a linear transformation from V to W. ∎

Observe that the linear transformation in Example A5.1 can be represented by matrix multiplication:

$$T(x, y, z) = \begin{pmatrix} 1 & -1 & 0 \\ 0 & 2 & 1 \\ -1 & 1 & 0 \\ 1 & 0 & 1 \end{pmatrix} \cdot \begin{pmatrix} x \\ y \\ z \end{pmatrix} .$$

This property is no accident. *Any* linear transformation of finite dimensional vector spaces can be represented by matrix multiplication, once one has selected a basis. We illustrate this idea with an example.

Example A5.3 Let V be the vector space of all polynomials in the variable x of degree less than or equal to three. Let $T : V \to V$ be given by

$$T(f) = f' ,$$

where the prime $'$ denotes the derivative. We can express T with a matrix as follows. Let

$$\begin{aligned} \mathbf{v}_1 &= 1 \\ \mathbf{v}_2 &= x \\ \mathbf{v}_3 &= x^2 \\ \mathbf{v}_4 &= x^3 \end{aligned}$$

be a basis for V. Then T maps

$$\begin{aligned} \mathbf{v}_1 &\longmapsto 0 \\ \mathbf{v}_2 &\longmapsto \mathbf{v}_1 \\ \mathbf{v}_3 &\longmapsto 2\mathbf{v}_2 \\ \mathbf{v}_4 &\longmapsto 3\mathbf{v}_3 . \end{aligned}$$

Now we construct a matrix, thinking of

$$\begin{aligned}\mathbf{v}_1 &\leftrightarrow (1,0,0,0)\\ \mathbf{v}_2 &\leftrightarrow (0,1,0,0)\\ \mathbf{v}_3 &\leftrightarrow (0,0,1,0)\\ \mathbf{v}_4 &\leftrightarrow (0,0,0,1)\,.\end{aligned}$$

In fact the matrix

$$\mathcal{T} = \begin{pmatrix} 0 & 1 & 0 & 0 \\ 0 & 0 & 2 & 0 \\ 0 & 0 & 0 & 3 \\ 0 & 0 & 0 & 0 \end{pmatrix}$$

will do the job. As an instance, notice that

$$\mathcal{T}\begin{pmatrix} 0 \\ 0 \\ 1 \\ 0 \end{pmatrix} = 2\begin{pmatrix} 0 \\ 1 \\ 0 \\ 0 \end{pmatrix},$$

which just says that the transformation maps $\mathbf{v}_3$ to $2\mathbf{v}_2$. The reader may check that this matrix multiplication also sends $\mathbf{v}_1, \mathbf{v}_3$, and $\mathbf{v}_4$ to the right targets. ∎

A6. Eigenvalues and Eigenvectors

Let A be an $n \times n$ matrix (i.e., n rows and n columns). We say that the number λ is an *eigenvalue* of the matrix A if there exists a nonzero column vector $\mathbf{v}$ such that $A\mathbf{v} = \lambda\mathbf{v}$. For such an eigenvalue λ, the corresponding nontrivial $\mathbf{v}$ which satisfy the equation are called *eigenvectors.*

Let I be the $n \times n$ identity matrix. Then we may write the defining equation for eigenvalues and eigenvectors as

$$(A - \lambda I)\mathbf{v} = 0\,.$$

Such an equation will have a nontrivial solution precisely when

$$\det\left(A - \lambda I\right) = 0\,. \qquad (*)$$

We call $(*)$ the *characteristic equation* associated with the matrix A.

Example A6.1 Let us find the eigenvalues and eigenvectors for the matrix

$$A = \begin{pmatrix} 4 & -1 \\ 2 & 1 \end{pmatrix}.$$

The characteristic equation is

$$0 = \det[A - \lambda I] = \det \begin{pmatrix} 4-\lambda & -1 \\ 2 & 1-\lambda \end{pmatrix}.$$

Calculating the determinant, we find that

$$\lambda^2 - 5\lambda + 6 = 0.$$

Thus $\lambda = 2, 3$. For the eigenvalue $\lambda = 2$, the defining condition for an eigenvector $\mathbf{v} = (v_1, v_2)$ is[1]

$$A\mathbf{v} = 2\mathbf{v}$$

or

$$\begin{pmatrix} 4 & -1 \\ 2 & 1 \end{pmatrix} \begin{pmatrix} v_1 \\ v_2 \end{pmatrix} = 2 \begin{pmatrix} v_1 \\ v_2 \end{pmatrix}.$$

This leads to the equations

$$\begin{aligned} 4x - y &= 2x \\ 2x + y &= 2y \end{aligned}$$

or

$$\begin{aligned} 2x - y &= 0 \\ 2x - y &= 0. \end{aligned}$$

We see, therefore, that $(1, 2)$, or any multiple of $(1, 2)$, is an eigenvector for $\lambda = 2$. (It will be typical that, corresponding to any given eigenvalue, there will be a *space* of eigenvectors.) Likewise, when $\lambda = 3$, we may calculate that the corresponding eigenvectors are all multiples of $(1, 1)$.

Observe that the eigenvectors that we have found—$(1, 2)$ and $(1, 1)$—form a basis for $\mathbb{R}^2$. ∎

A very important and useful fact is that, if an $n \times n$ matrix A has n distinct real eigenvalues $\lambda_1, \lambda_2, \ldots, \lambda_n$ then there is a nonsingular $n \times n$ matrix U, whose columns are the eigenvectors of A, such that

$$U^{-1}AU = \begin{pmatrix} \lambda_1 & 0 & \cdots & 0 \\ 0 & \lambda_2 & 0 & \cdots \\ & \vdots & & \\ 0 & 0 & \cdots & \lambda_n \end{pmatrix}.$$

[1]Throughout this discussion, we switch back and forth freely between row vectors and column vectors.

In other words, the theory of eigenvalues gives us, in this case, a device for diagonalizing the given matrix.

Example A6.2 Let us use the device of eigenvalues to diagonalize the matrix

$$A = \begin{pmatrix} -1 & -3 & 3 \\ -6 & 2 & 6 \\ -3 & 3 & 5 \end{pmatrix}.$$

We have

$$A - \lambda I = \begin{pmatrix} -1-\lambda & -3 & 3 \\ -6 & 2-\lambda & 6 \\ -3 & 3 & 5-\lambda \end{pmatrix}$$

and the characteristic polynomial is $-\lambda^3 + 6\lambda^2 + 24\lambda - 64$. It is not difficult to determine that the roots of this polynomial, i.e., the eigenvalues of A, are $2, -4, 8$. Solving as usual, we find that the corresponding eigenvectors are $(1,0,1)$, $(1,1,0)$, and $(0,1,1)$.

Now we form the matrix U by using the eigenvectors as columns. Thus

$$U = \begin{pmatrix} 1 & 1 & 0 \\ 0 & 1 & 1 \\ 1 & 0 & 1 \end{pmatrix}$$

and one readily calculates that

$$U^{-1} = \begin{pmatrix} \frac{1}{2} & -\frac{1}{2} & \frac{1}{2} \\ \frac{1}{2} & \frac{1}{2} & -\frac{1}{2} \\ -\frac{1}{2} & \frac{1}{2} & \frac{1}{2} \end{pmatrix}.$$

Finally, the diagonalized form of A is obtained by calculating

$$U^{-1}AU = \begin{pmatrix} 2 & 0 & 0 \\ 0 & -4 & 0 \\ 0 & 0 & 8 \end{pmatrix}. \qquad \blacksquare$$

Bibliography

[ALM] F. J. Almgren, *Plateau's Problem: An Invitation to Varifold Geometry*, Benjamin, New York, 1966.

[BLK] B. E. Blank and S. G. Krantz, *Calculus with Analytic Geometry*, Key Curriculum Press, Emeryville, CA, 2005.

[DER] J. Derbyshire, *Prime Obsession: Bernhard Riemann and the Greatest Unsolved Problem in Mathematics*, Joseph Henry Press, Washington, DC, 2003.

[EDD] A. Eddington, *The Expanding Universe*, Cambridge University Press, London, 1952.

[FOU] J. Fourier, *The Analytical Theory of Heat*, G. E. Stechert & Co., New York, 1878.

[GAK] T. Gamelin and D. Khavinson, The isoperimetric inequality and rational approximation, *Am. Math. Monthly* 96(1989), 18–30.

[GRK] R. E. Greene and S. G. Krantz, *Function Theory of One Complex Variable*, 2nd ed., American Mathematical Society, Providence, RI, 2002.

[HIT] S. Hildebrandt and A. J. Tromba, *The Parsimonious Universe*, Copernicus Press, New York, 1996.

[IXA] L. G. Ixaru, *Numerical Methods for Differential Equations and Applications*, D. Reidel Publishing Co., Dordrecht, Holland, 1984.

[JEA] J. Jeans, *The Astronomical Horizon*, Oxford University Press, London, 1945.

[KBO] A. C. King, J. Billingham, and S. R. Otto, *Differential Equations: Linear, Nonlinear, Ordinary, Partial*, Cambridge University Press, Cambridge, 2003.

[KNO] K. Knopp, *Elements of the Theory of Functions*, Dover, New York, 1952.

[KRA2] S. G. Krantz, *Real Analysis and Foundations*, CRC Press, Boca Raton, FL, 1992.

[KRA3] S. G. Krantz, *A Panorama of Harmonic Analysis*, Mathematical Association of America, Washington, DC, 1999.

[KRP1] S. G. Krantz and H. R. Parks, *A Primer of Real Analytic Functions*, 2nd ed., Birkhäuser, Boston, 2002.

[LAN] R. E. Langer, *Fourier Series: The Genesis and Evolution of a Theory*, Herbert Ellsworth Slaught Memorial Paper I, *Am. Math. Monthly* 54(1947).

[LUZ] N. Luzin, The evolution of "Function", Part I, Abe Shenitzer, ed., *Am. Math. Monthly* 105(1998), 59–67.

[MOR] F. Morgan, *Geometric Measure Theory: A Beginner's Guide*, Academic Press, Boston, 1988.

[OSS] R. Osserman, The isoperimetric inequality, *Bull. AMS* 84(1978), 1182–1238.

[RUD] W. Rudin, *Functional Analysis*, 2nd ed., McGraw-Hill, New York, 1991.

[SAB] K. Sabbagh, *The Riemann Hypothesis: The Greatest Unsolved Problem in Mathematics*, Farrar, Straus and Giroux, New York, 2003.

[STE] J. Stewart, *Calculus. Concepts and Contexts*, Brooks/Cole Publishing, Pacific Grove, CA, 2001.

[THO] G. B. Thomas (with Ross L. Finney), *Calculus and Analytic Geometry*, 7th ed., Addison-Wesley, Reading, MA, 1988.

[TIT] E. C. Titchmarsh, *Introduction to the Theory of Fourier Integrals*, The Clarendon Press, Oxford, 1948.

[WAT] G. N. Watson, *A Treatise on the Theory of Bessel Functions*, 2nd ed., Cambridge University Press, Cambridge, 1958.

Index